Handbook of Nanomaterials

Handbook of Nanomaterials

Emmanuel Craig

NY RESEARCH
P R E S S

New York

Published by NY Research Press
118-35 Queens Blvd., Suite 400,
Forest Hills, NY 11375, USA
www.nyresearchpress.com

Handbook of Nanomaterials
Emmanuel Craig

International Standard Book Number: 978-1-63238-871-1 (Hardback)

Cataloging-in-Publication Data

Handbook of nanomaterials / Emmanuel Craig.
    p. cm.
Includes bibliographical references and index.
ISBN 978-1-63238-871-1
1. Nanostructured materials. 2. Nanotechnology. 3. Microstructure. I. Craig, Emmanuel.
TA418.9.N35 H36 2022
620.5--dc23

Contents

Preface

Nanomaterials are the materials of which the size of a single unit is between 1 to 1000 nm. Research in nanomaterials uses material science as the basis to approach nanotechnology. Materials which have configurations at nanoscale often have unique electronic, optical or mechanical properties. On the basis of the sources, nanomaterials are classified into three types- engineered, incidental and natural. The nanomaterials which have been intentionally engineered and manufactured by humans to have specific properties are known as engineered nanomaterials. Incidental nanomaterials are produced during mechanical or industrial processes as a byproduct. The natural and functional nanomaterials which are found in biological systems are known as natural nanomaterials. This book brings forth some of the most innovative concepts and elucidates the unexplored aspects of nanomaterials. It will provide comprehensive knowledge to the readers. Coherent flow of topics, student-friendly language and extensive use of examples make this book an invaluable source of knowledge.

A detailed account of the significant topics covered in this book is provided below:

Chapter 1- The materials whose size lies between 1 and 100 nanometers are known as nanomaterials. Some of these are nanorods, nanoshells, nanofibers, silicene, nanosheets and nanomesh. Dendrimers, carbon-based nanomaterials, nanoporous material, etc. are some of the other types that are studied within it. This is an introductory chapter which will briefly introduce all the significant aspects of nanomaterials.

Chapter 2- Nanomaterials possess different physical, mechanical, electrical and optical properties. Structure, field emission, hardness, elastic modulus, adhesion, conductivity, scattering, absorption, etc. fall under the domain of these properties. This chapter has been carefully written to provide an easy understanding of the various properties of nanomaterials.

Chapter 3- Particles between the size of 1 and 100 nanometers are referred to as nanoparticles. Some of the processes associated with them are nanoparticle production, endocytosis and exocytosis of nanoparticles, etc. Common types of nanoparticles are magnetic nanoparticles, shell nanoparticles, superparamagnetic iron oxide nanoparticles, etc. All these related concepts of nanoparticles have been carefully analyzed in this chapter.

Chapter 4- The fibers that are generated from different polymers and have radius in nanometer range are referred to as nanofibers. Electrospinning, self-assembly, template synthesis, and thermal-induced phase separation are some of the methods used to make nanofibers. This chapter closely examines the key concepts of nanofibers to provide an extensive understanding of the subject.

Chapter 5- Nanocomposites are multiphase soild materials that have nano-scale repeat distances between their different phases. They improve the properties of mechanical strength, toughness and electrical, and thermal conductivity. Ceramic Matrix nanocomposites, metal matrix nanocomposites and polymer matrix nanocomposites are a few of its types. The topics elaborated in this chapter will help in gaining a better perspective about the subject of nanocomposites.

It gives me an immense pleasure to thank our entire team for their efforts. Finally in the end, I would like to thank my family and colleagues who have been a great source of inspiration and support.

Emmanuel Craig

Nanomaterials: An Introduction

The materials whose size lies between 1 and 100 nanometers are known as nanomaterials. Some of these are nanorods, nanoshells, nanofibers, silicene, nanosheets and nanomesh. Dendrimers, carbon-based nanomaterials, nanoporous material, etc. are some of the other types that are studied within it. This is an introductory chapter which will briefly introduce all the significant aspects of nanomaterials.

Nanomaterials are cornerstones of nanoscience and nanotechnology. Nanostructure science and technology is a broad and interdisciplinary area of research and development activity that has been growing explosively worldwide in the past few years. It has the potential for revolutionizing the ways in which materials and products are created and the range and nature of functionalities that can be accessed. It is already having a significant commercial impact, which will assuredly increase in the future.

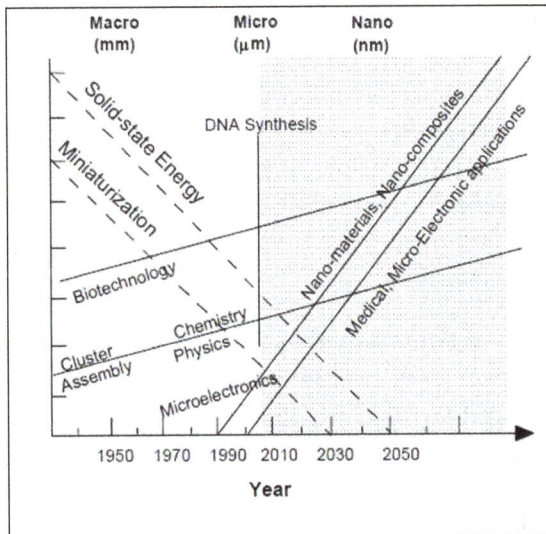

Evolution of science and technology and the future.

Nanoscale materials are defined as a set of substances where at least one dimension is less than approximately 100 nanometers. A nanometer is one millionth of a millimeter - approximately 100,000 times smaller than the diameter of a human hair. Nanomaterials are of interest because at this scale unique optical, magnetic, electrical, and other properties emerge. These emergent properties have the potential for great impacts in electronics, medicine, and other fields.

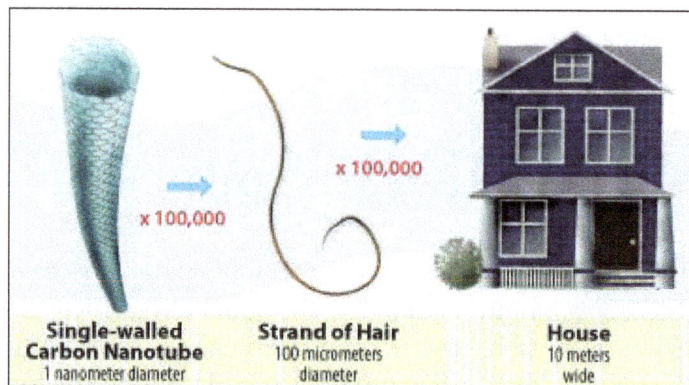

Nanomaterial (For example: Carbon nanotube).

Location of Nanomaterials

Some nanomaterials occur naturally, but of particular interest are engineered nanomaterials (EN), which are designed for, and already being used in many commercial products and processes. They can be found in such things as sunscreens, cosmetics, sporting goods, stain-resistant clothing, tires, electronics, as well as many other everyday items, and are used in medicine for purposes of diagnosis, imaging and drug delivery.

Engineered nanomaterials are resources designed at the molecular (nanometre) level to take advantage of their small size and novel properties which are generally not seen in their conventional, bulk counterparts. The two main reasons why materials at the nano scale can have different properties are increased relative surface area and new quantum effects. Nanomaterials have a much greater surface area to volume ratio than their conventional forms, which can lead to greater chemical reactivity and affect their strength. Also at the nano scale, quantum effects can become much more important in determining the materials properties and characteristics, leading to novel optical, electrical and magnetic behaviours.

Nanomaterials are already in commercial use, with some having been available for several years or decades. The range of commercial products available today is very broad, including stain-resistant and wrinkle-free textiles, cosmetics, sunscreens, electronics, paints and varnishes. Nanocoatings and nanocomposites are finding uses in diverse consumer products, such as windows, sports equipment, bicycles and automobiles. There are novel UV-blocking coatings on glass bottles which protect beverages from damage by sunlight, and longer-lasting tennis balls using butylrubber/nano-clay composites. Nanoscale titanium dioxide, for instance, is finding applications in cosmetics, sun-block creams and self-cleaning windows, and nanoscale silica is being used as filler in a range of products, including cosmetics and dental fillings.

Advances in Nanomaterials

The history of nanomaterials began immediately after the big bang when Nanostructures were formed in the early meteorites. Nature later evolved many other Nanostructures like seashells, skeletons etc. Nanoscaled smoke particles were formed during the use of fire by early humans. The scientific story of nanomaterials however began much later. One of the first scientific report is the colloidal gold particles synthesised by Michael Faraday as early as 1857. Nanostructured catalysts have also been investigated for over 70 years. By the early 1940's, precipitated and fumed silica nanoparticles were being manufactured and sold in USA and Germany as substitutes for ultrafine carbon black for rubber reinforcements.

Nanosized amorphous silica particles have found large-scale applications in many every-day consumer products, ranging from non-diary coffee creamer to automobile tires, optical fibers and catalyst supports. In the 1960s and 1970's metallic nanopowders for magnetic recording tapes were developed. In 1976, for the first time, nanocrystals produced by the now popular inert- gas evaporation technique was published by Granqvist and Buhrman. Recently it has been found that the Maya blue paint is a nanostructured hybrid material. The origin of its color and its resistance to acids and biocorrosion are still not understood but studies of authentic samples from Jaina Island show that the material is made of needle-shaped palygorskite (clay) crystals that form a superlattice with a period of 1.4 nm, with intercalates of amorphous silicate substrate containing inclusions of metal (Mg) nanoparticles. The beautiful tone of the blue color is obtained only when both these nanoparticles and the superlattice are present, as has been shown by the fabrication of synthetic samples.

Today nanophase engineering expands in a rapidly growing number of structural and functional materials, both inorganic and organic, allowing to manipulate mechanical, catalytic, electric, magnetic, optical and electronic functions. The production of nanophase or cluster-assembled materials is usually based upon the creation of separated small clusters which then are fused into a bulk-like material or on their embedding into compact liquid or solid matrix materials, e.g., nanophase silicon, which differs from normal silicon in physical and electronic properties, could be applied to macroscopic semiconductor processes to create new devices. For instance, when ordinary glass is doped with quantized semiconductor "colloids", it becomes a high performance optical medium with potential applications in optical computing.

Classification of Nanomaterials

Nanomaterials have extremely small size which having at least one dimension 100 nm or less. Nanomaterials can be nanoscale in one dimension (eg. surface films), two dimensions (eg. strands or fibres), or three dimensions (eg. particles). They can exist in single, fused, aggregated or agglomerated forms with spherical, tubular, and irregular shapes. Common types of nanomaterials include nanotubes, dendrimers, quantum dots and fullerenes. Nanomaterials have applications in the field of nano technology, and displays different physical chemical characteristics from normal chemicals (i.e., silver nano, carbon nanotube, fullerene, photocatalyst, carbon nano, silica).

According to Siegel, Nanostructured materials are classified as Zero dimensional, one dimensional, two dimensional, three dimensional nanostructures.

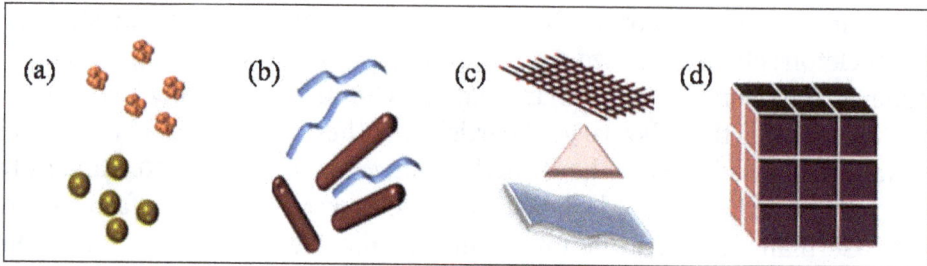

Classification of Nanomaterials (a) 0D spheres and clusters, (b) 1D nanofibers, wires, and rods, (c) 2D films, plates, and networks, (d) 3D nanomaterials.

Nanomaterials are materials which are characterized by an ultra fine grain size (< 50 nm) or by a dimensionality limited to 50 nm. Nanomaterials can be created with various modulation dimensionalities as defined by Richard W. Siegel: zero (atomic clusters, filaments and cluster assemblies), one (multilayers), two (ultrafine-grained overlayers or buried layers), and three (nanophase materials consisting of equiaxed nanometer sized grains) as shown in the above figure.

Importance of Nanomaterials

These materials have created a high interest in recent years by virtue of their unusual mechanical, electrical, optical and magnetic properties. Some examples are given below:

- Nanophase ceramics are of particular interest because they are more ductile at elevated temperatures as compared to the coarse-grained ceramics.

- Nanostructured semiconductors are known to show various non-linear optical properties. Semiconductor Q-particles also show quantum confinement effects which may lead to special properties, like the luminescence in silicon powders and silicon germanium quantum dots as infrared optoelectronic devices. Nanostructured semiconductors are used as window layers in solar cells.

- Nanosized metallic powders have been used for the production of gas tight materials, dense parts and porous coatings. Cold welding properties combined with the ductility make them suitable for metal-metal bonding especially in the electronic industry.

- Single nanosized magnetic particles are mono-domains and one expects that also in magnetic nanophase materials the grains correspond with domains, while boundaries on the contrary to disordered walls. Very small particles have special atomic structures with discrete electronic states, which give rise to special properties in addition to the superparamagnetism behaviour. Magnetic nanocomposites have been used for mechanical force transfer (ferrofluids), for high density information storage and magnetic refrigeration.

- Nanostructured metal clusters and colloids of mono- or plurimetallic composition have a special impact in catalytic applications. They may serve as precursors for new type of heterogeneous catalysts (Cortex-catalysts) and have been shown to offer substantial advantages concerning activity, selectivity and lifetime in chemical transformations and electrocatalysis (fuel cells). Enantioselective catalysis was also achieved using chiral modifiers on the surface of nanoscale metal particles.

• Nanostructured metal-oxide thin films are receiving a growing attention for the realization of gas sensors (NOx, CO, CO_2, CH4 and aromatic hydrocarbons) with enhanced sensitivity and selectivity. Nanostructured metal-oxide (MnO_2) finds application for rechargeable batteries for cars or consumer goods. Nanocrystalline silicon films for highly transparent contacts in thin film solar cell and nano-structured titanium oxide porous films for its high transmission and significant surface area enhancement leading to strong absorption in dye sensitized solar cells.

• Polymer based composites with a high content of inorganic particles leading to a high dielectric constant are interesting materials for photonic band gap structure.

Examples of Nanomaterials

Nanomaterials (gold, carbon, metals, meta oxides and alloys) with variety of morphologies (shapes) are depicted in figure.

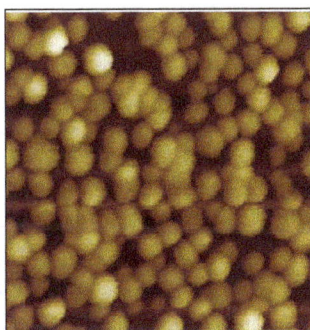

Au nanoparticle.

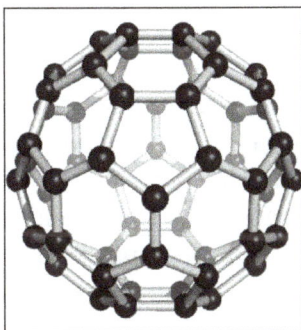

Buckminsterfullerene.

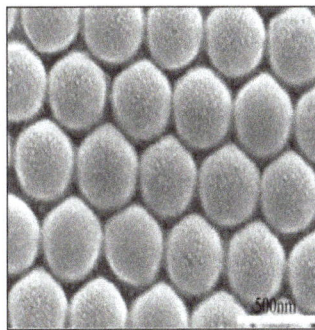

FePt nanosphere.

Titanium nanoflowe.

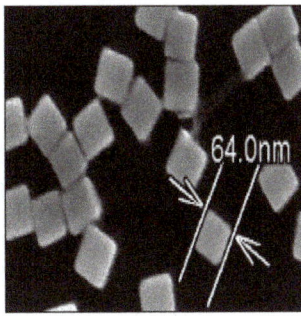

Silver nanocubes.

SnO_2 nanoflower.

Nanomaterials with a variety of morphologies.

Nanomaterial - Synthesis and Processing

Nanomaterials deal with very fine structures: a nanometer is a billionth of a meter. This indeed allows us to think in both the 'bottom up' or the 'top down' approaches to synthesize nanomaterials, i.e. either to assemble atoms together or to dis-assemble (break, or dissociate) bulk solids into finer pieces until they are constituted of only a few atoms. This domain is a pure example of interdisciplinary work encompassing physics, chemistry, and engineering upto medicine.

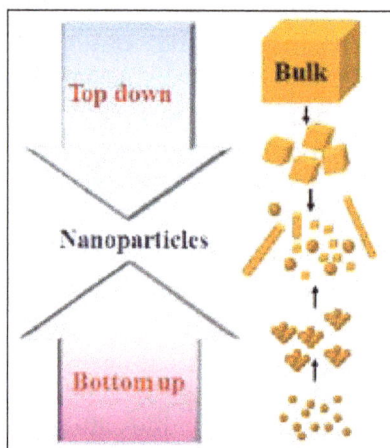

Schematic illustration of the preparative methods of nanoparticles.

Methods for Creating Nanostructures

There are many different ways of creating nanostructures: of course, macromolecules or nanoparticles or buckyballs or nanotubes and so on can be synthesized artificially for certain specific materials. They can also be arranged by methods based on equilibrium or near-equilibrium thermodynamics such as methods of self-organization and self-assembly (sometimes also called bio-mimetic processes). Using these methods, synthesized materials can be arranged into useful shapes so that finally the material can be applied to a certain application.

Mechanical Grinding

Mechanical attrition is a typical example of 'top down' method of synthesis of nanomaterials, where the material is prepared not by cluster assembly but by the structural decomposition of coarser-grained structures as the result of severe plastic deformation. This has become a popular method to make nanocrystalline materials because of its simplicity, the relatively inexpensive equipment needed, and the applicability to essentially the synthesis of all classes of materials. The major advantage often quoted is the possibility for easily scaling up to tonnage quantities of material for various applications. Similarly, the serious problems that are usually cited are;

• Contamination from milling media and atmosphere.

- To consolidate the powder product without coarsening the nanocrystalline microstructure.

In fact, the contamination problem is often given as a reason to dismiss the method, at least for some materials. Here we will review the mechanisms presently believed responsible for formation of nanocrystalline structures by mechanical attrition of single phase powders, mechanical alloying of dissimilar powders, and mechanical crystallisation of amorphous materials. The two important problems of contamination and powder consolidation will be briefly considered.

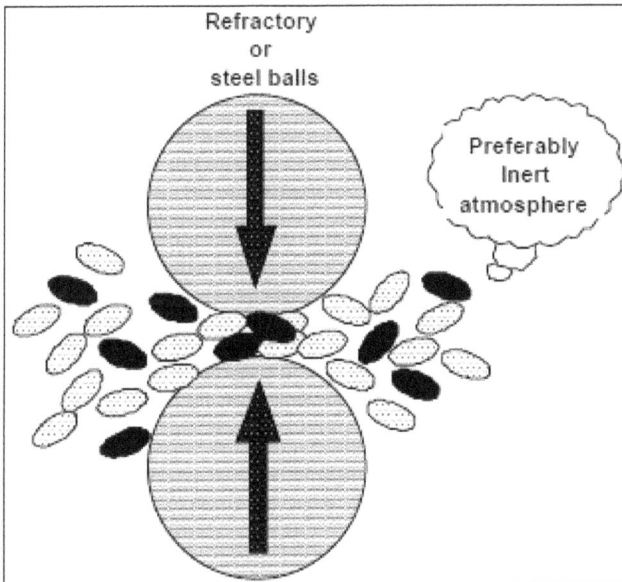

Schematic representation of the principle of mechanical milling.

Mechanical milling is typically achieved using high energy shaker, planetary ball, or tumbler mills. The energy transferred to the powder from refractory or steel balls depends on the rotational (vibrational) speed, size and number of the balls, ratio of the ball to powder mass, the time of milling and the milling atmosphere. Nanoparticles are produced by the shear action during grinding.

Milling in cryogenic liquids can greatly increase the brittleness of the powders influencing the fracture process. As with any process that produces fine particles, an adequate step to prevent oxidation is necessary. Hence this process is very restrictive for the production of non-oxide materials since then it requires that the milling take place in an inert atmosphere and that the powder particles be handled in an appropriate vacuum system or glove box. This method of synthesis is suitable for producing amorphous or nanocrystalline alloy particles, elemental or compound powders. If the mechanical milling imparts sufficient energy to the constituent powders a homogeneous alloy can be formed. Based on the energy of the milling process and thermodynamic properties of the constituents the alloy can be rendered amorphous by this processing.

Wet Chemical Synthesis of Nanomaterials

In principle we can classify the wet chemical synthesis of nanomaterials into two broad groups:

- The top down method: Where single crystals are etched in an aqueous solution for producing nanomaterials, For example, the synthesis of porous silicon by electrochemical etching.

- The bottom up method: Consisting of sol-gel method, precipitation etc. where materials containing the desired precursors are mixed in a controlled fashion to form a colloidal solution.

Sol-gel Process

The sol-gel process, involves the evolution of inorganic networks through the formation of a colloidal suspension (sol) and gelation of the sol to form a network in a continuous liquid phase (gel). The precursors for synthesizing these colloids consist usually of a metal or metalloid element surrounded by various reactive ligands. The starting material is processed to form a dispersible oxide and forms a sol in contact with water or dilute acid. Removal of the liquid from the sol yields the gel, and the sol/gel transition controls the particle size and shape. Calcination of the gel produces the oxide.

Sol-gel processing refers to the hydrolysis and condensation of alkoxide-based precursors such as $Si(OEt)_4$ (tetraethyl orthosilicate, or TEOS). The reactions involved in the sol-gel chemistry based on the hydrolysis and condensation of metal alkoxides M(OR) z can be described as follows:

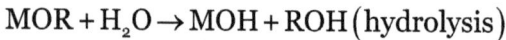

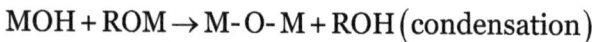

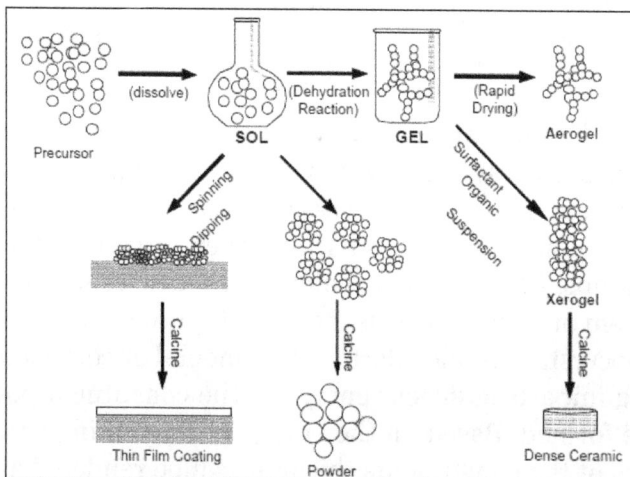

$$MOR + H_2O \rightarrow MOH + ROH \, (hydrolysis)$$

$$MOH + ROM \rightarrow M\text{-}O\text{-}M + ROH \, (condensation)$$

Schematic representation of sol-gel process of synthesis of nanomaterials.

Sol-gel method of synthesizing nanomaterials is very popular amongst chemists and is widely employed to prepare oxide materials. The sol-gel process can be characterized by a series of distinct steps:

- Formation of different stable solutions of the alkoxide or solvated metal precursor.

- Gelation resulting from the formation of an oxide- or alcohol- bridged network (the gel) by a polycondensation reaction that results in a dramatic increase in the viscocity of the solution.

- Aging of the gel (Syneresis), during which the polycondensation reactions continue until the gel transforms into a solid mass, accompanied by contraction of the gel network and expulsion of solvent from gel pores. Ostwald ripening (also referred to as coarsening, is the phenomenon by which smaller particles are consumed by larger particles during the growth process) and phase transformations may occur concurrently with syneresis. The aging process of gels can exceed 7 days and is critical to the prevention of cracks in gels that have been cast.

- Drying of the gel, when water and other volatile liquids are removed from the gel network. This process is complicated due to fundamental changes in the structure of the gel. The drying process has itself been broken into four distinct steps: (i) the constant rate period, (ii) the critical point, (iii) the falling rate period, (iv) the second falling rate period. If isolated by thermal evaporation, the resulting monolith is termed a xerogel. If the solvent (such as water) is extracted under supercritical or near super critical conditions, the product is an aerogel.

- Dehydration, during which surface- bound M-OH groups are removed, there by stabilizing the gel against rehydration. This is normally achieved by calcining the monolith at temperatures up to $800^{\circ}C$.

- Densification and decomposition of the gels at high temperatures (T>8000 C). The pores of the gel network are collapsed, and remaining organic species are volatilized.

The interest in this synthesis method arises due to the possibility of synthesizing non-metallic inorganic materials like glasses, glass ceramics or ceramic materials at very low temperatures compared to the high temperature process required by melting glass or firing ceramics.

The major difficulties to overcome in developing a successful bottom-up approach is controlling the growth of the particles and then stopping the newly formed particles from agglomerating. Other technical issues are ensuring the reactions are complete so that no unwanted reactant is left on the product and completely removing any growth aids that may have been used in the process. Also production rates of nano powders are very low by this process. The main advantage is one can get monosized nano particles by any bottom up approach.

Gas Phase Synthesis of Nanomaterials

The gas-phase synthesis methods are of increasing interest because they allow elegant way to control process parameters in order to be able to produce size, shape and chemical composition controlled nanostructures. Before we discuss a few selected pathways for gas-phase formation of nanomaterials, some general aspects of gas-phase synthesis needs to be discussed. In conventional chemical vapour deposition (CVD) synthesis, gaseous products either are allowed to react homogeneously or heterogeneously depending on a particular application.

- In homogeneous CVD, particles form in the gas phase and diffuse towards a cold surface due to thermophoretic forces, and can either be scrapped of from the cold surface to give nano-powders, or deposited onto a substrate to yield what is called 'particulate films'.

- In heterogeneous CVD, the solid is formed on the substrate surface, which catalyses the reaction and a dense film is formed.

In order to form nanomaterials several modified CVD methods have been developed. Gas phase processes have inherent advantages, some of which are noted here:

- An excellent control of size, shape, crystallinity and chemical composition.

- Highly pure materials can be obtained.

- Multicomonent systems are relatively easy to form.

- Easy control of the reaction mechanisms.

Most of the synthesis routes are based on the production of small clusters that can aggregate to form nano particles (condensation). Condensation occurs only when the vapour is supersaturated and in these processes homogeneous nucleation in the gas phase is utilised to form particles. This can be achieved both by physical and chemical methods.

Furnace

The simplest fashion to produce nanoparticles is by heating the desired material in a heatresistant crucible containing the desired material. This method is appropriate only for materials that have a high vapour pressure at the heated temperatures that can be as high as 2000°C. Energy is normally introduced into the precursor by arc heating, electronbeam heating or Joule heating. The atoms are evaporated into an atmosphere, which is either inert (e.g. He) or reactive (so as to form a compound). To carry out reactive synthesis, materials with very low vapour pressure have to be fed into the furnace in the form of a suitable precursor such as organometallics, which decompose in the furnace to produce a condensable material. The hot atoms of the evaporated matter lose energy by collision with the atoms of the cold gas and undergo condensation into small clusters via homogeneous nucleation. In case a compound is being synthesized, these precursors react in the gas phase and form a compound with the material that is separately injected in the reaction chamber. The clusters would continue to grow if they remain in the supersaturated region. To control their size, they need to be rapidly removed from the supersaturated environment by a carrier gas. The cluster size and its distribution are controlled by only three parameters:

- The rate of evaporation (energy input).

- The rate of condensation (energy removal).

- The rate of gas flow (cluster removal).

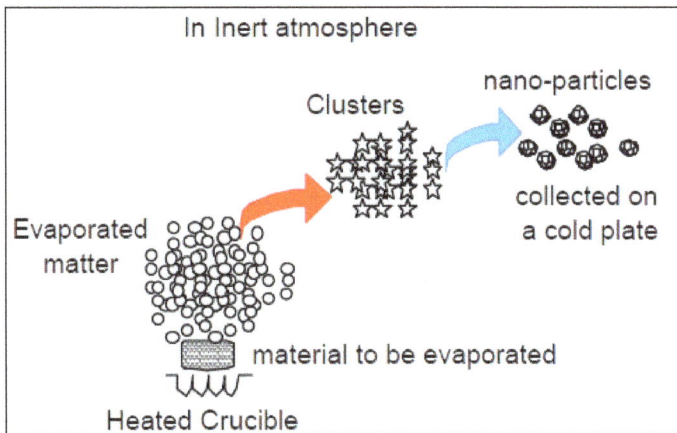

Schematic representation of gas phase process of synthesis of single phase nanomaterials from a heated crucible.

Because of its inherent simplicity, it is possible to scale up this process from laboratory (mg/day) to industrial scales (tons/day).

Flame Assisted Ultrasonic Spray Pyrolysis

In this process, precusrsors are nebulized and then unwanted components are burnt in a flame to get the required material, eg. ZrO_2 has been obtained by this method from a precursor of $Zr(CH_3 CH_2 CH_2O)_4$. Flame hydrolysis that is a variant of this process is used for the manufacture of fused silica. In the process, silicon tetrachloride is heated in an oxy-hydrogen flame to give a highly dispersed silica. The resulting white amorphous powder consists of spherical particles with sizes in the range 7-40 nm. The combustion flame synthesis, in which the burning of a gas mixture, e.g. acetylene and oxygen or hydrogen and oxygen, supplies the energy to initiate the pyrolysis of precursor compounds, is widely used for the industrial production of powders in large quantities, such as carbon black, fumed silica and titanium dioxide. However, since the gas pressure during the reaction is high, highly agglomerated powders are produced which is disadvantageous for subsequent processing. The basic idea of low pressure combustion flame synthesis is to extend the pressure range to the pressures used in gas phase synthesis and thus to reduce or avoid the agglomeration. Low pressure flames have been extensively used by aerosol scientists to study particle formation in the flame.

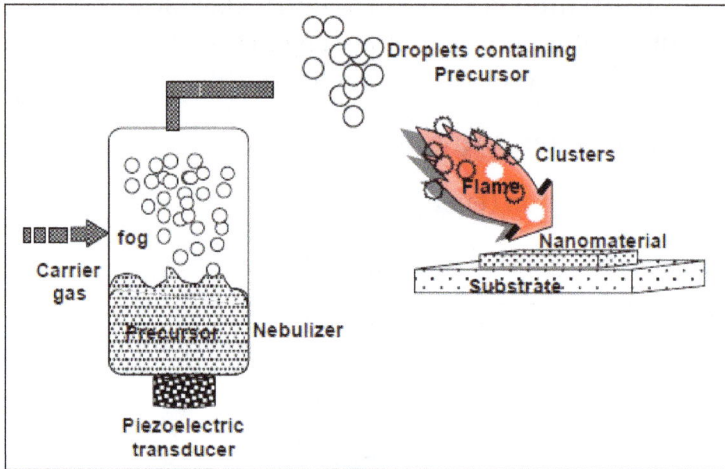

Flame assisted ultrasonic spray pyrolysis.

A key for the formation of nanoparticles with narrow size distributions is the exact control of the flame in order to obtain a flat flame front. Under these conditions the thermal history, i.e. time and temperature, of each particle formed is identical and narrow distributions result. However, due to the oxidative atmosphere in the flame, this synthesis process is limited to the formation of oxides in the reactor zone.

Gas Condensation Processing (GPC)

In this technique, a metallic or inorganic material, e.g. a suboxide, is vaporised using thermal evaporation sources such as crucibles, electron beam evaporation devices or sputtering sources in an atmosphere of 1-50 mbar He (or another inert gas like Ar, Ne, Kr). Cluster form in the vicinity of the source by homogenous nucleation in the gas phase and grow by coalescence and incorporation of atoms from the gas phase.

The cluster or particle size depends critically on the residence time of the particles in the growth system and can be influenced by the gas pressure, the kind of inert gas, i.e. He, Ar or Kr, and on the evaporation rate/vapour pressure of the evaporating material. With increasing gas pressure, vapour pressure and mass of the inert gas used the average particle size of the nanoparticles increases. Lognormal size distributions have been found experimentally and have been explained theoretically by the growth mechanisms of the particles. Even in more complex processes such as the low pressure combustion flame synthesis where a number of chemical reactions are involved the size distributions are determined to be lognormal.

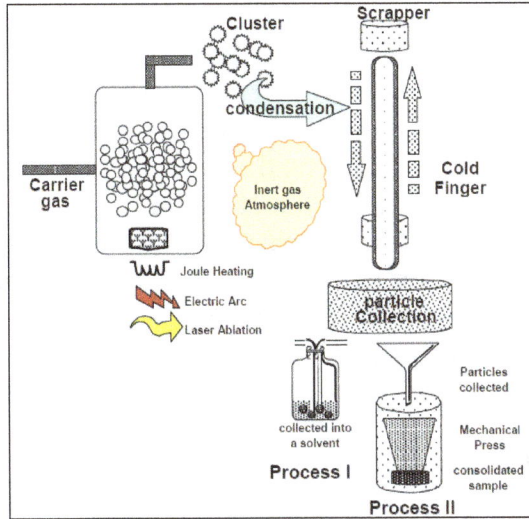

Schematic representation of typical set-up for gas condensation synthesis of nanomaterials followed by consolidation in a mechanical press or collection in an appropriate solvent media.

Originally, a rotating cylindrical device cooled with liquid nitrogen was employed for the particle collection: the nanoparticles in the size range from 2-50 nm are extracted from the gas flow by thermophoretic forces and deposited loosely on the surface of the collection device as a powder of low density and no agglomeration. Subsequenly, the nanoparticles are removed from the surface of the cylinder by means of a scraper in the form of a metallic plate. In addition to this cold finger device several techniques known from aerosol science have now been implemented for the use in gas condensation systems such as corona discharge, etc. These methods allow for the continuous operation of the collection device and are better suited for larger scale synthesis of nanopowders. However, these methods can only be used in a system designed for gas flow, i.e. a dynamic vacuum is generated by means of both continuous pumping and gas inlet via mass flow controller. A major advantage over convectional gas flow is the improved control of the particle sizes. It has been found that the particle size distributions in gas flow systems, which are also lognormal, are shifted towards smaller average values with an appreciable reduction of the standard deviation of the distribution. Depending on the flow rate of the He-gas, particle sizes are reduced by 80% and standard deviations by 18%.

The synthesis of nanocrystalline pure metals is relatively straightforward as long as evaporation can be done from refractory metal crucibles (W, Ta or Mo). If metals with high melting points or metals which react with the crucibles, are to be prepared, sputtering, i.e. for W and Zr, or laser or electron beam evaporation has to be used. Synthesis of alloys or intermetallic compounds by thermal evaporation can only be done in the exceptional cases that the vapour pressures of the elements are similar. As an alternative, sputtering from an alloy or mixed target can be employed. Composite materials such as Cu/Bi or W/Ga have been synthesised by simultaneous evaporation from two separate crucibles onto a rotating collection device. It has been found that excellent intermixing on the scale of the particle size can be obtained.

However, control of the composition of the elements has been difficult and reproducibility is poor. Nanocrystalline oxide powders are formed by controlled postoxidation of primary nanoparticles of a pure metal (e.g. Ti to TiO2) or a suboxide (e.g. ZrO to ZrO2). Although the gas condensation method including the variations have been widely employed to prepared a variety of metallic and ceramic materials, quantities have so far been limited to a laboratory scale. The quantities of metals are below 1 g/day, while quantities of oxides can be as high as 20 g/day for simple oxides such as CeO2 or ZrO2. These quantities are sufficient for materials testing but not for industrial production. However, it should be mentioned that the scale-up of the gas condensation method for industrial production of nanocrystalline oxides by a company called nanophase technologies has been successful.

Chemical Vapour Condensation (CVC)

As shown schematically in figure, the evaporative source used in GPC is replaced by a hot wall reactor in the Chemical Vapour Condensation or the CVC process. Depending on the processing parameters nucleation of nanoparticles is observed during chemical vapour deposition (CVC) of thin films and poses a major problem in obtaining good film qualities. The original idea of the novel CVC process which is schematically shown below where, it was intended to adjust the parameter field during the synthesis in order to suppress film formation and enhance homogeneous nucleation of particles in the gas flow. It is readily found that the residence time of the precursor in the reactor determines if films or particles are formed. In a certain range of residence time both particle and film formation can be obtained.

Adjusting the residence time of the precursor molecules by changing the gas flow rate, the pressure difference between the precursor delivery system and the main chamber occurs. Then the temperature of the hot wall reactor results in the fertile production of nanosized particles of metals and ceramics instead of thin films as in CVD processing. In the simplest form a metal organic precursor is introduced into the hot zone of the reactor using mass flow controller. Besides the increased quantities in this continuous process compared to GPC has been demonstrated that a wider range of ceramics including nitrides and carbides can be synthesised. Additionally, more complex oxides such as BaTiO3 or composite structures can be formed as well. Appropriate precursor compounds can be readily found in the CVD literature. The extension to production of nanoparticles requires the determination of a modified parameter field in order to promote particle formation instead of film formation. In addition to the formation of single phase nanoparticles by CVC of a single precursor the reactor allows the synthesis of:

- Mixtures of nanoparticles of two phases or doped nanoparticles by supplying two precursors at the front end of the reactor.

- Coated nanoparticles, i.e., $n-ZrO_2$ coated with $n-Al_2O_3$ or vice versa, by supplying a second precursor at a second stage of the reactor. In this case nanoparticles which have been formed by homogeneous nucleation are coated by heterogeneous nucleation in a second stage of the reactor.

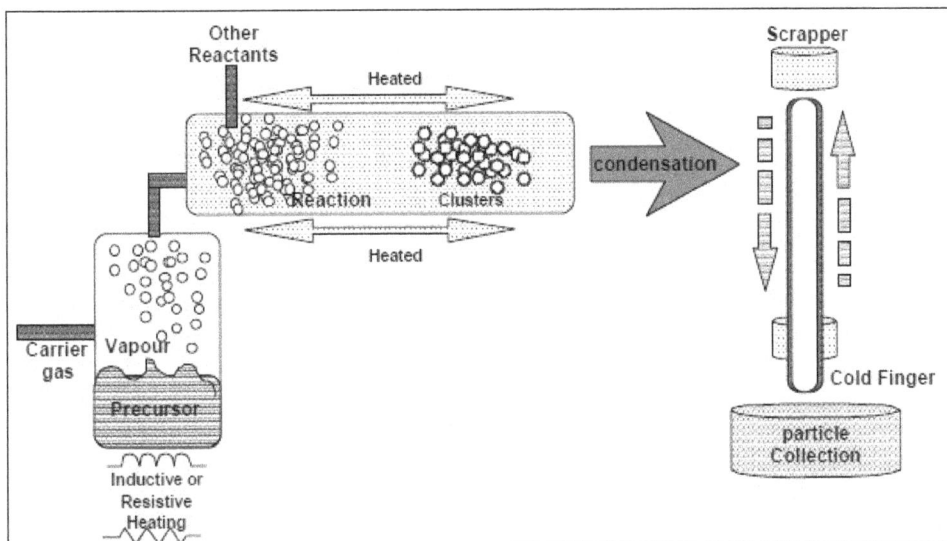

A schematic of a typical CVC reactor.

Because CVC processing is continuous, the production capabilities are much larger than in GPC processing. Quantities in excess of 20 g/hr have been readily produced with a small scale laboratory reactor. A further expansion can be envisaged by simply enlarging the diameter of the hot wall reactor and the mass flow through the reactor.

Sputtered Plasma Processing

In this method is yet again a variation of the gas-condensation method excepting the fact that the source material is a sputtering target and this target is sputtered using rare gases and the constituents are allowed to agglomerate to produce nano-material. Both dc (direct current) and rf (radio-frequency) sputtering has been used to synthesize nanoparticles. Again reactive sputtering or multitarget sputtering has been used to make alloys and oxides, carbides, nitrides of materials. This method is specifically suitable for the preparation of ultrapure and non-agglomerated nanoparticles of metal.

Microwave Plasma Processing

This technique is similar to the previously discussed CVC method but employs plasma instead of high temperature for decomposition of the metal organic precursors. The method uses microwave plasma in a 50 mm diameter reaction vessel made of quartz placed in a cavity connected to a microwave generator. A precursor such as a chloride compound is introduced into the front end of the reactor. Generally, the microwave cavity is designed as a single mode cavity using the TE10 mode in a WR975 waveguide with a frequency of 0.915 GHz. The major advantage of the plasma assisted pyrolysis in contrast to the thermal activation is the low temperature reaction which reduces the tendency for agglomeration of the primary particles. This is also true in the case of plasma-CVD processes. Additionally, it has been shown that by introducing another precursor into a second reaction zone of the tubular reactor, e.g. by splitting the microwave guide tubes, the primary particles can be coated with a second phase. For example, it has been demonstrated that ZrO_2 nanoparticles can be coated by Al_2O_3. In this case the inner ZrO_2 core is crystalline, while the Al_2O_3 coating is amorphous. The reaction sequence can be reversed with the result that an amorphous Al_2O_3 core is coated with crystalline ZrO_2. While the formation of the primary particles occurs by homogeneous nucleation, it can be easily estimated using gas reaction kinetics that the coating on the primary particles grows heterogeneously and that homogeneous nucleation of nanoparticles originating from the second compound has a very low probability. A schematic representation of the particle growth in plasma's is given below:

Particle Precipitation Aided CVD

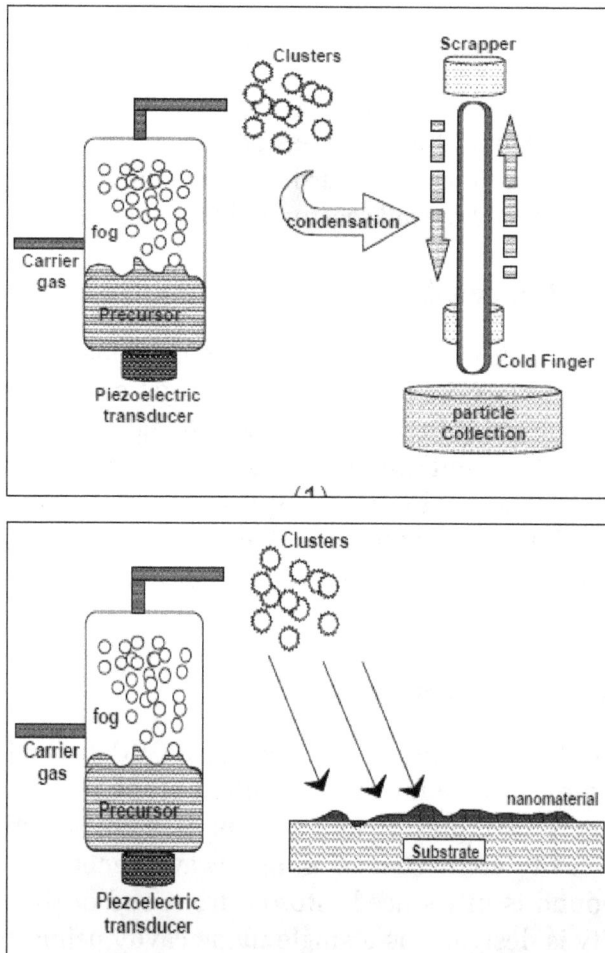

Schematic representation of (1) nanoparticle, and (2) particulate film formation.

In another variation of this process, colloidal clusters of materials are used to prepare nanoparticles. The CVD reaction conditions are so set that particles form by condensation in the gas phase and collect onto a substrate, which is kept under a different condition that allows heterogeneous nucleation. By this method both nanoparticles and particulate films can be prepared. An example of this method has been used to form nanomaterials eg. SnO2, by a method called pyrosol deposition process, where clusters of tin hydroxide are transformed into small aerosol droplets, following which they are reacted onto a heated glass substrate.

Laser Ablation

Laser ablation has been extensively used for the preparation of nanoparticles and particulate films. In this process a laser beam is used as the primary excitation source of ablation for generating clusters directly from a solid sample in a wide variety of applications. The small dimensions of the particles and the possibility to form thick films make this method quite an efficient tool for the production of ceramic particles and coatings and also an ablation source for analytical applications such as the coupling to induced coupled plasma emission spectrometry, ICP, the formation of the nanoparticles has been explained following a liquefaction process which generates an aerosol, followed by the cooling/solidification of the droplets which results in the formation of fog. The general dynamics of both the aerosol and the fog favours the aggregation process and micrometer-sized fractal-like particles are formed. The laser spark atomizer can be used to produce highly mesoporous thick films and the porosity can be modified by the carrier gas flow rate. ZrO_2 and SnO_2 nanoparticulate thick films were also synthesized successfully using this process with quite identical microstructure. Synthesis of other materials such as lithium manganate, silicon and carbon has also been carried out by this technique.

Properties of Nanomaterials

Nanomaterials have the structural features in between of those of atoms and the bulk materials. While most microstructured materials have similar properties to the corresponding bulk materials, the properties of materials with nanometer dimensions are significantly different from those of atoms and bulks materials. This is mainly due to the nanometer size of the materials which render them: (i) large fraction of surface atoms; (ii) high surface energy; (iii) spatial confinement; (iv) reduced imperfections, which do not exist in the corresponding bulk materials.

Due to their small dimensions, nanomaterials have extremely large surface area to volume ratio, which makes a large to be the surface or interfacial atoms, resulting in more "surface" dependent material properties. Especially when the sizes of nanomaterials are comparable to length, the entire material will be affected by the surface properties of nanomaterials. This in turn may enhance or modify the properties of the bulk materials. For example, metallic nanoparticles can be used as very active catalysts. Chemical sensors from nanoparticles and nanowires enhanced the sensitivity and sensor selectivity. The nanometer feature sizes of nanomaterials also have spatial confinement effect on the materials, which bring the quantum effects.

The energy band structure and charge carrier density in the materials can be modified quite differently from their bulk and in turn will modify the electronic and optical properties of the materials. For example, lasers and light emitting diodes (LED) from both of the quantum dots and quantum wires are very promising in the future opto-elections. High density information storage using quantum dot devices is also a fast developing area. Reduced imperfections are also an important factor in determination of the properties of the nanomaterials. Nanosturctures and Nanomaterials favors of a selfpurification process in that the impurities and intrinsic material defects will move to near the surface upon thermal annealing. This increased materials perfection affects the properties of nanomaterials. For example, the chemical stability for certain nanomaterials may be enhanced, the mechanical properties of nanomaterials will be better than the bulk materials. The superior mechanical properties of carbon nanotubes are well known. Due to their nanometer size, nanomaterials are already known to have many novel properties. Many novel applications of the nanomaterials rose from these novel properties have also been proposed.

Optical Properties

One of the most fascinating and useful aspects of nanomaterials is their optical properties. Applications based on optical properties of nanomaterials include optical detector, laser, sensor, imaging, phosphor, display, solar cell, photocatalysis, photoelectrochemistry and biomedicine.

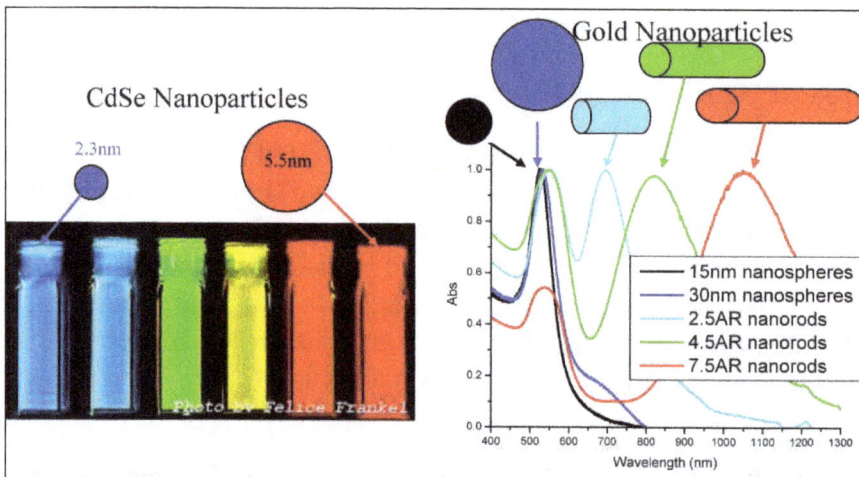

Fluorescence emission of (CdSe) ZnS quantum dots of various sizes
and absorption spectra of various sizes and shapes of gold nanoparticles.

The optical properties of nanomaterials depend on parameters such as feature size, shape, surface characteristics, and other variables including doping and interaction with the surrounding environment or other nanostructures. Likewise, shape can have dramatic influence on optical properties of metal nanostructures. Figure With the CdSe semiconductor nanoparticles, a simple change in size alters the optical properties of the nanoparticles. When metal nanoparticles are enlarged, their optical properties change only slightly as observed for the different samples of gold nanospheres in fig. However, when an anisotropy is added to the nanoparticle, such as growth of nanorods, the optical properties of the nanoparticles change dramatically.

Electrical Properties

"Electrical Properties of Nanoparticles" discuss about fundamentals of electrical conductivity in nanotubes and nanorods, carbon nanotubes, photoconductivity of nanorods, electrical conductivity of nanocomposites. One interesting method which can be used to demonstrate the steps in conductance is the mechanical thinning of a nanowire and measurement of the electrical current at a constant applied voltage. The important point here is that, with decreasing diameter of the wire, the number of electron wave modes contributing to the electrical conductivity is becoming increasingly smaller by well-defined quantized steps.

In electrically conducting carbon nanotubes, only one electron wave mode is observed which transport the electrical current. As the lengths and orientations of the carbon nanotubes are different, they touch the surface of the mercury at different times, which provides two sets of information: (i) the influence of carbon nanotube length on the resistance; and (ii) the resistances of the different nanotubes. As the nanotubes have different lengths, then with increasing protrusion of the fiber bundle an increasing number of carbon nanotubes will touch the surface of the mercury droplet and contribute to the electrical current transport.

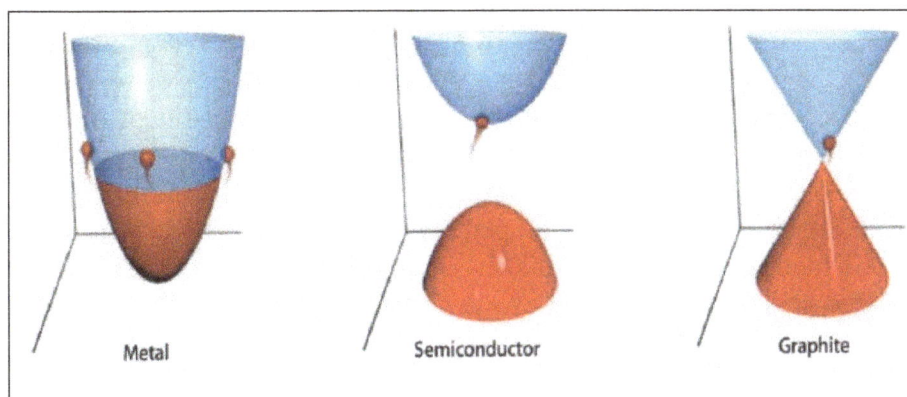

Metal   Semiconductor   Graphite

Electrical behavior of naotubes.

Mechanical Properties

"Mechanical Properties of Nanoparticles" deals with bulk metallic and ceramic materials, influence of porosity, influence of grain size, superplasticity, filled polymer composites, particle-filled polymers, polymer-based nanocomposites filled with plate-lets, carbon nanotube-based composites. The discussion of mechanical properties of nanomaterials is, in to some extent, only of quite basic interest, the reason being that it is problematic to produce macroscopic bodies with a high density and a grain size in the range of less than 100 nm. However, two materials, neither of which is produced by pressing and sintering, have attracted much greater interest as they will undoubtedly achieve industrial importance.

These materials are polymers which contain nanoparticles or nanotubes to improve their mechanical behaviors, and severely plastic-deformed metals, which exhibit aston-ishing properties. However, because of their larger grain size, the latter are generally not accepted as nanomaterials. Experimental studies on the mechanical properties of bulk nanomaterials are generally impaired by major experimental problems in produc-ing specimens with exactly defined grain sizes and porosities. Therefore, model calcu-lations and molecular dynamic studies are of major importance for an understanding of the mechanical properties of these materials.

Filling polymers with nanoparticles or nanorods and nanotubes, respectively, leads to significant improvements in their mechanical properties. Such improvements depend heavily on the type of the filler and the way in which the filling is conducted. The latter point is of special importance, as any specific advantages of a nanoparticulate filler may be lost if the filler forms aggregates, thereby mimicking the large particles. Particulate-filled polymer-based nanocomposites exhibit a broad range of failure strengths and strains. This depends on the shape of the filler, particles or platelets, and on the degree of agglomeration. In this class of material, polymers filled with silicate platelets exhibit the best mechanical properties and are of the greatest economic relevance. The larger the particles of the filler or agglomerates, the poorer are the properties obtained. Al-though, potentially, the best composites are those filled with nanofibers or nanotubes, experience teaches that sometimes such composites have the least ductility. On the oth-er hand, by using carbon nanotubes it is possible to produce composite fibers with ex-tremely high strength and strain at rupture. Among the most exciting nanocomposites are the polymerceramic nanocomposites, where the ceramic phase is platelet-shaped. This type of composite is preferred in nature, and is found in the structure of bones, where it consists of crystallized mineral platelets of a few nanometers thickness that are bound together with collagen as the matrix. Composites consisting of a polymer matrix and defoliated phyllosilicates exhibit excellent mechanical and thermal properties.

Magnetic Properties

Bulk gold and Pt are non-magnetic, but at the nano size they are magnetic. Surface at-oms are not only different to bulk atoms, but they can also be modified by interaction with other chemical species, that is, by capping the nanoparticles. This phenomenon opens the possibility to modify the physical properties of the nanoparticles by capping them with appropriate molecules. Actually, it should be possible that non-ferromagnetic bulk materials exhibit ferromagnetic-like behavior when prepared in nano range. One can obtain magnetic nanoparticles of Pd, Pt and the surprising case of Au (that is dia-magnetic in bulk) from non-magnetic bulk materials. In the case of Pt and Pd, the fer-romagnetism arises from the structural changes associated with size effects.

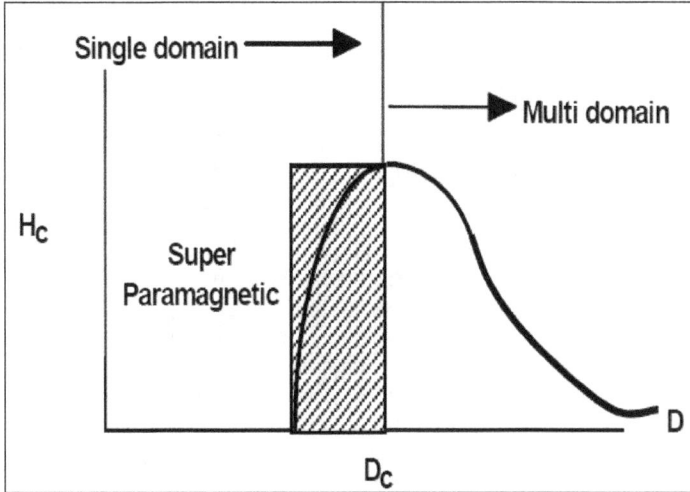

Magnetic properties of nanostrucutred materials.

However, gold nanoparticles become ferromagnetic when they are capped with appro-priate molecules: the charge localized at the particle surface gives rise to ferromagnet-ic-like behavior.

Surface and the core of Au nanoparticles with 2 nm in diameter show ferromagnetic and paramagnetic character, respectively. The large spin-orbit coupling of these noble metals can yield to a large anisotropy and therefore exhibit high ordering tempera-tures. More surprisingly, permanent magnetism was observed up to room temperature for thiol-capped Au nanoparticles. For nanoparticles with sizes below 2 nm the local-ized carriers are in the 5d band. Bulk Au has an extremely low density of states and becomes diamagnetic, as is also the case for bare Au nanoparticles. This observation suggested that modification of the d band structure by chemical bonding can induce ferromagnetic like character in metallic clusters.

Fuel Cells

A fuel cell is an electrochemical energy conversion device that converts the chemical energy from fuel (on the anode side) and oxidant (on the cathode side) directly into electricity. The heart of fuel cell is the electrodes. The performance of a fuel cell elec-trode can be optimized in two ways; by improving the physical structure and by using more active electro catalyst. A good structure of electrode must provide ample surface area, provide maximum contact of catalyst, reactant gas and electrolyte, facilitate gas transport and provide good electronic conductance. In this fashion the structure should be able to minimize losses.

Carbon Nanotubes - Microbial Fuel Cell

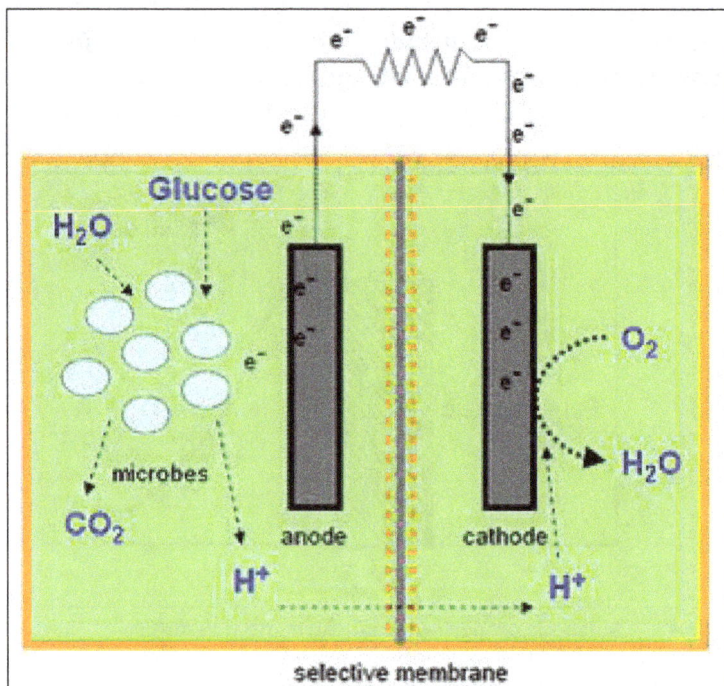

Schematic representation of microbial fuel cell.

Microbial fuel cell is a device in which bacteria consume water-soluble waste such as sugar, starch and alcohols and produces electricity plus clean water. This technology will make it possible to generate electricity while treating domestic or industrial wastewater. Microbial fuel cell can turn different carbohydrates and complex substrates present in wastewaters into a source of electricity. The efficient electron transfer between the microorganism and the anode of the microbial fuel cell plays a major role in the performance of the fuel cell. The organic molecules present in the wastewater posses a certain amount of chemical energy, which is released when converting them to simpler molecules like CO2. The microbial fuel cell is thus a device that converts the chemical energy present in water-soluble waste into electrical energy by the catalytic reaction of microorganisms.

Carbon nanotubes (CNTs) have chemical stability, good mechanical properties and high surface area, making them ideal for the design of sensors and provide very high surface area due to its structural network. Since carbon nanotubes are also suitable supports for cell growth, electrodes of microbial fuel cells can be built using of CNT. Due to three-dimensional architectures and enlarged electrode surface area for the entry of growth medium, bacteria can grow and proliferate and get immobilized. Multi walled CNT scaffolds could offer self-supported structure with large surface area through which hydrogen producing bacteria (e.g., E. coli) can eventually grow and proliferate. Also CNTs and MWCNTs have been reported to be biocompatible for different eukaryotic cells. The efficient proliferation of hydrogen producing bacteria throughout an electron conducting scaffold of CNT can form the basis for the potential application as electrodes in MFCs leading to efficient performance.

Catalysis

Higher surface area available with the nanomaterial counterparts, nano-catalysts tend to have exceptional surface activity. For example, reaction rate at nano-aluminum can go so high, that it is utilized as a solid-fuel in rocket propulsion, whereas the bulk aluminum is widely used in utensils. Nano-aluminum becomes highly reactive and supplies the required thrust to send off pay loads in space. Similarly, catalysts assisting or retarding the reaction rates are dependent on the surface activity, and can very well be utilized in manipulating the rate-controlling step.

Phosphors for High-definition TV

The resolution of a television, or a monitor, depends greatly on the size of the pixel. These pixels are essentially made of materials called "phosphors," which glow when struck by a stream of electrons inside the cathode ray tube (CRT). The resolution improves with a reduction in the size of the pixel, or the phosphors. Nanocrystalline zinc selenide, zinc sulfide, cadmium sulfide, and lead telluride synthesized by the sol-gel techniques are candidates for improving the resolution of monitors. The use of nano-phosphors is envisioned to reduce the cost of these displays so as to render highdefinition televisions (HDTVs) and personal computers affordable to be purchase.

Next-generation Computer Chips

The microelectronics industry has been emphasizing miniaturization, whereby the circuits, such as transistors, resistors, and capacitors, are reduced in size. By achieving a significant reduction in their size, the microprocessors, which contain these components, can run much faster, thereby enabling computations at far greater speeds. However, there are several technological impediments to these advancements, including lack of the ultrafine precursors to manufacture these components; poor dissipation of tremendous amount of heat generated by these microprocessors due to faster speeds; short mean time to failures (poor reliability), etc. Nanomaterials help the industry break these barriers down by providing the manufacturers with nanocrystalline starting materials, ultra-high purity materials, materials with better thermal conductivity, and longer-lasting, durable interconnections (connections between various components in the microprocessors).

Example: Nanowires for junctionless transistors.

Transistors are made so tiny to reduce the size of sub assemblies of electronic systems and make smaller and smaller devices, but it is difficult to create high-quality junctions. In particular, it is very difficult to change the doping concentration of a material over distances shorter than about 10 nm. Researchers have succeeded in making the junctionless transistor having nearly ideal electrical properties. It could potentially operate faster and use less power than any conventional transistor on the market today. The device consists of a silicon nanowire in which current flow is perfectly controlled by a silicon gate that is separated from the nanowire by a thin insulating layer. The entire silicon nanowire is heavily n-doped, making it an excellent conductor. However, the gate is p-doped and its presence has the effect of depleting the number of electrons in the region of the nanowire under the gate. The device also has near-ideal electrical properties and behaves like the most perfect of transistors without suffering from current leakage like conventional devices and operates faster and using less energy.

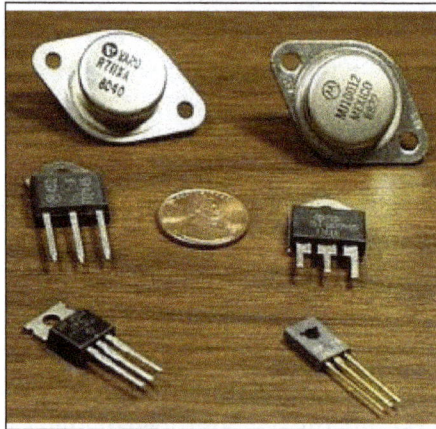

Silicon nanowires in junctionless transistors.

Elimination of Pollutants

Nanomaterials possess extremely large grain boundaries relative to their grain size. Hence, they are very active in terms of their chemical, physical, and mechanical properties. Due to their enhanced chemical activity, nanomaterials can be used as catalysts to react with such noxious and toxic gases as carbon monoxide and nitrogen oxide in automobile catalytic converters and power generation equipment to prevent environmental pollution arising from burning gasoline and coal.

Sun-screen Lotion

Prolonged UV exposure causes skin-burns and cancer. Sun-screen lotions containing nano-TiO_2 provide enhanced sun protection factor (SPF) while eliminating stickiness. The added advantage of nano skin blocks (ZnO and TiO_2) arises as they protect the skin by sitting onto it rather than penetrating into the skin. Thus they block UV radiation effectively for prolonged duration. Additionally, they are transparent, thus retain natural skin color while working better than conventional skin-lotions.

Sensors

Sensors rely on the highly active surface to initiate a response with minute change in the concentration of the species to be detected. Engineered monolayers (few Angstroms thick) on the sensor surface are exposed to the environment and the peculiar functionality (such as change in potential as the CO/anthrax level is detected) is utilized in sensing.

Disadvantages of Nanomaterials

- Instability of the particles: Retaining the active metal nanoparticles is highly challenging, as the kinetics associated with nanomaterials is rapid. In order to retain nanosize of particles, they are encapsulated in some other matrix. Nanomaterials are thermodynamically metastable and lie in the region of high-energy local-minima. Hence they are prone to attack and undergo transformation. These include poor corrosion resistance, high solubility, and phase change of nanomaterials. This leads to deterioration in properties and retaining the structure becomes challenging.

- Fine metal particles act as strong explosives owing to their high surface area coming in direct contact with oxygen. Their exothermic combustion can easily cause explosion.

- Impurity: Because nanoparticles are highly reactive, they inherently interact with impurities as well. In addition, encapsulation of nanoparticles becomes necessary when they are synthesized in a solution (chemical route). The stabilization of nanoparticles occurs because of a non-reactive species engulfing the reactive nano-entities. Thereby, these secondary impurities become a part of the synthesized nanoparticles, and synthesis of pure nanoparticles becomes highly difficult. Formation of oxides, nitrides, etc can also get aggravated from the impure environment/surrounding while synthesizing nanoparticles. Hence retaining high purity in nanoparticles can become a challenge hard to overcome.

- Biologically harmful: Nanomaterials are usually considered harmful as they become transparent to the cell-dermis. Toxicity of nanomaterials also appears predominant owing to their high surface area and enhanced surface activity. Nanomaterials have shown to cause irritation, and have indicated to be carcinogenic. If inhaled, their low mass entraps them inside lungs, and in no way they can be expelled out of body. Their interaction with liver/blood could also prove to be harmful (though this aspect is still being debated on).

- Difficulty in synthesis, isolation and application: It is extremely hard to retain the size of nanoparticles once they are synthesized in a solution. Hence, the nanomaterials have to be encapsulated in a bigger and stable molecule/material. Hence free nanoparticles are hard to be utilized in isolation, and they have to be interacted for intended use via secondary means of exposure. Grain growth is inherently present in nanomateirals during their processing. The finer grains tend to merge and become bigger and stable grains at high temperatures and times of processing.

- Recycling and disposal: There are no hard-and-fast safe disposal policies evolved for nanomaterials. Issues of their toxicity are still under question, and results of exposure experiments are not available. Hence the uncertainty associated with affects of nanomaterials is yet to be assessed in order to develop their disposal policies.

Synthesis of Nanomaterials

Materials scientists are conducting research to develop novel materials with better properties, more functionality and lower cost than the existing one. Several physical, chemical methods have been developed to enhance the performance of nanomaterials displaying improved properties with the aim to have a better control over the particle size, distribution.

Methods to Synthesis of Nanomaterials

In general, top-down and bottom-up are the two main approaches for nanomaterials synthesis:

- Top-down: Size reduction from bulk materials.

- Bottom-up: Material synthesis from atomic level.

Top-down routes are included in the typical solid state processing of the materials. This route is based with the bulk material and makes it smaller, thus breaking up larger particles by the use of physical processes like crushing, milling or grinding. Usually this route is not suitable for preparing uniformly shaped materials, and it is very difficult to realize very small particles even with high energy consumption. The biggest problem with top-down approach is the imperfection of the surface structure. Such imperfection would have a significant impact on physical properties and surface chemistry of nanostructures and nanomaterials. It is well known that the conventional top-down technique can cause significant crystallographic damage to the processed patterns.

Bottom-up approach refers to the build-up of a material from the bottom: atom-by-atom, molecule-by-molecule or cluster-by-cluster. This route is more often used for preparing most of the nano-scale materials with the ability to generate a uniform size, shape and distribution. It effectively covers chemical synthesis and precisely controlled the reaction to inhibit further particle growth. Although the bottom-up approach is nothing new, it plays an important role in the fabrication and processing of nanostructures and nanomaterials.

Synthesis of nanoparticles to have a better control over particles size distribution, morphology, purity, quantity and quality, by employing environment friendly economical processes has always been a challenge for the researchers. The choice of synthesis technique can be a key factor in determining the effectiveness of the photovoltaic as studies. There are many methods of synthesizing titanium dioxide, such as hydro-thermal, combustion synthesis, gas-phase methods, microwave synthesis and sol-gel processing.

Hydrothermal Synthesis

Hydrothermal synthesis is typically carried out in a pressurised vessel called an auto-clave with the reaction in aqueous solution. The temperature in the autoclave can be raised above the boiling point of water, reaching the pressure of vapour saturation. Hydrothermal synthesis is widely used for the preparation of TiO_2 nanoparticles which can easily be obtained through hydrothermal treatment of peptised precipitates of a titanium precursor with water. The hydrothermal method can be useful to control grain size, particle morphology, crystalline phase and surface chemistry through regulation of the solution composition, reaction temperature, pressure, solvent properties, addi-tives and aging time.

Solvothermal Method

The Solvothermal method is identical to the hydrothermal method except that a variety of solvents other than water can be used for this process. This method has been found to be a versatile route for the synthesis of a wide variety of nanoparticles with narrow size distributions, particularly when organic solvents with high boiling points are chosen. The solvothermal method normally has better control of the size and shape distributions and the crystallinity than the hydrothermal method, and has been employed to synthesize TiO_2 nanoparticles and nanorods with/without the aid of surfactants.

Chemical Vapor Deposition (CVD)

This process is often used in the semiconductor industry to produce high-purity, high-performance thin films. In a typical CVD process, the substrate is exposed to volatile precursors, which react and decompose on the substrate surface to produce the desired film. Frequently, volatile by products that are produced are removed by gas flow through the reaction chamber. The quality of the deposited materials strong-ly depends on the reaction temperature, the reaction rate, and the concentration of the precursors. Cao et al. prepared Sn_4^+-doped TiO_2 nanoparticles films by the CVD method and found that more surface defects were present on the surface due to dop-ing with Sn. Gracia et al. synthesized M (Cr, V, Fe, Co)-doped TiO_2 by CVD and found that TiO_2 crystallized into the anatase or rutile structures depending on the type and amount of cation present in the synthesis process. Moreover, upon annealing, partial segregation of the cations in the form of M_2O_n was observed. The advantages of this method include the uniform coating of the nanoparticles or nano film. However, this process has limitations including the higher temperatures required, and it is difficult to scaleup.

Thermal Decomposition and Pulsed Laser Ablation

Pure and doped metal nanomaterials can be synthesized via decomposing metal alkoxides and salts by applying high energy using heat or electricity. However, the properties of the produced nanomaterials strongly depend on the precursor concentrations, the flow rate of the precursors and the environment. Kim et al. synthesized TiO_2 nanoparticles with a diameter less than 30 nm via the thermal decomposition of titanium alkoxide or $TiCl_4$ at 1200°C. Liang et al produced TiO_2 nanoparticles with a diameter ranging from 3 to 8 nm by pulsed laser ablation of a titanium target immersed in an aqueous solution of surfactant or deionized water. Nagaveni et al prepared W, V, Ce, Zr, Fe, and Cu ion-doped anatase TiO_2 nanoparticles by a solution combustion method and found that the solid solution formation was limited to a narrow range of concentrations of the dopant ions. However, the drawbacks of these methods are high cost and low yield, and difficulty in controlling the morphology of the synthesized nanomaterials.

Templating

The synthesis of nanostructure materials using the template method has become extremely popular during the last decade. In order to construct materials with a similar morphology of known characterized materials (templates); this method utilizes the morphological properties with reactive deposition or dissolution. Therefore, it is possible to prepare numerous new materials with a regular and controlled morphology on the nano and microscale by simply adjusting the morphology of the template material. A variety of templates have been studied for synthesizing titania nanomaterials. This method has some disadvantages including the complicated synthetic procedures and, in most cases, templates need to be removed, normally by calcinations, leading to an increase in the cost of the materials and the possibility of contamination.

Combustion

Combustion synthesis leads to highly crystalline particles with large surface areas. The process involves a rapid heating of a solution containing redox groups. During combustion, the temperature reaches approximately 650°C for one or two minutes making the material crystalline. Since the time is so short, the transition from anatase to rutile is inhibited.

Gas Phase Methods

Gas phase methods are ideal for the production of thin films. Gas phase can be carried out chemically or physically. Chemical Vapour Deposition (CVD) is a widely used industrial technique that can coat large areas in a short space of time. During the procedure, titanium dioxide is formed from a chemical reaction or decomposition of a precursor in the gas phase. Physical vapour deposition (PVD) is another thin film deposition technique. Films are formed from the gas phase but without a chemical transition from precursor to product. For TiO_2 thin films, a focused beam of electrons heats the titanium dioxide material. The electrons are produced from a tungsten wire heated by a current. This is known as Electron beam (E-beam) evaporation. Titanium dioxide films deposited with E-beam evaporation have superior characteristics over CVD grown films such as, smoothness, conductivity, presence of contaminations and crystallinity. Reduced TiO_2 powder (heated at 900°C in a hydrogen atmosphere) is necessary for the required conductance needed to focus an electron beam on the TiO_2.

Microwave Synthesis

Various TiO_2 materials have been synthesised using microwave radiation. Microwave techniques eliminate the use of high temperature calcination for extended periods of time and allow for fast, reproducible synthesis of crystalline TiO_2 nanomaterials. Corradi et al prepared colloidal TiO_2 nanoparticle suspensions within 5 minutes using microwave radiation. High quality rutile rods were developed combining hydrothermal and microwave synthesis, while TiO_2 hollow, open ended nanotubes were synthesised through reacting anatase and rutile crystals in NaOH solution.

Conventional Sol-gel Method

The sol-gel method is a versatile process used for synthesizing various oxide materials. This synthetic method generally allows control of the texture, the chemical, and the morphological properties of the solid. This method also has several advantages over other methods, such as allowing impregnation or coprecipitation, which can be used to introduce dopants. The major advantages of the sol-gel technique includes molecular scale mixing, high purity of the precursors, and homogeneity of the solgel products with a high purity of physical, morphological, and chemical properties. In a typical sol-gel process, a colloidal suspension, or a sol, is formed from the hydrolysis and polymerization reactions of the precursors, which are usually inorganic metal salts or metal organic compounds such as metal alkoxides. A general flowchart for a complete sol-gel process is shown in figure.

Any factor that affects either or both of these reactions is likely to impact the properties of the gel. These factors, generally referred to as sol-gel parameters, includes type of precursor, type of solvent, water content, acid or base content, precursor concentration, and temperature. These parameters affect the structure of the initial gel and, in turn, the properties of the material at all subsequent processing steps.

After gelation, the wet gel can be optionally aged in its mother liquor, or in another solvent, and washed. The time between the formation of a gel and its drying, known as aging, is also an important parameter. A gel is not static during aging but can continue to undergo hydrolysis and condensation.

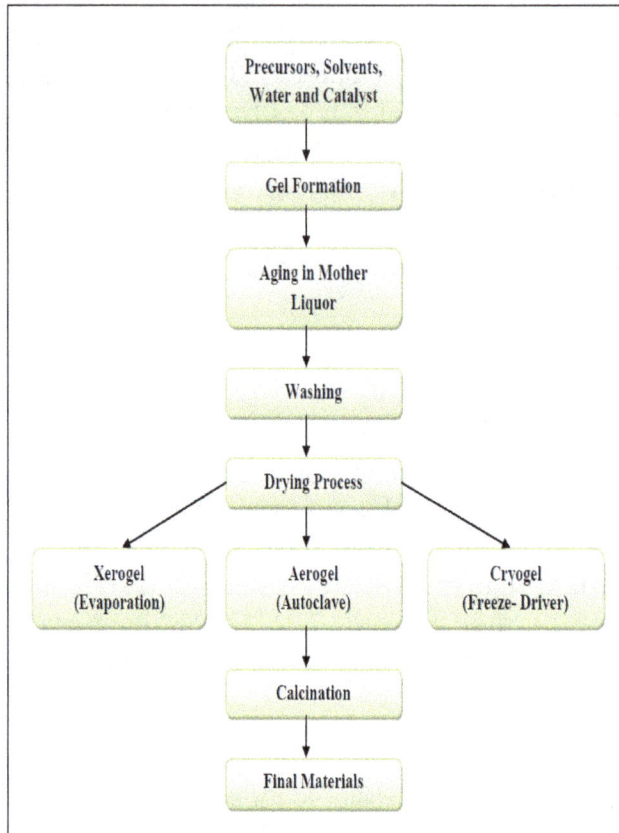

Sol-Gel and Drying Flowchart.

Table: Important Parameters in the Various Steps of a Sol- Gel Process.

Step	Purpose	Important Parameters
Solution Chemistry	To form gel.	Type of Precursor, Type of Solvent, Water Content, Precursor Concentration, Temperature, pH.
Aging	To allow a gel undergo changes in properties.	Time, Temperature, Composition of the pore liquid, Aging environment.
Drying	To remove solvent from a gel.	Drying method (evaporative super critical and freeze drying),Temperature and heating rate, Pressure and pressurization rate, Time.
Calcination	To change the physical /chemical properties of the solid, often resulting in crystallization and densification.	Temperature and heating rate, Time, Gaseous environment (inert, reactive gases).

Furthermore, syneresis, which is the expulsion of solvent due to gel shrinkage, and coarsening, which is the dissolution and reprecipitation of particles, can occur. These phenomena can affect both the chemical and structural properties of the gel after its initial formation. Then it must be dried to remove the solvent. Table showed a summary of the key steps in a sol-gel process which includes the aim of each step along with experimental parameters that can be manipulated.

Aerogel

One important parameter that affects a sol-gel product is the drying condition. Due to the surface tension of the liquid, a capillary pressure gradient is present in the pore walls and this may be able to collapse most of the pore volume when solvent is removed. One convenient way to avoid pore collapse is to remove the liquid from the pores above the critical temperature (T_c) and critical pressure (P_c) of the fluid, namely, supercritical drying . Under supercritical conditions, there is no longer a distinction between the liquid and vapor phases: the densities become equal; there is no liquid-vapor interface and no capillary pressure. This type of drying prevents the formation a liquid-vapor meniscus which recedes during the emptying of the pores in the wet gels. The resulting dried gel, called an aerogel, has a pore volume similar to that of the wet gel.

Xerogel

Conventional evaporative drying induces capillary pressure associated with the liquid vapor interface within a pore, causing shrinkage of the gel network. In a sample with a distribution of pore sizes, the resultant differential capillary pressure often collapses the porous network during drying. The dried sample often has low surface area and pore volume.

Cryogel

Another way of avoiding the presence of liquid-vapor interface is to freeze the pore liquid and sublime the resulting solid under vacuum. In this method, the gel liquid is first frozen and thereafter dried by sublimation. Therefore, the formation of a liquid-vapor meniscus is prevented. The materials obtained are then also termed cryogels. Their surface area and mesopore volume tend to be smaller than those of aerogels, although they remain significant. However, freeze-drying does not permit the preparation of monolithic gels. The reason is that the growing crystals reject the gel network, pushing it out of the way until it is stretched to the breaking point. It is this phenomenon that allows gels to be used as hosts for crystal growth: the gel is so effectively excluded that crystals nucleated in the pore liquid are not contaminated with the gel phase; the crystals can grow up to a size of a few millimetres before the strain is so great that macroscopic fractures appear in the gel. Nevertheless, the gel network may eventually be destroyed by the nucleation and growth of solvent crystals, which tend to produce very large pores. To attenuate this event, a rapid freeze process known as flash freezing has been developed. It is also important that the solvent has a low expansion coefficient and a high pressure of sublimation.

Applications of Sol-gel Method

Applications for sol-gel process derive from the various special shapes obtained directly from the gel state (monoliths, films, fibers, and monosized powders) combined with compositional and microstructural control and low processing temperatures. Compared with other methods, such as the solid-state method, the advantages of using sol-gel process include:

- The use of synthetic chemicals rather than minerals enables high purity materials to be synthesized.

- It involves the use of liquid solutions as mixtures of raw materials. Since the mixing is with low viscosity liquids, homogenization can be achieved at a molecular level in a short time.

- Since the precursors are well mixed in the solutions, they are likely to be equally well-mixed at the molecular level when the gel is formed; thus on heating the gel, chemical reaction will be easy and at a low temperature.

- Changing physical characteristics such as pore size distribution and pore volume can be achieved.

- Incorporating multiple components in a single step can be achieved.

- Producing different physical forms of samples is manageable.

Experimental Procedures and Characterization Techniques

Chemicals Used

Most of the chemicals used in the research are standard chemicals that are normally available in the laboratory. Special materials for DSSC are mostly bought from Solaronix. Table shows the list of materials used in this research.

Table: List of materials used.

Chemicals	Manufacturer	Purity	Usage
Titanium (IV)isopropoxide	Sigma-Aldrich	98%	Nanoparticles
Ethanol	Changshu Yangyuan Chemical	99.9%	Solvent
Acetic acid	Merck	99 %	Catalysts
DI-water	-	-	Hydrolysis
Acetone	Merk	97 %	Clean substrate
Methanol	Molychem	99.8%	Clean substrate
Isopropanol	Fisher Scientific	99 %	Clean substrate
Acetylacetone	Sigma - Aldrich	39.5%	Binder
FTO-glass	Solaronix, Spektron	-	Substrate

DMF Solvent	Merck	98 %	Solvent
Triton X-100	Sigma-Aldrich	-	Surfactant
Polyethylene glycol(600)	Merck	-	Surfactant
N719 dye	Solaronix	-	Sensitizer
Rhodamine dye	Sigma - Aldrich	-	Sensitizer
Coumarin 30 B	Sigma - Aldrich	-	Sensitizer
Plasitol	Solaronix	-	electrolyte
Idolyte TG 50	Solaronix	-	electrolyte
Aluminium nitrate	Loba Chemie	98 %	Dopant Material
Silver nitrate	Qualigens	99.9 %	Dopant Material
Magnesium nitrate	Loba Chemie	98 %	Dopant Material
Nickel nitrate	Merck	97 %	Dopant Material
Chromium nitrate	Himedia	98 %	Dopant Material

Synthesis of Titanium Dioxide Nanopowders

The nano-TiO_2 powder was prepared with titanium isopropoxide solution as the raw material. In a typical experiment, 60 ml of deionized water and 5 ml of glacial acetic acid were dissolved at room temperature to obtain solution A. 14 ml titanium isopropoxide was dissolved in 40 ml of anhydrous ethanol with constant stirring to form solution B. Then, the solution B was added drop-wise into the solution A within 60 min under vigorous stirring. Subsequently, the obtained sol wasstirred continuously for 2 h and aged for 48 h at room temperature. As-prepared TiO_2 gels were dried for 10h at 80°C. The obtained solids were ground and finally calcinated at 450°C for 2 h (heating rate = 3°C/min).

Substrate Cleaning

Coated glass with highly F-doped Transparent Conducting Oxide (TCO) usually serves as a support for the dye-sensitized oxide. It allows light transmission while providing good conductivity for current collection. The substrates are first dipped into Acetone with ultrasonic bath for 15 minutes to dissolve unwanted organic materials and to remove dust and contamination material that are left on the substrates post manufacture. Another 15 minutes of ultrasonic bath in methanol is followed in order to remove the acetone and materials that are not cleansed or dissolved by acetone. Finally, a 10 minute ultrasonic bath in isopropoemal was needed to further remove the residual particles on the substrates.

TiO_2 Photoanode Deposition on FTO

It is very important to work with a fingerprint free Transparent Conducting Oxide (TCO), always gloves were used and TCO was cleaned with alcohol prior to use. TCO was heated to 50°C at the beginning of the process to increase the adhesion and Scotch 3M adhesive tape were applied on the edges of the conductive side of the TCO glass. The reason for applying tapes was preparing a mould such that nonsintered TiO_2 has always same area and thickness for all samples. A certain proportions of nano-TiO_2 powder with ethanol, acetylacetone (A.R), polyethylene glycol and triton(X-100) were mixing for 30 minutes in agate mortar. Then TiO_2 colloidal was dropped on the conductive side of the TCO after the conductive side of the TCO was checked by the multimeter. Then, the TiO_2 paste was uniformly distributed over the TCO by Doctor Blade method. Doctor blade means a film smoothing method using any steel, rubber, plastic, or other type of blade used to apply or remove a liquid substance from another surface.

The term "doctor blade" is derived from the name of a blade used in conjunction with the ductor roll on the letter press. The term "ductor blade" eventually mutates into the term "doctor blade".

Heat Treatment for Photoanode

The tapes were removed from the glass and plates were sintered at 450°C for 30 min in air above the TiO_2 coated TCO was required after TiO_2 material deposition. Colour of TiO_2 becomes brown in the middle of the sintering process and then its colour changes to the brownish-white. This colour remained till the end of the sintering process. This is to ensure that the polymer or macromolecules in TiO_2 colloid such as Acetylacetone can be removed, leaving tiny holes in TiO_2 layers, resulting in better dye absorption and better contact between TiO_2 particles. In consequence, it optimizes the chances of electrons being excited by the photons and increases the amount of excited electrons entering into the TiO_2 conduction band.

Preparation of Dye Solution

The dye solution used in this research was N719, Rhodamine B and Coumarin 30B dyes are commonly used in DSSC laboratories. The material is normally available in powder form from commercial companies and dissolved in chemicals before use. The samples need to be fully covered with dye solution for 24 hours for the dye particles to be fully absorbed. Due to the fact that dyes are the light absorbing material, it needs to be store in the dark, preventing the loss of functionality and the samples need to be rinsed with different solvent medium.

Heat Treatment of FTO Coated TiO2 (a) Before and (b) After.

Preparation of Counter Electrode

The platinization procedure given by Solaronix was applied because the material was taken from Solaronix. Actually, this method is simply called thermal decomposition which is most widely used platinization procedure. Plasitol was applied on the surface by using a brush. All TCO glasses were sintered at 400°C for 5 minutes for decomposition which was the minimum required calcination condition according to the procedure.

Nanocrystalline Dye Sensitized Solar Cell Assembly

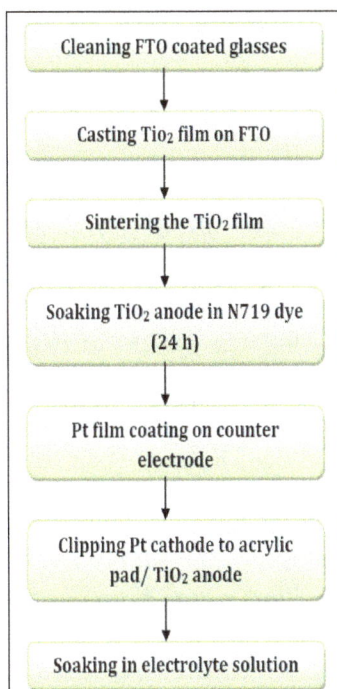

Cleaning FTO coated glasses

Casting Tio₂ film on FTO

Sintering the TiO₂ film

Soaking TiO₂ anode in N719 dye (24 h)

Pt film coating on counter electrode

Clipping Pt cathode to acrylic pad/ TiO₂ anode

Soaking in electrolyte solution

Preparation of Dye Sensitized Solar Cell.

Sensitized TiO_2 photo-anode and the counter electrode were stacked together face to face and the liquid electrolyte, Idolyte TG 50 solution drop penetrated into the working space and counter electrode via capillary action. The two electrodes were held with binder clips. The flow chart of preparation of dye sensitized solar cell and schematic diagram was shown in figure.

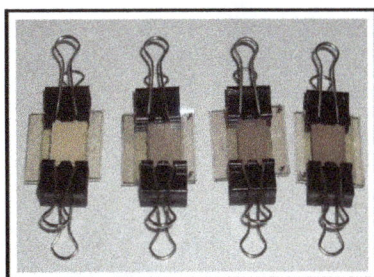

Schematic Diagram of Dye Sensitized Solar Cells.

Characterization

The synthesized nanomaterials are characterized by the following analytical tools which are described in detailed:

- XRD analysis.

- UV-Vis spectroscopy.

- Field Emission Scanning Electron Microscopy.

- Energy Dispersive X-Ray Spectroscopy.

- Photoluminescence spectroscopy.

XRD Analysis

Solid materials are formed by atoms or atomic group arranged in certain way. When an x-ray beam is injected into the material, it would be scattered by atoms. If two or more x-ray beams scattered by the atoms that have some phase differences are superimposed onto each other, diffraction is occurred. The x-ray diffraction instrument (SHIMAD-ZU-6000 Model) is used to collect the intensities of the scattered signals to get the diffraction pattern of the measured sample. This pattern is normally as the signal intensity versus the phase angle. When such a pattern is used in the crystal surface calibration process for the sample, the material's crystalline structure, such as the orientation and phase angles, can be obtained. Identification of the phases was made with the help of the Joint Committee on Powder Diffraction Standards (JCPDS) files. An advantage for using X- ray diffraction measurement is that it can analyse the material without causing damages on the material.

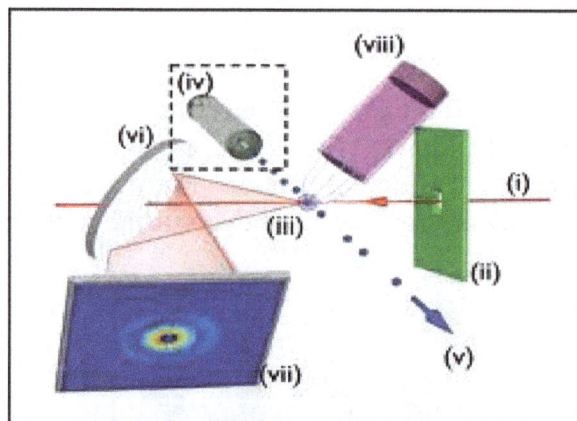

Photograph of a Typical XRD Diffractometer.

UV-Vis Spectroscopy

Many molecules absorb ultraviolet (UV) or visible light. The absorption of UV or visible radiation is caused by the excitation of outer electrons, from their ground state to an excited state. The Bouguer-Lambert-Beer law forms the mathematical physical basis for the light absorption measurements on gas and in solution. According to this law, absorbance is directly proportional to the path length l, and the concentration of the absorbing substance, c, and can be expressed as A = εbc, where ε is a constant of proportionality, called the absorbtivity. In addition, absorption strongly depends on the types of samples, and the environment of the sample. For instance, molecules absorb radiation of various wavelengths depending on the structural groups present within the molecules, and show a number of absorption bands in the absorption spectrum. The solvent in which the absorbing species is dissolved also has an effect on the spectrum of the species. Moreover, the size of the particle is also important. If the size of the particle d>>λ, light interacts with the samples instead of absorption, with parts of the light scattered and reflected.

When dealing with solid samples, light penetrates into the sample; undergoes numerous reflections, refractions and diffraction and emerges finally diffusely at the surface. The Bouguer Lambert-Beer law cannot handle solid samples, which is based on the assumption that the light intensity is not lost by scattering and reflection processes.

Diffuse reflectance measurements are usually analyzed on the basis of the Kubelka-Munk equation:

$$F\left(R_\infty\right)=\frac{k}{s}\left(1-R\right)^2/2R$$

where k and s are absorption and scattering coefficients respectively, and R is the reflectance at the front face. $F\left(R\infty\right)$ is termed the Kubelka-Munk function and is proportional to the concentration of the adsorbate molecules. From the onset of the plot of Kubelka-Munk function vs wavelength or photoenergy, the band gap energy of a semiconductor can be easily calculated. However, to measure a diffuse reflectance spectrum, the diffusely reflected light must be collected with an integrated sphere, avoiding secularly reflected light, and using a reference standard ($BaSO_4$ or white standard).

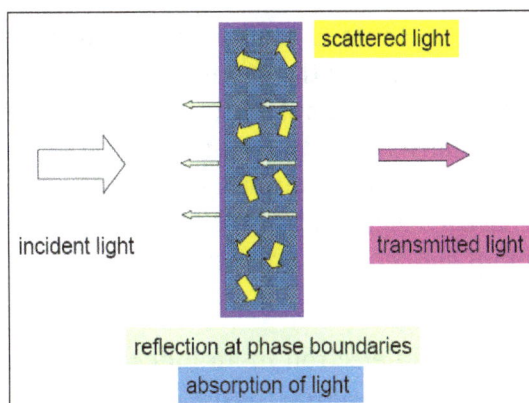

Interaction of Light with Solid Sample.

Field Emission Scanning Electron Microscopy (FESEM)

Electron micrograph images were taken on a FEI QUANTA 200F with a Schottky electron gunas shown in figure. Measurements were carried out at an accelerating voltage range of 5 – 15 kV. Powdered samples were evenly distributed on a mounted carbon tape surface. Loose powdered sample was removed with canned air spray. The Field Emission Scanning Electron Microscope (FESEM) is a type of electron microscope that images the sample surface by scanning it with a high-energy beam of electrons in a raster scan pattern. The electrons interact with the shells in atoms that make up the sample producing signals that contain information about the sample's surface topography, composition and other properties such as electrical conductivity. The types of signals produced by an SEM include Secondary Electrons (SE), Back Scattered Electrons (BSE), characteristic X-rays, light (cathodoluminescence), Specimen current and Transmitted Electrons (STEM). Generally the most common or standard detection mode is SE imaging. The spot size in a Field Emission SEM is smaller than in conventional SEM and can therefore produce very high-resolution images, revealing details in the range of 1 to 5 nm in size.

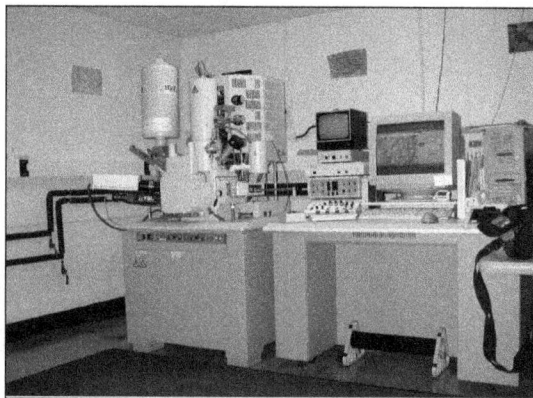

FEI QUANTA 200F.

Energy Dispersive X-ray Spectroscopy (EDS or EDX)

EDS or EDX is an analytical technique used for the elemental analysis or chemical characterization of a sample. It is one of the variants of X-ray fluorescence spectroscopy which relies on the investigation of a sample through interactions between electromagnetic radiation and matter, analyzing X-rays emitted by the matter in response to being hit with charged particles. Its characterization capabilities are due in large part to the fundamental principle that each element has a unique atomic structure allowing X-rays that are characteristic of an element's atomic structure to be identified uniquely from one another.

The incident beam may excite an electron in an inner shell, ejecting it from the shell while creating an electron hole. An electron from an outer, higher-energy shell then fills the hole, and the difference in energy between the higher-energy shell and the lower energy shell may be released in the form of an X-ray. The number and energy of the X-rays emitted from a specimen can be measured by an energy dispersive spectrometer. As the energy of the X-ray are characteristic of the difference in energy between the two shells, and of the atomic structure of the element from which they were emitted, this allows the elemental composition of the specimen to be measured as shown in the figure.

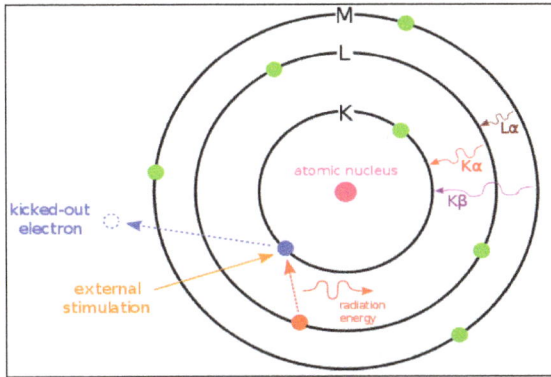

Principle of EDS.

Photoluminescence Spectroscope

Photoluminescence (PL) is the spontaneous emission of light from a material under optical excitation. PL measurement is a kind of powerful and non destructive technique, which has been carried out on most of semiconductors. To date, there are many different type lasers have been widely used in the PL setup, for example, He-Cd laser with 325 nm, Ar⁺ laser with 316 nm/514 nm/488 nm, Nd:YAG pulsed laser with 266 nm, tunable solid state lasers and so on.

When we use pump laser to provide pulsed excitation, the lifetime information of excited state can be obtained. Then the setup will be called Time-Resolved PL (TRPL). When light of sufficient energy is illuminated a material, photons are absorbed and excitations are created. These excited carriers relax and emit a photon. Then PL spectrum can be collected and analyzed. However, only the energy of photons is equal to or higher than the bandgap, the absorption can happen in materials.

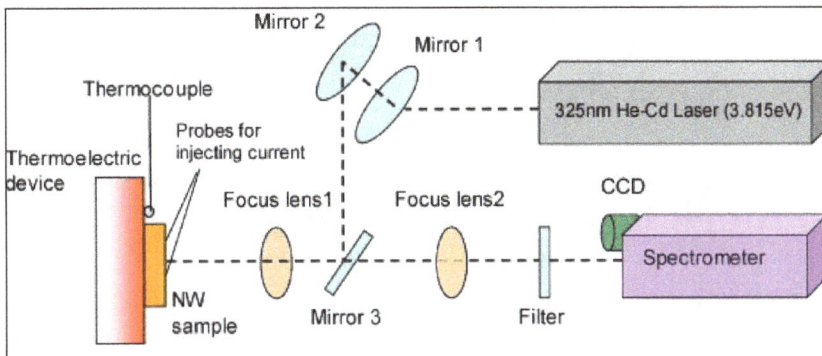

Typical Experimental Set-Ups for PL Measurements.

Therefore, we have to choose different excitation source to do the measurements according to different material with different electronic band structure. The PL peak positions reveal transition energies and the PL intensity implicates the relative rates of radiative and non-radiative recombination. We also can change other external parameters during the PL measurement, such as temperature, excitation power and applied external perturbation such as magnetic field and electrical field and pressure, which can help us further understand the electronic states and bands.

Solar Cell Efficiency Analysis

Cell Placement on Solar Simulator.

The solar cell efficiency is determined by its current-voltage (JV) characteristics under standard illumination conditions (Keithley, model 2400). A standard solar spectrum of air mass 1.5 (AM 1.5) with an intensity of 100 W/m² also referred to as 1 sun, is used for solar cell characterization. The AM 1.5 spectrum corresponds to sunlight that has path through the atmosphere 1.5 times longer than when the sun is directly overhead. The sunlight will be attenuated differently by the earth atmosphere dependent on the incident radiation angle. In the lab, the illumination conditions are provided by a calibrated lamp source. The current-voltage characteristics are monitored under illumination by varying an external load from zero load (short-circuit condition) to infinite load (open-circuit condition). The cell is placed on the simulator stage as shown in figure, where the positive plug is connected to the Pt side and the negative probe is connected to the other substrate with TiO_2.

Nanomaterials and Cell Interactions

Understanding the interactions between nanomaterials and the cell cannot be over-emphasized in the areas of nanotoxicology and nanomedicine. Nanotoxicology, a branch of bionanoscience focuses on the study of the hazardous interactions between nanomaterials and the ecosystem and further ascertaining its consequent implications. Nanomaterials are unambiguously defined as materials having particles or constituents with external dimension in the nanoscale be tween 1 - 100 nm. They can either originate from combustion, manufacturing or naturally occurring processes. In recent times, there has been a steady exponential increase in the production of nanomaterials by combustion and manufacturing processes as a result of its exploitation in applications cutting across several industrial sectors and disciplines. A 30-fold increase in nano-based products between 2011 and 2015 and an approximate market sale of $1 trillion globally in 2015 have been reported. Furthermore, there is a projected increase of total production of nanomaterials from an approximate 2300 tons in 2006 to 58,000 tons by the end of 2020. Consequently, there is a current elevation in the exposure of living things in the ecosystem to nanomaterials and there is prediction for continued increase. This has intensified the concerns of the potential toxicity arising from these exposures. The cell being the basic functional unit of living things would best be used for the study to ensure less complexity and maximal understanding of the mechanisms involved and the role of metabolism. However, due to the complexity of nanomaterial and cell interactions, the mechanism of action of nanomaterial toxicity is not yet fully understood but a proposed and accepted mechanism by which nanomaterials may induce cytotoxicity is through inducing reactive oxygen species (ROS) which can initiate oxidative stress which could subsequently lead to cytotoxicity, DNA damage, and other adverse effects. To get the basics, the composition of their physical, chemical and other properties that influence their toxicity must first be discussed and understood. Finding answers to questions such as: Is the toxicity reversible or irreversible? What is the relationship to exposure (threshold of effect)? Which sub-populations are susceptible? What are the target organs? Is the effect species specific? Are there trans-generational effects? What is the role of metabolism? etc. is also essential.

Determining Factors in Nanomaterial-cell Interactions

Inorganic nanomaterials otherwise called nanoparticles/nanocrystals are composed of specific and unique physico-chemical properties which are factors that influence their complex interactions with cells. These include: 1) particle size and distribution; 2) surface area, charge and coatings; 3) shape/structure of particle; and 4) dissolution and aggregation. Other properties such as magnetic, optical, electronic, thermal and mechanical make them widely used in several applications and consumer products. These properties enhance their cell permeating ability and penetration of other biological barriers into living organisms.

Particle Size and Distribution

Research has shown higher degree of toxicity of nanomaterials in relation to their larger bulk particles thereby leading to the assumption that nano particles are more effective in causing damage. A direct correlation between nanoparticles size, its distribution in tissues and consequent toxicity has been reported.

Surface Charge and Coatings

Surface charge is a major determinant of nano particle dispersing features which plays an important role in binding to cell membrane, the absorption of ions and subsequent cellular uptake. Research reports enhanced toxicity due to increased surface charge of Iron Oxide (FeO) and Silver nanoparticles ($AgNO_3$) respectively while surface coatings indirectly affect aggregation and dissolution properties thereby enhancing the surface charge.

Shape and Structure of Particle

The morphology and shape of a nanoparticle are very important factors that influence their toxicity. Morphology, i.e. spheres, rods, truncated triangles, particles, cubes, wires, fields and coatings etc. affects the kinetics, transport and subsequent cellular uptake of nanoparticles. To buttress this fact, inhibition of Escherichia coli has been shown to be greater by triangular nanoplates in comparison to spherical- or rod-shaped Ag nano particles which could be due to high atom density of the triangular nano particles.

Dissolution and Aggregation

These properties are important in governing nanoparticles behavior and toxicity. Due to the fact that nanoparticles are not found isolated in nature, taking into perspective the added presence of other environmental stressors. Waste nanoparticles are released as aggregates and soluble ions into the environment. Dissolution and aggregation are processes that are largely influenced by size, surface properties and colloidal stability of which the later is in turn influenced by environmental stressors which include temperature, pH, and ionic strength thereby increasing exposure levels and subsequent toxicity. A study by showed silver nanoparticles exhibited high and rapid aggregation in media at high ionic strength.

Routes of Exposure, Transport and Fate of Nanomaterials

Synthesized nanomaterials are fast becoming a part of our everyday life due to our daily use of cosmetics, food packs, drugs, biosensors etc. to enhance drug delivery systems and odor-combating properties. This has spiked the rate of exposure to nanomaterials and their supposed toxic effects. It is therefore of essence to investigate and deeply understand the different routes of the body's exposure to these particles, their transport and their eventual fate and behavior in the body which influences their toxicity.

Nanomaterials can be released into the environment by intention or unintentionally through manufacturing processes such as atmospheric emissions and waste streams from production industries. Environmental exposure to nanoparticles in clothes, sunscreen, cosmetics, and health care products is directly related to their usage. Nanoparticles emitted settle on land and water and potentially contaminate ground and surface waters, soil and potentially become toxic to aquatic life and plant products.

Nanoparticles intentionally released into the environment by technological applications, diffuse releases from wear and spillage also greatly increase exposure.

Although nanomaterials from engineered processes are minimal, airborne particles increase inhalation exposure and could undergo aggregation into larger particles or chains thereby changing their composition and potential effect on entrance into the body system. In the respiratory system, due to its high surface area/activity, unusual morphology and small diameters, enhanced toxicity based on nanostructure occurs. Nanoparticles have been found to have higher deposition rates in lungs of individuals with asthma/chronic obstructive pulmonary diseases than in healthy individuals.

It is suggested that on inhalation, nanoparticles deposit haphazardly on the alveolar surface, likely leading to a scattered chemo-attractant signal which results in reduced recognition and macrophage responses. It has also been reported that there is decreased clearance of less than 25% of 50 - 100 nm particles within the first 24 hr after inhalation.

In relation to the skin, exposure could either be intentional or non-intentional. Use of lotions, cosmetics, wound dressing, detergents and clothes containing nanomaterials constitute intentional exposure e.g. use of sunscreen containing Nano TiO_2 and ZnO materials. Diffuse release from wear and abrasion of clothes when worn and washed also contribute to dermal exposure.

Nanomaterials Toxicity

The potential toxic effects of nanomaterials on the ecosystem is influenced by its physiochemical properties such as type, size, surface area/coatings, charge etc. Toxicities that occur to the living systems are consequences of various causes/dysfunctions such as ROS production, loss of membrane integrity, releases of toxic metal ions that bind with specific cell receptors and undergo certain conformations that inhibit normal cell function resulting in cytotoxicity, genotoxicity and possible cell necrosis.

With respect to these observations, nanomaterials toxicity will be discussed based on the types and classes of nanomaterials, e.g. metallic, metal oxides, carbon nanotubes and quantum dots.

Metal and Metal Oxide Nanomaterials

Metallic nanomaterials and their oxides turn out to be the most used in industries and technological applications such as health, textiles and cosmetics. Gold nanoparticles, according to research, have been reported to be safe (not cytotoxic on cellular uptake). The group investigated cellular uptake and potential cytotoxicity of gold nanoparticles in human leukaemic cells and reported that these spherical shaped gold particles paired with different surface coatings were not toxic to human cells on exposure and uptake. However, although not found to be cytotoxic, gold nanoparticles have the capacity to cause cellular damage as shown by a study on citrate-capped gold nanospheres. The group reported their ability to aid the formation of abnormal act in filaments which resulted in reduced cell proliferation, adhesion and impaired motility.

Silver nanoparticles which are widely used for therapeutics due to their potent antimicrobial activity readily undergo ionization thereby enhancing their toxicity. Exposure to high levels of silver consistently over a long period causes argyria, breathing problems, lungs and throat irritation, stomach pains and skin allergic reactions. A group study reported that with increasing doses (10 - 75 μg/ml) of silver nanoparticles (15 - 30 nm) there was a decrease in cell viability over a period of 24 h and this was found to be likely mediated by oxidative stress due to more than a 10-fold increase of ROS levels in cells exposed to 50 μg/ml silver (15 nm) particles.

Another study examined the capacity of various nanoparticles and nanotubes to be cytotoxic and cause DNA damage and oxidative stress centering on metal oxide nanoparticles (CuO, TiO_2, ZnO, $CuZnFe_2O_4$, Fe_3O_4, Fe_2O_3). Results indicated a wide difference in the level of different metal oxide nanoparticles cytotoxicity. CuO nanoparticles were the most cytotoxic and genotoxic, TiO_2 was responsible for DNA damage, ZnO had adverse effects on cell viability and DNA, $CuZnFe_2O_4$ induced DNA lesions while iron oxide particles showed little or no cytotoxicity. CeO_2 particles are seen as non-cytotoxic and non-inflammatory to cells on uptake, but suppress ROS production and induce cellular resistance to external source of oxidative stress.

Carbon Nanotubes

Carbon nanotubes are widely used in applications for commercial products due to their exceptional nanostructure and properties, thereby increasing human and environmental exposure. This has engineered the investigation of the toxicity of these carbon nanotubes by several studies, especially the multi-walled carbon nanotubes (MWCNTs) mostly used due to its relative low cost. Some of these studies have reported the capacity of carbon nanotubes to cause inflammatory and apoptosis responses in human T cells.

One study reported MWCNTs activation of genes involved in cellular transport, metabolism, cell cycle regulation, and stress response in human skin fibroblasts. On examining the response of mouse embryonic stem cells DNA to MWCNTs uptake, another group reported that on accumulation, nanotubes had the ability to induce apoptosis and activate the tumor suppressor protein p53 within 2 h of uptake.

Quantum Dots

These are nanocrystals consisting 1000 to 100,000 atoms and are capable of emitting "quantum effects", i.e. prolonged fluorescence. Their exceptional optical and electrical properties make them valuable in the biomedical applications such as biomedical imaging, labeling neoplastic cells, DNA, and cell membrane receptors.

The skin is the most susceptible route of exposure for quantum dots to the body system and subsequently cells. A group study supporting this fact, reported the susceptibility of rat skin to quantum dots penetration is essentially limited to the topmost stratum corneum layers of unbroken skin. A similar study also reported cytotoxicity of CdTe quantum dots with cysteamine and mercaptopropionic acid coatings on uptake by pheochromocytoma cells. Cell necrosis was found to arise due to membrane bleeding and chromatin condensation. These studies have shown that quantum dots cytotoxicity can be minimized by regulating processing parameters during synthesis such as surface coatings, and UV light exposure.

Mechanisms of Nanomaterial Toxicity

In response to the recent and proposed increase in the toxic exposure of cells to nano-materials and particles, researchers have undertaken studies to determine and under-stand the different mechanisms through which these particles and their bulk coun-terparts interact with and affect the cells adversely. Due to their uniquely different properties, different nanomaterials exhibit different toxic potentials as seen in a study which examined (CuO, TiO_2, ZnO, $CuZnFe_2O_4$, Fe_3O_4, Fe_2O_3) nanometal oxides and re-ported CuO as the most potent genotoxic and cytotoxic nanometal oxide.

Results from in vitro assessment of nanoparticle toxicity have reported adverse effects at different levels of the cell structure. Certain endpoints measured include malforma-tion, oxidative stress, stagnant growth/development, and gene expression. However, this is a sequential process whereby reactive oxyen species (ROS) and free radicals are generated which induce oxidative stress, lipid peroxidation, DNA damage and subse-quently cell necrosis.

The Concept of Cellular Uptake

Cellular uptake is usually a two step process that includes binding to membrane recep-tors and transport/internalization. Certain factors such as nanoparticle charge, size, type of nanoparticle and the surface charge of the cell membrane play important roles in cellular uptake. The positive charge of nanoparticles has greatly influenced their up-take due to the fact that their electrostatic interactions with the negatively charged cell membrane are favourable.

However, recent findings prove that there has been cellular uptake of negatively charged particles which invariably suggests that electrostatic interactions only partly influence cellular uptake of nanoparticles. This therefore brings into view the role of the pro-tein corona as a fundamental element in nanoparticle/cell interactions and subsequent toxicity. The formation and composition of the protein corona depends on the physi-co-chemical properties of the nano material/particle involved as it varies per particle. The protein corona influences cellular uptake of nanoparticles by creating an interface through modifying/masking the surface properties of a nanoparticle. Having discussed these basic surface in teractions resulting in cellular uptake, let's attempt to decipher the different mechanisms through which the potential toxic effects of nanoparticles occur on internalization.

Oxidative Stress

Oxidative stress is generally defined as a disproportion between the rate of production of ROS and the cells ability to reduce or mop it up, which may be either as a result of increased ROS production/a decrease in cells disease mechanism or both. When there is uncontrolled/over production of ROS, the results include generation of protein radicals, induction of lipid peroxidation, Breakage of DNA strand and nuclei acids modification, modulation of gene expression and subsequent cell necrosis and genotoxicity.

Several authors have investigated the potential role of oxidative stress in Ag nanoparticles toxicity. A concentration-dependent increase in ROS production and oxidative stress was reported following a 7-day exposure to Ag nanoparticles (100 nm) at 10 and 100 μg/ml. This was evaluated using antioxidant enzyme activity. A similar report showed concentration-dependent enzyme activities following the investigation of superoxide $\left(O_2^-\right)$ and stimulation of antioxidant defense mechanisms. A third study reported instability of lysosomes resulting from initiation of apoptosis by Ag nanoparticles.

Genotoxicity

The in vitro genotoxic assessments of different nanoparticles have been reported. They include chromosomal fragmentation, DNA strands breaks, point mutuations, oxidative DNA adducts, alterations in gene expression profiles, potential mutagenesis and carcinogenesis. Research has reported genotoxicity mediated by direct interactions of nanoparticles with DNA and excess ROS production.

Regarding particle size, a group study reported that smaller sized nanoparticles were more genotoxic compared to their bulk counterparts. Surface coatings of positively charged nanoparticles have also been found to enhance genotoxicity.

Nanocatalysis

Nanocatalysis is a rapidly growing field which involves the use of nanomaterials as catalysts for a variety of homogeneous and heterogeneous catalysis applications. Heterogeneous catalysis represents one of the oldest commercial practices of nanoscience; nanoparticles of metals, semiconductors, oxides, and other compounds have been widely used for important chemical reactions.

Although surface science studies have contributed significantly to our fundamental understanding of catalysis, most commercial catalysts, are still produced by "mixing, shaking and baking" mixtures of multi-components; their nanoscale structures are not well controlled and the synthesis-structure-performance relationships are poorly understood. Due to their complex physico-chemical properties at the nanometer scale, even characterization of the various active sites of most commercial catalysts proves to be elusive.

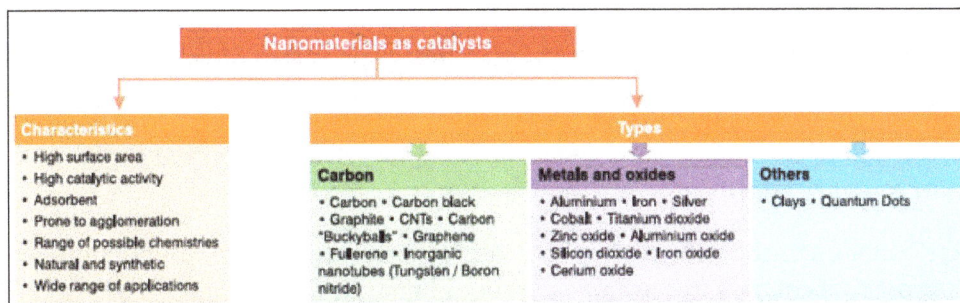

A key objective of nanocatalysis research is to produce catalysts with 100% selectivity, extremely high activity, low energy consumption, and long lifetime. This can be achieved only by precisely controlling the size, shape, spatial distribution, surface composition and electronic structure, and thermal and chemical stability of the individual nanocomponents.

The field of nanocatalysis (the use of nanoparticles to catalyze reactions) has undergone an explosive growth during the past decade, both in homogeneous and heterogeneous catalysis. Since nanoparticles have a large surface-to-volume ratio compared to bulk materials, they are attractive candidates for use as catalysts.

In homogeneous catalysis, transition metal nanoparticles in colloidal solutions are used as catalysts. In this type of catalysis, the colloidal transition metal nanoparticles are finely dispersed in an organic or aqueous solution, or a solvent mixture. The colloidal nanoparticle solutions must be stabilized in order to prevent aggregation of the nanoparticles and also to be good potential recyclable catalysts. Metal colloids are very efficient catalysts because a large number of atoms are present on the surface of the nanoparticles. The method that is used in synthesizing transition metal nanoparticles in colloidal solutions is very important for catalytic applications. The reduction method employed controls the size and the shape of the transition metal nanoparticles that are formed, which are very important in catalytic applications.

Common Synthesis Methods for Colloidal Nanoparticles

Chemical Reduction Method

Important Characteristics:

- Reduction of a metal salt in solution using reducing agents like alcohol, sodium borohydride, etc.

- Precursor transition metal salts are reduced to form the transition metal nanoparticles.

Thermal, Photochemical and Sonochemical Reduction Method

Important Characteristics:

- Decomposition of the precursor Organometallic salt to the zerovalent form.

- Reduction of precursor metal salt or degradation of an Organometallic complex by radiation X.ray Of gamna-ray and also UV-visible radiation by use of xenon or mercury lamp.

- Reduction of the precursor metal salt by anacoustic cavitation phenomena and growth of colloids in the sonicated liquid medium.

Ligand Displacement Method

Important Characteristics:

- Displacement of ligand in the organometallic complex (e.g. amine ligands are displaced by thiol ligands).

Condensation of Metal Vapor

Important Characteristics:

- Evaporation of transition metal vapors at reduced pressure and subsequent co-condensation of these metals at low temperature with organic vapors.

- Nanoparticles are formed by nucleation and growth when the frozen metal organic mixture is warmed to the point of melting.

- No precise control on size of the nanoparticles.

Electrochemical Reduction Method

Important Characteristics:

- Precursor metal ions are reduced at the cathode using a sacrificial anode as the metal source.

- The metal at anode is oxidized in presence of a quaternary ammonium salt which acts as both the electrolyte and the stabilizer.

The different reduction methods that have been used to synthesize colloidal transition metal nanoparticles for homogeneous catalysis are summarized below. Chemical reduction of the precursor transition metal salt is the most widely used method of synthesizing transition metal nanocatalysts in colloidal solution. There are four other synthetic methods to prepare colloidal transition metal nanocatalysts that are not as commonly used. These synthetic methods include:

- Thermal, photochemical, or sonochemical reduction of the precursor transition metal salts.

- Ligand reduction and displacement from organometallic precursors.

- Metal vapor synthesis.

- Electrochemical reduction of transition metal precursor salts. Stabilization of nanoparticles in the solution is necessary in order to prevent agglomeration and aggregation. For catalytic applications, the choice of stabilizers plays an important role in determining the reactivity of nanoparticles.

A good stabilizer is one that protects the nanoparticles during the catalytic process, but does not neutralize the surface of nanoparticles resulting in loss of catalytic activity. The choice of a stabilizer to be used for capping the nanoparticles is usually a balancing act between passivation of the nanoparticle surface and the fraction of available sites for catalysis, and also affects the size and shape of the nanoparticles formed.

Heterogeneous metal nanocatalysts are prepared by adsorption of nanoparticles on to supports, which involves functionalization of supports to adsorb nanoparticles on to them and, fabrication of nanostructures on the supports by lithographic techniques.

Benefits of Nanocatalysts in the Chemical Industry

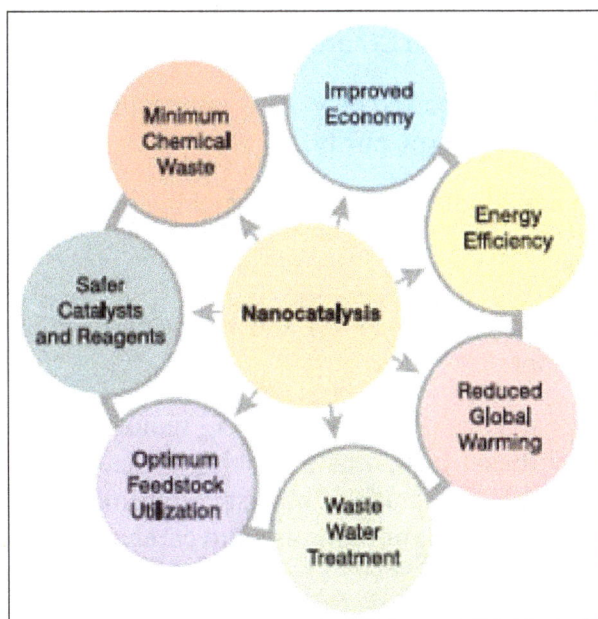

Benefits of nanocatalysis.

- Increased selectivity and activity of catalysts by controlling pore size and particle characteristics.

- Replacement of precious metal catalysts by catalysts tailored at the nanoscale and use of base metals, thus improving chemical reactivity and reducing process costs.

- Catalytic membranes by design that can remove unwanted molecules from gases or liquids by controlling the pore size and membrane characteristics In view of the numerous potential benefits that can accrue through their use, nanostructured catalysts have been the subject of considerable research attention in recent times. Many applications and patents have also been realized adopting such nanostructured catalysts leading to significant process improvements as exemplified below.

Important Applications of Nanocatalysis

Sector: Biomass

Application: Biomass gasification to produce high syn gas and biomass pyrolysis for production of bio-oil.

Process Improvements:

- Novel Al_2O_3 supported NiO catalyst reduces tar yield significantly and increases tar removal efficiency to 99% • Significant increase in gas yield.

- Lighter fractions of H2 and CO are increased in the syn gas composition while heavier fractions of CH_4 and CO are reduced, thus improving syn gas quality.

Catalyst: Nano NiO catalyst supported on γ- Al_2O_3 microspheres of 3 mm size (Johnson Mathey Company, greater than 99% purity).

Application: Production of biodiesel from waste cooking oil.

Process Improvements:

- Esterification of fatty acids (FFAs) and transesterification of triglycerides to biodiesel in one pot.

- Solid acid nanocatalysis of $Al_{0.9}H_{0.3}PW_{12}O_{40}$ nanotubes with double acid sites yield 96% of biodiesel from waste cooking oil as compared to 42.6% with conventional $H_3PW_{12}O_{40}$ catalyst Catalyst: Aluminium dodeca-tungsto-phosphate ($Al_{0.9}H_{0.3}PW_{12}O_{40}$) nanotubes as solid catalysts with surface area of 278 m^2/g.

Application: Green diesel production using Fischer-Tropsch Synthesis (FTS).

Process Improvements:

- Improving the FTS technology for production of high molecular weight waxes, followed by their hydrocracking to generate liquid fuels.

- Improved efficiency of slurry and fixed-bed reactors, used in FTS from biosyngas.

- Produce long, linear-chain paraffin waxes in fixed bed and slurry FTS reactors.

Catalyst:

- Nano Fe and Co powders (10-50 nm) are used as FTS catalysts in slurry reactors, promoted by other metals like Mn, Cu and alkalis.

- Produced by thermal plasma chemical vapor deposition (TPCVD) and cluster spray techniques.

- Minimize liquid-solid diffusion resistance.

- Multi-walled carbon nanofilaments (MWCNF), produced by CO_2 sequestration via dry reforming for gas-to-liquid FTS, with the iron carbide content rendering catalytic activity.

Sector: Oil, Gas and Fossil Fuels.

- Paraffin Dehydrogenation.

- Naphtha Reforming.

- Selective Hydrogenation.

- Hydrodesufurization.

Application: Improved economic catalytic combustion of JP-10 aviation fuel using hydrocarbon fuel soluble nano catalyst.

Process Improvements:

- 50 ppm addition of catalyst in JP-10 reduces the ignition temperature required to initiate combustion by about 240°C.

Catalyst: Hexanethiol monolayer protected Palladium clusters < 1.5 nm.

Application: Hydrogen production by steam reforming of ethanol over nanostructured indium oxide catalysts.

Process Improvements:

- At 623K, 99% conversion with mesoporous In_2O_3/KIT-6 catalyst exhibit high production rates from ethyl alcohol at low-temperatures and yield low concentration of CO impurity in comparison with other reported catalysts.

Catalyst: Mesoporous In_2O_3 prepared using Mobil Composition of Matter No. 41 (MCM-41) silica catalyst as templates with particle size of 2-3 nm and surface area of 107 m2/g to 173 m²/g.

Application: Adsorptive desulfurization and bio desulfurization of fossil oils.

Process Improvements: In situ coupling desulfurization using assembly of nano adsorbents (nano γ- Al2O3) onto surfaces of Pseudomonas de lafieldii.

Catalyst: Nano γ- Al_2O_3 (10 nm in width and 100-200 nm in length) with specific surface area of 339 m²/g.

Application: Hydrodesulfurization of diesel.

Process Improvements: Hydrodesulfurization of dibenzothiophene increased by 20% using SDM NiMo/Al-HMS nanocatalyst at 330°C as compared to commercial catalysts.

Catalyst: Synthesis of new NiMo/Al hexagonal, mesoporous structured nanocomposite catalyst by supercritical deposition method.

Sector: Fuel Cells.

Application: Core-shell nanocatalysts for fuel cell applications.

Process Improvements:

- Pt atoms are placed at the surface of other metal nanoparticles.

- All the Pt atoms are available for catalytic reactions at the surface.

- Pt clusters on ruthenium nanoparticles produce high activity per unit of Pt mass.

Catalyst: Smooth and compact Pt shell for better oxygen reduction reactions in fuel cell applications.

Application: In situ hydrogen production by reaction of ammonia and nanocatalysts.

Process Improvements:

- Ammonia is stored as a coordination complex with a transition metal compound in solid composition.

- It acts as the hydrogen fuel precursor for a vehicle internal combustion engine that is operated to use hydrogen or a combination of hydrogen and gasoline as fuel.

- Ammonia dissociation catalyst tube containing a catalyst bed and maintained at 750°C is used to dissociate ammonia into nitrogen and hydrogen atoms.

Catalyst: The dissociation catalyst is a mixture of nanometer size particles of Co-NiO-Cu-Zr catalyst deposited on high surface area of TiO_2 and 2% Pt deposited on alumina particles.

Dendrimers

Dendrimers are nano-sized, radially symmetric molecules with well-defined, homogeneous, and monodisperse structure consisting of tree-like arms or branches. These hyperbranched molecules were first discovered by Fritz Vogtle in 1978, by Donald Tomalia and co-workers in the early 1980s, and at the same time, but independently by George R. Newkome. The second group called synthesized macromolecules 'arborols. Dendrimers might also be called 'cascade molecules', but this term is not as much established as 'dendrimers'. Dendrimers are nearly monodisperse macromolecules that contain symmetric branching units built around a small molecule or a linear polymer core. 'Dendrimer' is only an architectural motif and not a compound. Polyionic dendrimers do not have a persistent shape and may undergo changes in size, shape, and flexibility as a function of increasing generations. Dendrimers are hyperbranched macromolecules with a carefully tailored architecture, the end-groups (i.e., the groups reaching the outer periphery), which can be functionalized, thus modifying their physicochemical or biological properties. Dendrimers have gained a broad range of applications in supramolecular chemistry, particularly in host-guest reactions and self-assembly processes. Dendrimers are characterized by special features that make them promising candidates for a lot of applications. Dendrimers are highly defined artificial macromolecules, which are characterized by a combination of a high number of functional groups and a compact molecular structure. The emerging role of dendritic macromolecules for anticancer therapies and diagnostic imaging is remarkable. The advantages of these well-defined materials make them the newest class of macromolecular nano-scale delivery devices. Dendritic macromolecules tend to linearly increase in diameter and adopt a more globular shape with increasing dendrimer generation. Therefore, dendrimers have become an ideal delivery vehicle candidate for explicit study of the effects of polymer size, charge, and composition on biologically relevant properties such as lipid bilayer interactions, cytotoxicity, internalization, blood plasma retention time, biodistribution, and filtration.

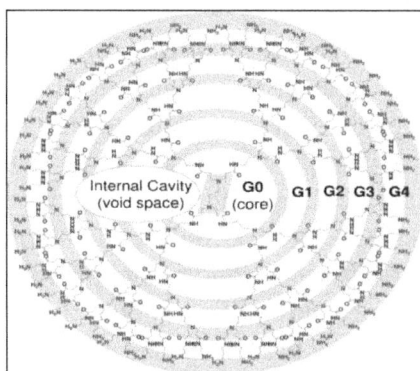

Schematic representation of a generation G4 dendrimer with 64 amino groups at the periphery. This dendrimer starts from an ethylene diamine core; the branches or arms were attached by exhaustive Michael addition to methyl acrylate followed by exhaustive aminolysis of the resulting methyl ester using ethylene diamine

Structure and Chemistry

The structure of dendrimer molecules begins with a central atom or group of atoms labeled as the core. From this central structure, the branches of other atoms called 'dendrons' grow through a variety of chemical reactions. There continues to be a debate about the exact structure of dendrimers, in particular whether they are fully extended with maximum density at the surface or whether the end-groups fold back into a densely packed interior. Dendrimers can be prepared with a level of control not attainable with most linear polymers, leading to nearly monodisperse, globular macromolecules with a large number of peripheral groups as seen in figure, the structure of some dendrimer repeat units, for example, the 1,3-diphenylacetylene unit developed by Moore.

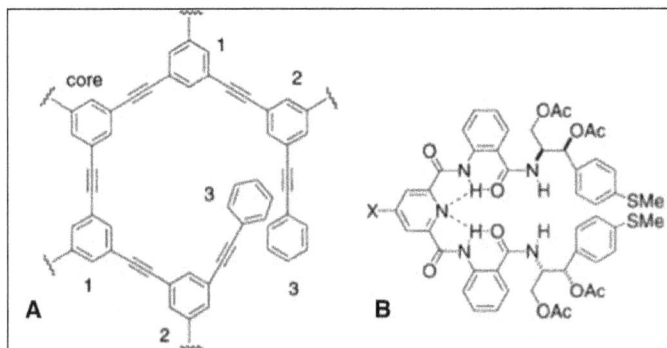

Types of dendrimers. (A) More type dendrimers consisting of phenyl acetylene subunits at the third-generation different arms may dwell in the same space, and the fourth-generation layer potential overlaps with the second-generation layer. (B) Parquette-type dendrons are chiral, non-racemic, and with intramolecular folding driven by hydrogen bonding.

Dendrimers are a new class of polymeric belongings. Their chemistry is one of the most attractive and hastily growing areas of new chemistry. Dendrimer chemistry, as other specialized research fields, has its own terms and abbreviations. Furthermore, a more brief structural nomenclature is applied to describe the different chemical events taking place at the dendrimer surface. Dendrigrafts are a class of dendritic polymers like dendrimers that can be constructed with a well-defined molecular structure, i.e., being monodisperse. The unique structure of dendrimers provides special opportunities for host-guest chemistry and is especially well equipped to engage in multivalent interactions. At the same time, one of the first proposed applications of dendrimers was as container compounds, wherein small substrates are bound within the internal voids of the dendrimer. Experimental evidence for unimolecular micelle properties was established many years ago both in hyperbranched polymers and dendrimers.

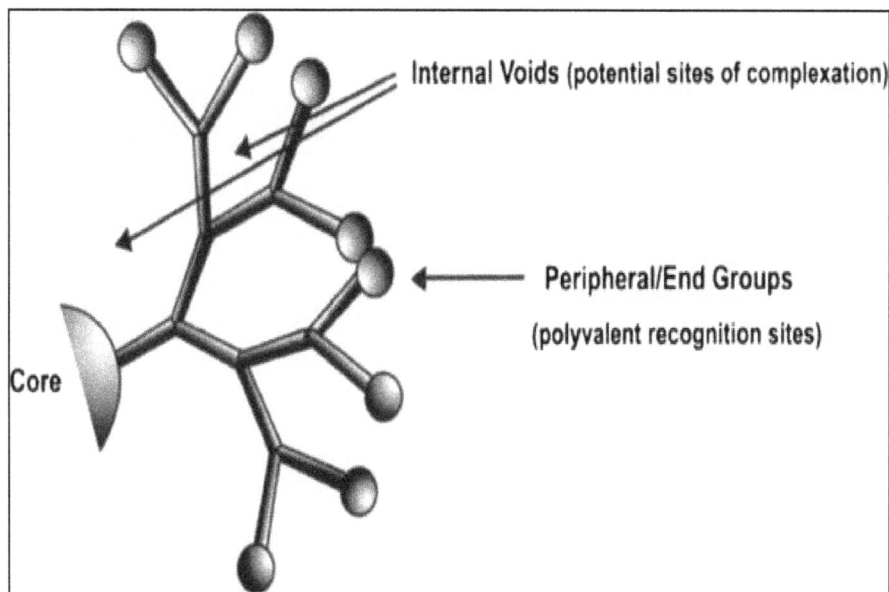

Three main parts of a dendrimer: the core, end-groups, and subunits linking the two molecules.

Dendrimers are just in between molecular chemistry and polymer chemistry. They relate to the molecular chemistry world by virtue of their step-by-step controlled synthesis, and they relate to the polymer world because of their repetitive structure made of monomers. The three traditional macromolecular architectural classes (i.e., linear, cross-linked, and branched) are broadly recognized to generate rather polydisperse products of different molecular weights. In contrast, the synthesis of dendrimers offers the chance to generate monodisperse, structure-controlled macromolecular architectures similar to those observed in biological systems. Dendrimers are generally prepared using either a divergent method or a convergent one. In the different methods, dendrimer grows outward from a multifunctional core molecule. The core molecule reacts with monomer molecules containing one reactive and two dormant groups, giving the first-generation dendrimer. Then, the new periphery of the molecule is activated for reactions with more monomers.

Cascade reactions are the Foundation of Dendrimer Synthesis

The basic cascade or iterative methods that are currently employed for synthesis were known to chemists much earlier. For example, similar schemes form the basis of solid-phase peptide synthesis. In turn, biology has long exploited similar iterative strategies in biochemical synthetic pathways; one example is provided by fatty acid biosynthesis.

The synthesis of dendrimers follows either a divergent or convergent approach.

Dendrimers can be synthesized by two major approaches. In the divergent approach, used in early periods, the synthesis starts from the core of the dendrimer to which the arms are attached by adding building blocks in an exhaustive and step-wise manner. In the convergent approach, synthesis starts from the exterior, beginning with the molecular structure that ultimately becomes the outermost arm of the final dendrimer. In this strategy, the final generation number is pre-determined, necessitating the synthesis of branches of a variety of requisite sizes beforehand for each generation.

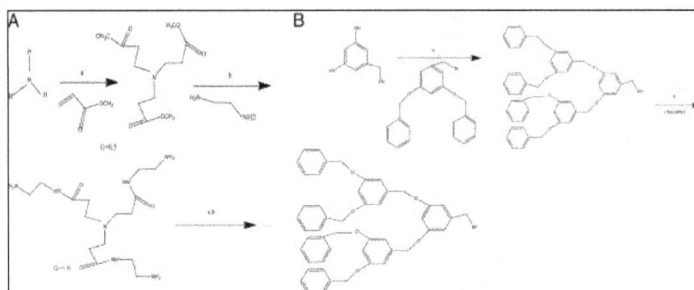

Approaches for the synthesis if dendrimers. (A) Divergent approach: synthesis of radially symmetric polyamidoamine (PAMAM)dendrimers using ammonia as the trivalent core; the generations are added at each synthetic cycle (two steps), leading to an exponential increase in the number of surface functional groups. (B) Convergent approach: synthesis of dendrons or wedges or branches that will become the periphery of the dendrimer when coupled to a multivalent core in the last step of the synthesis.

Properties of Dendrimers

When comparing dendrimers with other nanoscale synthetic structures (e.g., traditional polymers, buck balls, or carbon nanotubes), these are either highly non-defined or have limited structural diversity.

Pharmacokinetic Properties

Pharmacokinetic properties are one of the most significant aspects that need to be considered for the successful biomedical application of dendrimers, for instance, drug delivery, imaging, photodynamic therapy, and neutron capture therapy. The diversity of potential applications of dendrimers in medicine results in increasing interest in this area. For example, there are several modifications of dendrimers' peripheral groups which enable to obtain antibody-dendrimer, peptide-dendrimer conjugates or dendritic boxes that encapsulate guest molecules.

Covalent Conjugation Strategies

The strategy of coupling small molecules to polymeric scaffolds by covalent linkages to improve their pharmacological properties has been under experimental test for over three decades. In most cases, however, the conjugated dendritic assembly functions as 'pro-drug' where, upon internalization into the target cell, the conjugate must be liberated to activate the drug.

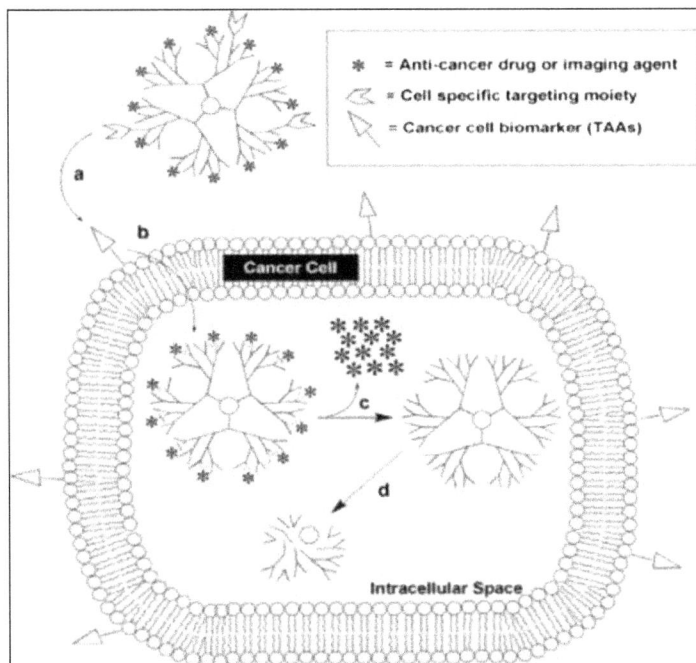

Requirements for dendrimer-based, cancer-targeted drug delivery. (A) Dendrimers with multiple surface functional groups can be directed to cancer cells by tumor-targeting entities that include folate or antibodies specific for tumor-associated antigens (TAAs). (B) The next step is ingestion into the cell which, in the case of folate targeting, occurs by membrane receptor-mediated endocytosis. (C) Once inside the cell, the drug generally must be released from the dendrimer, which, for the self-immolative method, results in the simultaneous disintegration of the dendritic scaffold (D).

Polyvalency

Polyvalency is useful as it provides for versatile functionalization; it is also extremely important to produce multiple interactions with biological receptor sites, for example, in the design of antiviral therapeutic agents.

Self-assembling Dendrimers

Another fascinating and rapidly developing area of chemistry is that of self-assembly. Self-assembly is the spontaneous, precise association of chemical species by specific, complementary intermolecular forces. Recently, the self-assembly of dendritic structures has been of increasing interest. Because dendrimers contain three distinct structural parts (the core, end-groups, and branched units connecting the core and periphery), there are three strategies for self-assembling dendrimers. The first is to create dendrons with a core unit that is capable of recognizing itself or a ditopic or polytopic core structure, therefore leading to spontaneous formation of a dendrimer.

A self-assembling dendrimer using pseudorotaxane formation as the organizing force was reported by Gibson and coworkers.

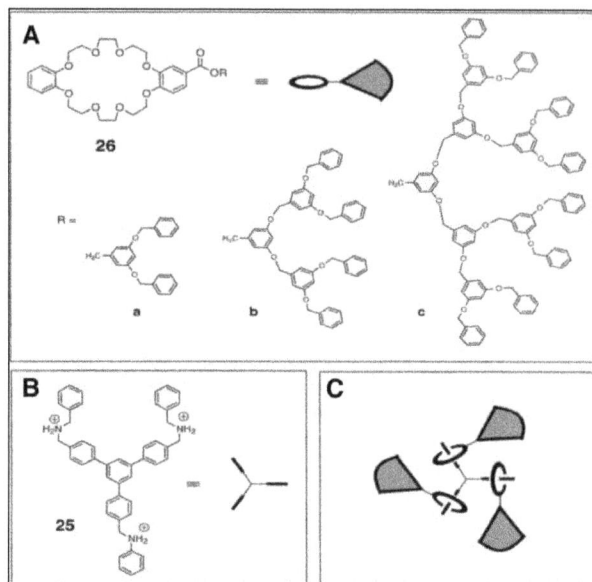

Gibson's self-assembling dendrimers using pseudorotaxane formation. (A) Crown ethers with dendritic substituents. (B) Triammonium ion core. (C) Schematic of tridendron formed by triple pseudorotaxane self-assembly.

Electrostatic Interactions

Molecular recognition events at dendrimer surfaces are distinguished by the large number of often identical end-groups presented by the dendritic host. When these groups are charged, the surface may have as a polyelectrolyte and is likely to electrostatically attract oppositely charged molecules. One example of electrostatic interactions between polyelectrolyte dendrimers and charged species include the aggregation of methylene blue on the dendrimer surface and the binding of EPR probes such as copper complexes and nitroxide cation radicals.

Applications

Today, dendrimers have several medicinal and practical applications.

Dendrimers in Biomedical Field

Dendritic polymers have advantage in biomedical applications. These dendritic polymers are analogous to protein, enzymes, and viruses, and are easily functionalized. Dendrimers and other molecules can either be attached to the periphery or can be encapsulated in their interior voids . Modern medicine uses a variety of this material as potential blood substitutes, e.g., polyamidoamine dendrimers.

Anticancer Drugs

Perhaps the most promising potential of dendrimers is in their possibility to perform controlled and specified drug delivery, which regards the topic of nanomedicine. One of the most fundamental problems that are set toward modern medicine is to improve pharmacokinetic properties of drugs for cancer. Drugs conjugated with polymers are characterized by lengthened half-life, higher stability, water solubility, decreased immunogenicity, and antigenicity. Unique pathophysiological traits of tumors such as extensive angiogenesis resulting in hypervascularization, the increased permeability of tumor vasculature, and limited lymphatic drainage enable passive targeting, and as a result, selective accumulation of macromolecules in tumor tissue. This phenomenon is known as 'enhanced permeation and retention' (EPR). The drug-dendrimer conjugates show high solubility, reduced systemic toxicity, and selective accumulation in solid tumors. Different strategies have been proposed to enclose within the dendrimer structure drug molecules, genetic materials, targeting agents, and dyes either by encapsulation, complexation, or conjugation.

Dendrimers in Drug Delivery

In 1982, Maciejewski proposed, for the first time, the utilization of these highly branched molecules as molecular containers. Host-guest properties of dendritic polymers are currently under scientific investigation and have gained crucial position in the field of supramolecular chemistry. Host-guest chemistry is based on the reaction of binding of a substrate molecule (guest) to a receptor molecule (host).

Transdermal Drug Delivery

Clinical use of NSAIDs is limited due to adverse reactions such as GI side effects and renal side effects when given orally. Transdermal drug delivery overcomes these bad effects and also maintains therapeutic blood level for longer period of time. Transdermal delivery suffers poor rates of transcutaneous delivery due to barrier function of the skin. Dendrimers have found applications in transdermal drug delivery systems. Generally, in bioactive drugs having hydrophobic moieties in their structure and low water solubility, dendrimers are a good choice in the field of efficient delivery system.

Gene Delivery

The primary promise that the combination of understanding molecular pathways of disease and the complete human genome sequence would yield safer and more efficient medicines and revolutionize the way we treat patients has not been fulfilled to date. However, there is little doubt that genetic therapies will make a significant contribution to our therapeutic armamentarium once some of the key challenges, such as specific and efficient delivery, have been solved. The ability to deliver pieces of DNA to the required parts of a cell includes many challenges. Current research is being performed to find ways to use dendrimers to traffic genes into cells without damaging or deactivating the DNA. To maintain the activity of DNA during dehydration, the dendrimer/ DNA complexes were encapsulated in a water soluble polymer and then deposited on or sandwiched in functional polymer films with a fast degradation rate to mediate gene transfection. Based on this method, PAMAM dendrimer/DNA complexes were used to encapsulate functional biodegradable polymer films for substrate-mediated gene delivery. Research has shown that the fast-degrading functional polymer has great potential for localized transfection.

Dendrimers as Magnetic Resonance Imaging Contrast Agents

Dendrimer-based metal chelates act as magnetic resonance imaging contrast agents. Dendrimers are extremely appropriate and used as image contrast media because of their properties.

Dendritic Sensors

Dendrimers, although are single molecules, can contain high numbers of functional groups on their surfaces. This makes them striking for applications where the covalent connection or close proximity of a high number of species is important. Balzani and coworkers investigated the fluorescence of a fourth-generation poly (propylene amine) dendrimer decorated with 32 dansyl units at the periphery. Since the dendrimer contains 30 aliphatic amine units in the interior, suitable metal ions are able to coordinate. It was observed that when a Co^{2+} ion is incorporated into the dendrimer, the strong fluorescence of all the dansyl units is quenched. Low concentrations of Co^{2+} ions (4.6×10^{-7} M) can be detected using a dendrimer concentration of 4.6×10^{-6} M. The many fluorescent groups on the surface serve to amplify the sensitivity of the dendrimer as a sensor.

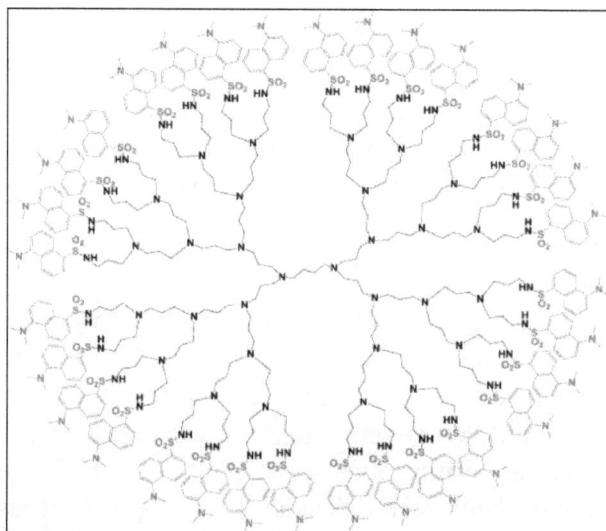

Poly (propylene amine) dendrimer, containing 32 dansyl units at its periphery.

Dendrimers used for Enhancing Solubility

PAMAM dendrimers are expected to have potential applications in enhancing solubility for drug delivery systems. Dendrimers have hydrophilic exteriors and interiors, which are responsible for its unimolecular micelle nature. Dendrimer-based carriers offer the opportunity to enhance the oral bioavailability of problematic drugs. Thus, dendrimer nano carriers offer the potential to enhance the bioavailability of drugs that are poorly soluble and substrates for efflux transporters.

Photodynamic Therapy

Photodynamic therapy (PDT) relies on the activation of a photosensitizing agent with visible or near-infrared (NIR) light. Upon excitation, a highly energetic state is formed which, upon reaction with oxygen, affords a highly reactive singlet oxygen capable of inducing necrosis and apoptosis in tumor cells. Dendritic delivery of PDT agents has been investigated within the last few years in order to improve upon tumor selectivity, retention, and pharmacokinetics.

Miscellaneous Dendrimer Applications

Clearly, there are many other areas of biological chemistry where application of dendrimer systems may be helpful. Cellular delivery using carrier dendritic polymers is used in the purification of water dendrimer-based product in cosmetics contaminated by toxic metal ion and inorganic solute, and dendrimer-based commercial products organic solutes.

Carbon-based Nanomaterials

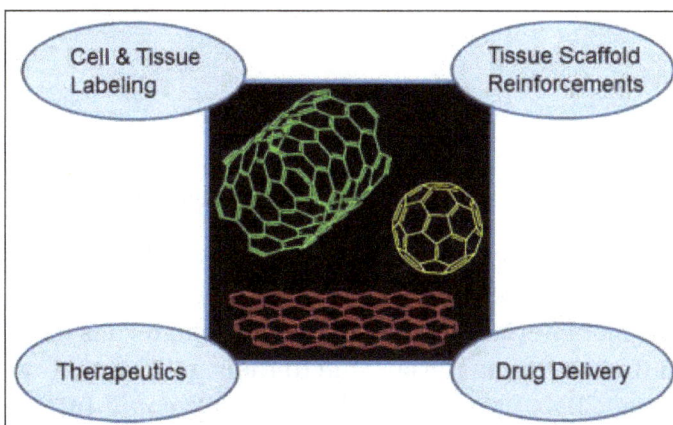

Functional carbon-based nanomaterials (CBNs) have become important due to their unique combinations of chemical and physical properties (i.e., thermal and electrical conductivity, high mechanical strength, and optical properties), extensive research efforts are being made to utilize these materials for various industrial applications, such as high-strength materials and electronics. These advantageous properties of CBNs are also actively investigated in several areas of biomedical engineering.

Graphite is one of the oldest and most widely used natural materials. More traditionally known as the main ingredient of pencil lead, from which the name "graphite" originated, it is now more widely used in several large-scale industrial applications, such as carbon raising in steelmaking, battery electrodes, and industrial-grade lubricants.[1] Due to its high demand, the consumption of synthetic graphite has significantly increased in recent years. Extensive scientific investigation into graphite has revealed that its unique combination of physical properties stems from its macromolecular structure, which consists of stacked layers of hexagonal arrays of sp^2 carbon.

With the deeper appreciation and development of nanofabrication techniques and nanomaterials that have progressed within the last two decades, graphite is now being actively used as a starting material to engineer various types of carbon-based nanomaterials (CBNs), including single or multi-walled nanotubes, fullerenes, nanodiamonds, and grapheme. These CBNs possess excellent mechanical strength, electrical and thermal conductivity, and optical properties; much of the research efforts have been focused on utilizing these advantageous properties for various applications, such as high-strength composite materials and electronics.

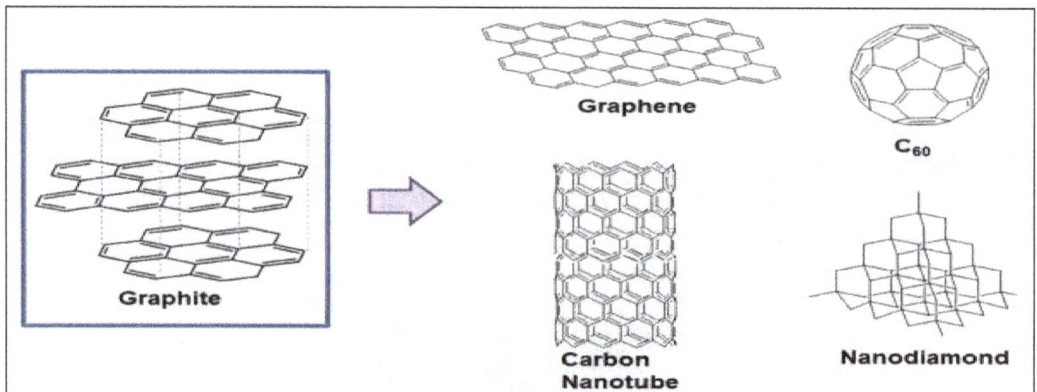

Various types of carbon-based nanomaterials.

The field of biomedical engineering has also embraced the growing popularity and influence of CBNs in recent years, because many of its applications rely heavily on the performance of biomaterials. Carbon-based nanomaterials have been widely regarded as highly attractive biomaterials due to their multi-functional nature. In addition, incorporating CBNs into existing biomaterials could further augment their functions. Therefore, CBNs have found their way into many areas of biomedical research, including drug delivery systems, tissue scaffold reinforcements, and cellular sensors.

Carbon Nanotubes

Ever since their discovery, carbon nanotubes (CNTs) have become the most widely used CBNs. Carbon nanotubes are commonly synthesized by arc discharge or chemical vapor deposition of graphite. They have a cylindrical carbon structure, and possess a wide range of electrical and optical properties stemming not only from their extended sp^2-carbon, but also from their tunable physical properties (e.g., diameter, length, single-walled vs. multi-walled, surface functionalization, and chirality). Due to the diverse array of their useful properties, CNTs have been explored for use in many industrial applications. For example, CNTs are well known for their superb mechanical strength: their measured rigidity and flexibility are greater than that of some commercially available high-strength materials (e.g., high tensile steel, carbon fibers, and Kevlar). Thus, they have been utilized as reinforcing elements for composite materials such as plastics and metal alloys, which have already led to several commercialized products. However, the possibility of CNT-incorporated composites as super high-strength load-bearing materials has not been met with satisfactory results, mostly due to their poor interaction with the surrounding matrices, which leads to inefficient load transfer from the matrices to the CNTs.

Many recent research efforts have been geared toward incorporating CNTs into various materials to utilize their multi-functional nature (i.e., electrical and thermal conductivity, and optical properties) rather than focusing purely on composite mechanical strength. For example, the excellent electrical properties of CNTs coupled with their nanoscale dimensions are of great interest in electronics for the construction of nanoscale electronic circuitry. In addition, CNTs are known to have low threshold electric fields for field emission, as compared with other common field emitters. Thus, CNTs are actively explored in high-efficiency electron emission devices such as electron microscopes, flat display panels, and gas-discharge tubes. Carbon nanotubes also display strong luminescence from field emission, which could be used in lighting elements.

Carbon Nanotubes in Biomedical Engineering

There is considerable interest in using CNTs for various biomedical applications. The physical properties of CNTs, such as mechanical strength, electrical conductivity, and optical properties, could be of great value for creating advanced biomaterials. Carbon nanotubes can also be chemically modified to present specific moieties (e.g., functional groups, molecules, and polymers) to impart properties suited for biological applications, such as increased solubility and biocompatibility, enhanced material compatibility, and cellular responsiveness. Their applications include cell and tissue labeling agents, injectable drug delivery systems, and biomaterial reinforcements.

Cell and Tissue Labeling and Imaging

The possibility of using CNTs as labeling and imaging agents has been discussed since their discovery due to their unique optical properties. Carbon nanotubes have optical transitions in the near-infrared (NIR) region, which has been shown to be useful in biological tissue because NIR has greater penetration depth and lower excitation scattering. In addition, fluorescence in the NIR region displays much lower autofluorescence than do the ultraviolet or visible ranges. These properties make CNTs potent imaging agents with higher resolution and greater tissue depth for NIR fluorescence microscopy and optical coherence tomography. For example, Cherukuri et al. successfully monitored CNTs in phagocytic cells and those intravenously administered into mice using NIR fluorescence.

Raman spectroscopy has been used extensively to characterize the structural features of CNTs, as they are highly sensitive to Raman scattering because of their extensive symmetric carbon bonds. The characteristic Raman signatures of CNTs have also been utilized as cellular probes. For example, Liu et al. detected CNTs in various tissues after intravenous delivery into mice using Raman spectroscopy.

Drug Delivery Systems

Drug delivery has benefited greatly from the advances in nanotechnology by using a variety of nanomaterials (i.e., liposomes, polymersomes, microspheres, and polymer conjugates) as vehicles to deliver therapeutic agents. Carbon nanotubes have also been investigated extensively as drug delivery systems, since CNTs have been shown to interact with various biomacromolecules (i.e., proteins and DNA) by physical adsorption. In addition, several chemical modification schemes have been developed to conjugate therapeutic molecules or targeting moieties covalently to CNTs.

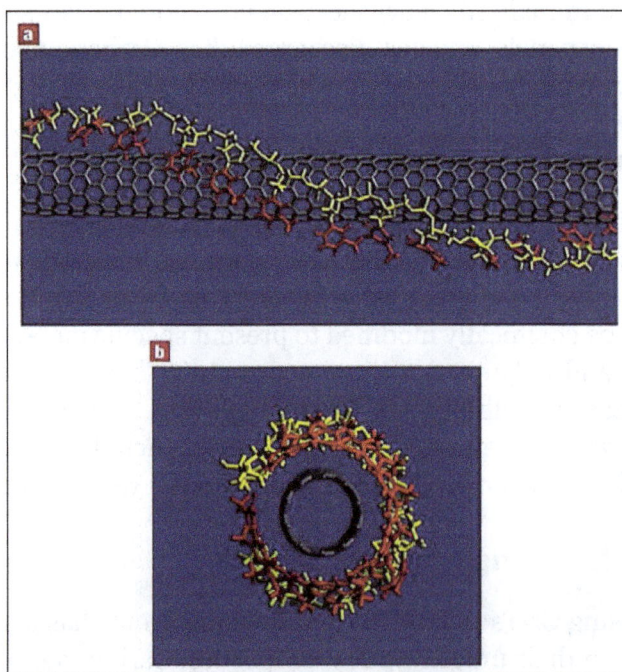

Binding model of a carbon nanotube (CNT) with a poly(T) DNA sequence. The DNA wraps around the CNT in a right-handed helical structure. The bases (red) orient to stack with the surface of the nanotube, and extend away from the sugar-phosphate backbone (yellow). The DNA wraps to provide a tube within which the CNT can reside, hence converting it into a water-soluble object.

It provided an important insight into the interactions between CNTs and DNA molecules. Carbon nanotubes were shown to be effectively dispersed in aqueous media in the presence of single-stranded DNA (ssDNA). Spectroscopic and microscopic analyses provided evidence of strong interactions between DNA molecules and CNTs, resulting in individual dispersion. Molecular dynamics modeling showed that the base of ssDNA interacted with the surface of CNT via $\pi-\pi$ stacking, resulting in helical wrapping of ssDNA chains around the CNT. This research highlights the potential use of CNTs for gene delivery, as well as DNA-specific separation techniques for molecular electronics.

Reinforcing Tissue Engineering Scaffolds

The use of CNTs as composite reinforcements for tissue engineering scaffolds to date has primarily been focused on enhancing their mechanical properties. Commonly used scaffold materials, such as hydrogels and fibrous scaffolds, are inherently soft in order to mimic the stiffness of natural tissues and thus often lack structural strength and support. Incorporating CNTs into these materials has been shown to enhance their mechanical properties. For example, Shin et al. demonstrated that incorporating CNTs into photocrosslinkable gelatin hydrogels resulted in significant increases in their tensile strengths. Sen et al. also demonstrated improvement in the tensile strengths of CNT-reinforced polystyrene and polyurethane fibrous membranes.

More recently, researchers have turned their attention to utilizing the multi-functional nature of CNTs in engineering tissue scaffolds. Most notably, CNTs have been incorporated to fabricate electrically conductive scaffolds. Most of the biomaterials used for tissue engineering applications are electrically insulating, as they are made from non-conductive polymers. However, certain applications, such as neural and cardiac tissues, would benefit greatly from conductive scaffolds that can effectively propagate electrical signals across the tissue constructs for proper electrophysiological functions. For example, Kam et al. applied electrical stimulation to neural stem cells (NSCs) grown on CNT-laminin composite films, and demonstrated improved action potentials of NSCs and differentiation into functional neural networks. Shin et al. cultured cardiomyocytes on CNT-reinforced gelatin hydrogel and observed their enhanced electrophysiological activities, and ultimately developed functional cardiac tissue. These works highlight the ability of CNTs to impart electrical conductivity successfully to otherwise non-conductive biomaterials.

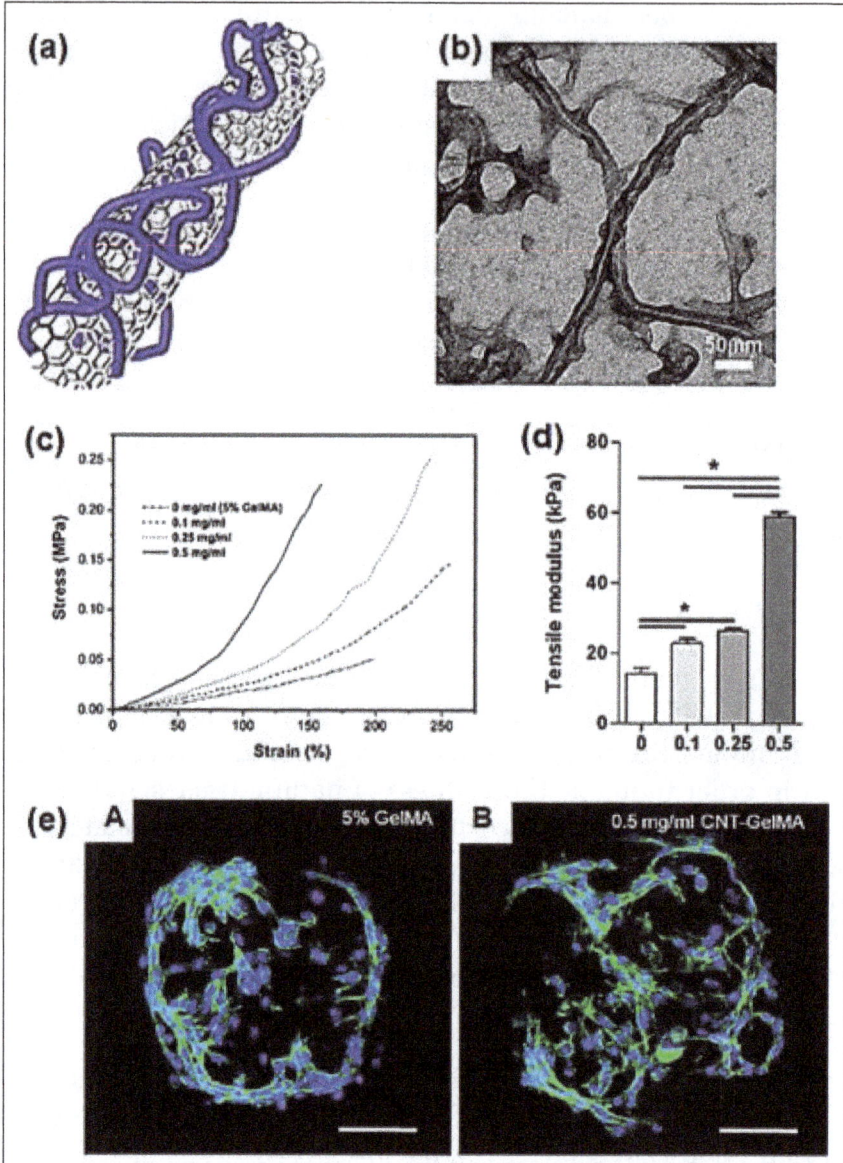

(a) Schematic description of gelatin methacrylate (GelMA) coated onto a carbon nanotube (CNT). (b) High resolution TEM image of CNT-GelMA. (c) Stress-strain curves of CNT-GelMA hydrogels with various concentrations of CNTs. (d) Tensile moduli of CNT-GelMA hydrogels. (e) 3T3 fibroblasts cultured on GelMA hydrogel (A) and CNT-GelMA hydrogel (B).(Scale bar: 100 μm).

Cytotoxicity of CNTs

Despite many successful applications in biomedical engineering, there is a growing concern for safety with CNTs. Some recent in vitro studies have reported increased cytotoxicity of CNTs due to their cellular uptake, agglomeration, and induced oxidative stress. These conflicting results regarding the biocompatibility of CNTs largely stem from the variability of CNTs (i.e., size, surface properties, and functionalization) and testing subjects (i.e., in vitro vs. in vivo, types of cells, tissues, and animals tested). In addition, increased cytotoxicity has often been attributed to incomplete removal of metal catalysts used to prepare CNTs. Most in vivo studies using CNTs have shown that they did not cause significant toxicity and reported renal clearance from the body, although small portions of CNTs have been shown to accumulate in certain organs, such as lungs, liver, and spleen, and may cause inflammation. However, the cytotoxicity seems to be more highly variable and more pronounced at the cellular level, based on several in vitro cell culture studies.

With the continued growth of CNTs in various biomedical fields, more systematic biological evaluations of CNTs having various chemical and physical properties are warranted in order to determine their precise pharmacokinetics, cytotoxicity, and optimal dosages. Nevertheless, many studies have shown that toxicity can effectively be minimized by functionalizing the CNTs with biocompatible polymers or surfactants to prevent aggregation.

Graphene

Graphene is the latest nanomaterial to burst onto the scene. The ground-breaking work by Geim and Novoselov provided a simple method for extracting graphene from graphite viaexfoliation and explored its unique electrical properties. Graphene and CNTs possess similar electrical, optical, and thermal properties, but the two-dimensional atomic sheet structure of graphene enables more diverse electronic characteristics; the existence of quantum Hall effect and massless Dirac fermions help explain the low-energy charge excitation at room temperature and the optical transparency in infrared and visible range of the spectrum.[34] In addition, graphene is structurally robust yet highly flexible, which makes it attractive for engineering thin, flexible materials.

Graphene in Biomedical Engineering

Research into utilizing graphene for biomedical applications has been limited to date, as graphene research itself is still in its infancy. Graphene oxide (GO), produced by oxidation of graphite under acidic conditions, is more commonly used, as it offers several advantages over using pure graphene. First, GO is dispersible in aqueous media, which is essential for biological applications. Second, GO presents hydrophilic functional groups that enable chemical functionalization. Third, GO has broader ranges of physical properties than pure graphene due to its structural heterogeneity. Several biomedical applications including injectable cellular labeling agents, drug delivery systems, and scaffold reinforcements have been explored using GO, as has similarly been done with CNTs. For example, Zhang et al. functionalized GO with folic acid (FA) as a cancer targeting molecule and loaded doxorubicin and camptothecin, known cancer drugs, onto the large surface area of GO via $\pi-\pi$ stacking . The drug-loaded FA-GO showed improved cancer targeting capability and anti-cancer activity as compared with drugs delivered alone or drugs carried with unmodified GO. Sun et al. also demonstrated that poly(ethylene glycol)-conjugated GO with targeting molecules could be used as a cellular sensor by utilizing the intrinsic photoluminescence property of GO at the NIR region. In another study, Zhang et al.incorporated GO into poly(vinyl alcohol) hydrogels to improve their mechanical strength.

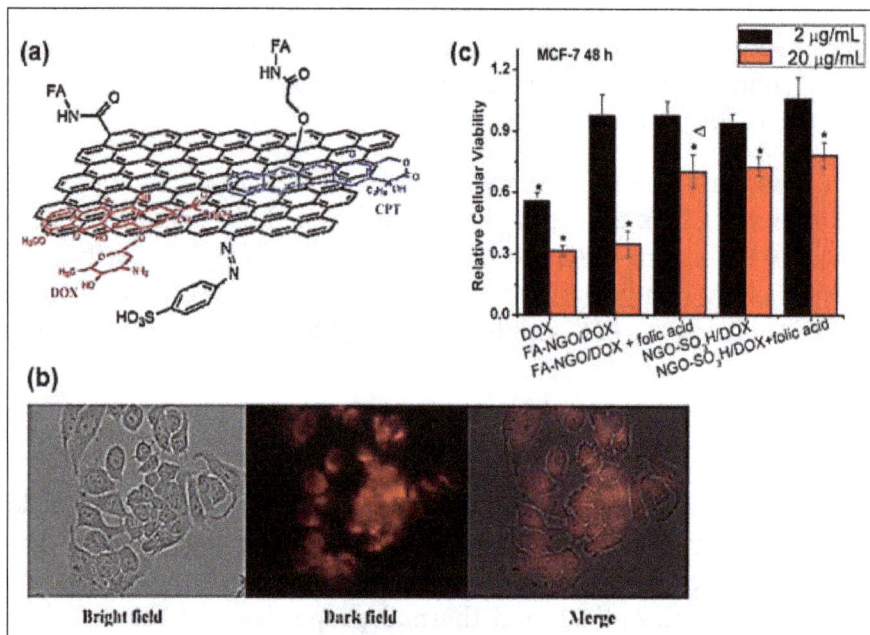

(a) Graphene oxide (GO) was funtionalized with folic acid (FA) as a targeting molecule (FA-NGO), then loaded with anti-cancer drug, doxorubicin (DOX) or camptothecin (CPT). (b) FA-NGO was localized into MCF-7 (human breast cancer) cells, identified with fluorescent labeling with rhodamine. (c) Greater anti-cancer activity of DOX was observed when FA-NGO was used as a drug carrier.

Other Carbon-based Nanomaterials

Buckminsterfullerene (C_{60}), also commonly known as the buckyball, is a spherical closed-cage structure (truncated icosahedron) made of sixty sp^2 carbon. Its discovery in 1985 and subsequent investigation led to the uncovering of electronic properties, stemming from its highly symmetrical structure, and potential applications, culminating in a Nobel Prize in 1996.[42] It can be argued that the scientific pursuit of CBNs and their potential applications began with the discovery of C_{60}. The popularity of C_{60} has somewhat diminished in recent years with the rise of more scalable and practical CBNs such as CNTs and graphene. However, its uniform size and shape as well as availability for chemical modification led many scientists to develop C_{60} derivatives for therapeutic purposes. Perhaps the most fascinating and highly promising aspect of C_{60} is its anti-human immunodeficiency virus (HIV) activity. Schinazi et al. first discovered a group of water-soluble C_{60} derivatives capable of inhibiting HIV protease activity by binding to its active site, due to their unique molecular structure and hydrophobicity. Various C_{60} derivatives have since been developed that display anti-HIV activity by targeting other important HIV enzymes, such as reverse transcriptase. These results demonstrate that C_{60} p derivatives may become a potent group of AIDS therapeutics in the future.

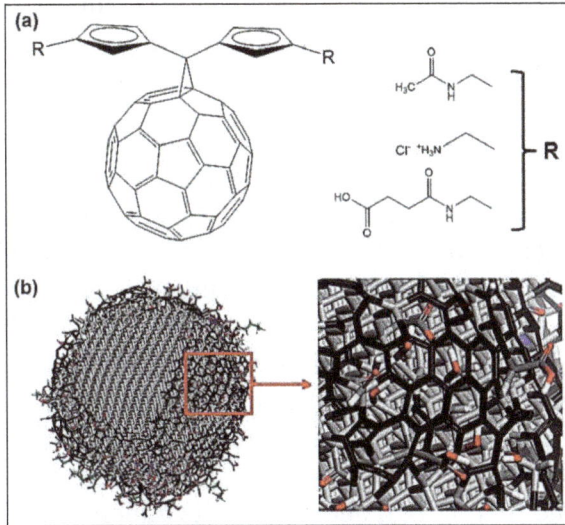

(a) A class of C_{60} derivatives that display anti-HIV activity. (b) Molecular structure of a 5 nm-diameter nanodiamond (ND). The ND is mostly made up of sp^3 carbon, but the outer layer is functionalized with sp^2 carbon and other functional groups.

Nanodiamond (ND) has also generated interest in the field of biomedical engineering in recent years .Nanodiamonds are synthesized by high energy treatment of graphite, most commonly via detonation, and are smaller than 10 nm. They have similar physical properties as bulk diamond, such as fluorescence and photoluminescence, as well as biocompatibility. Unlike other CBNs, NDs are made up mostly of tetrahedral clusters of sp^3-carbon. The surface of nanodiamonds, however, is functionalized with various functional groups or sp^2-carbon for colloidal stability, which enables chemical modification for targeted drug and gene delivery and tissue labeling. For example, Lien et al. recently used fluorescent and magnetic NDs for cell labeling. polyethyleneimine (PEI)-conjugated NDs were highly effective as gene carriers, without the cytotoxicity associate with PEI alone.

Nanoporous Material

Nanoporous materials consist of a regular organic or inorganic framework supporting a regular, porous structure. The size of the pores is generally 100 nm or smaller. Most nanoporous materials can be classified as bulk materials or membranes. Activated carbon and zeolites are two examples of bulk nanoporous materials, while cell membranes can be thought of as nanoporous membranes. A porous medium or a porous material is a material containing pores (voids). The skeletal portion of the material is often called the "matrix" or "frame". The pores are typically filled with a fluid (liquid or gas). There are many natural nanoporous materials, but artificial materials can also be manufactured. One method of doing so is to combine polymers with different melting points, so that upon heating, one polymer degrades. A nanoporous material with consistently sized pores has the property of letting only certain substances pass through while blocking others.

Nanoporous materials are of great interest due to their applications in filtration, extraction, separation, and catalysis. Due to the small pores in these materials, discrimination between molecules and ions based on sizes and shapes is possible, while the confined environment enhances chemical reactions. For making small pores, the self-organization of liquid crystals and their polymers is very appealing. For this goal, smectic phases can be used to prepare nanoporous membranes. Thin nanoporous films based on hydrogen bonded smectic liquid crystalline acrylates having straight pores with a very narrow size distribution have been fabricated. The anionic pores have a 2D geometry with the integrity of the film maintained by the presence of a small concentration of a covalent smectic crosslinker. This entity determines to what extent the smectic layers may separate and their length determines the dimension of pore that is formed, typically in the range of 1 nm.

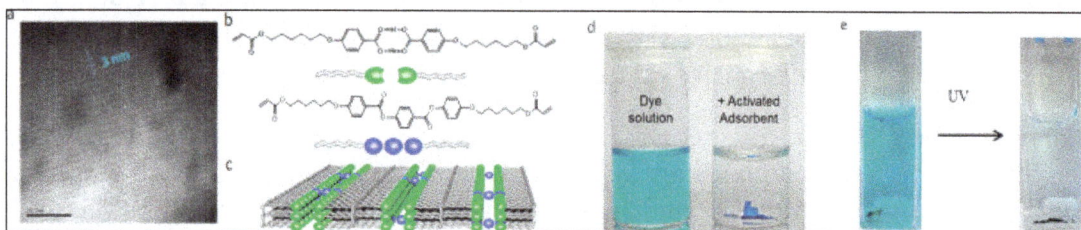

(a) Cryo-TEM picture showing nanopores in a smectic LC network after filling with barium cations for contrast, (b) chemical structure of the hydrogen bonded dimer (green) and cross linker (blue), and (c) schematic representation of the nanoporous polymer after polymerization. A two-dimensional porous structure is formed after breaking the hydrogen bonds in an alkaline solution. The layers are held together by covalent smectic crosslinkers. (d) Adsorption of a cationic dye in the nanoporous adsorbent. Vials with the initial dye solution and the dye solution with the adsorbent. (e) Exposure of photoresponsive smectic LC polymer film to UV light results in the adsorption of a cationic dye.

These porous systems can be used as absorbents which are able to efficiently selectively adsorb cationic dyes over anionic, zwitterionic, and larger cations. These materials show extraordinary adsorption capacities of nearly 1 g dye for 1 g adsorbentand a high adsorption rate constant, which are very competitive with other adsorbents.

The adsorbate can be released by acid treatment, which makes the reuse of adsorbate possible. Such pH responsive LC adsorbates could also be interesting for the recovery of valuable materials from waste streams or seawater, for example. Photoresponsive nanoporous polymer filmshave also been fabricated by adding a photoresponsive azobenzene cross linker to the hydrogen bonded smectic liquid crystalline polymer network. Upon exposure to UV light a decrease in the smectic layer spacing was observed, suggesting a decrease in pore size. In addition, the binding sites in the material were changed with light, leading to light induced adsorption of cations and cationic dyes. Recently, the use of these porous systems as reaction medium and printable pH responsive hydrogels having asymmetric swelling properties have also been reported.

Applications of Nanomaterials

As nanomaterials have exclusive, advantageous physical, chemical, and mechanical properties, they can be applied in a host of applications.

Next-generation Computer Chips

The microelectronics sector has paid special attention to miniaturization, which involves reducing the size of circuits like transistors, capacitors, and resistors. A considerable reduction in their size enables microprocessors developed using these parts, to operate much faster, thus allowing computations at much greater speeds.

However, there are a number of technical obstacles to achieving these advancements, such as the lack of ultrafine precursors to make these parts, inadequate dissipation of huge amounts of heat generated by these microprocessors because of faster speeds, poor mean time to failures (poor reliability), etc.

Nanomaterials help the industry to overcome these obstacles by offering manufacturers materials with better thermal conductivity, nanocrystalline starting materials, ultra-high-purity materials, and longer-lasting, durable interconnections (connections between different parts in the microprocessors).

Kinetic Energy Penetrators with Improved Lethality

The Department of Defense (DoD) has been using depleted uranium (DU) projectiles (penetrators) for its battle against hardened targets and armored vehicles of enemies. But DU has residual radioactivity; therefore, it is explosive, harmful (carcinogenic), and lethal to personnel who use them. However, some of the key reasons for the sustained use of DU penetrators are that they have an exclusive self-sharpening mechanism on impact with a target, as well as the lack of an appropriate non-explosive, non-toxic alternative for DU.

Nanocrystalline tungsten heavy alloys can be used for such self-sharpening mechanisms due to their exclusive deformation characteristics, for example, grain-boundary sliding. Therefore, nanocrystalline tungsten heavy alloys and composites are being assessed for use as alternative DU penetrators.

Better Insulation Materials

Nanocrystalline materials manufactured by the sol-gel method give rise to foam-like structures known as "aerogels". In spite of being extremely lightweight and porous, these aerogels can hold loads equal to 100 times their weight. Aerogels are made up of continuous 3D networks of particles with air (or any other fluid, such as a gas) trapped at their interstices.

Since aerogels are porous and include air trapped at the interstices, they are used for insulation in homes, offices, etc. This considerably reduces the cooling and heating bills, thus saving power and decreasing the associated environmental pollution.

They are also being employed as materials for "smart" windows, which darken when the sun is very bright (same as in changeable lenses in sunglasses and prescription spectacles), and lighten when the sun is not shining very brightly.

Phosphors for High-definition TV

The resolution of a monitor or television is subject to the size of the pixel. These pixels are fundamentally composed of materials known as "phosphors," which glow when struck by a stream of electrons within the cathode ray tube (CRT). The resolution enhances with a reduction in the pixel size or the phosphors.

Nanocrystalline zinc selenide, cadmium sulfide, zinc sulfide, and lead telluride prepared through the sol-gel methods are potential materials for enhancing the resolution of monitors. The use of nanophosphors is intended to lower the cost of these displays to make personal computers and high-definition televisions (HDTVs) affordable for an average household in the United States.

Low-cost Flat-panel Displays

In the laptop (portable) computer industry, the demand for flat-panel displays is high. Japan is leading in this area, mainly due to its RandD efforts on the materials for these displays.

The resolution of these display devices can be significantly improved by synthesizing nanocrystalline phosphors, while considerably bringing down the manufacturing costs. Furthermore, the flat-panel displays manufactured using nanomaterials have far higher contrast and brightness compared to the traditional ones due to their improved magnetic and electrical properties.

Tougher and Harder Cutting Tools

Cutting tools made of nanocrystalline materials like carbides of tantalum, tungsten, and titanium, are a lot harder, much more erosion-resistant and wear-resistant, and last longer than their traditional (large-grained) equivalents. They also allow the manufacturer to machine several materials much faster, thereby boosting productivity and largely minimizing manufacturing costs.

Moreover, miniaturizing microelectronic circuits necessitates microdrills (drill bits having diameters lesser than the thickness of an average human hair [100 μm]) with improved edge retention and much better wear resistance. Nanocrystalline carbides are being used in these micro drills since they are much harder, stronger, and wear-resistant.

Elimination of Pollutants

Nanocrystalline materials have very large grain boundaries corresponding to their grain size. Therefore, they are very active with regards to their physical, chemical, and mechanical properties. Owing to their improved chemical activity, nanomaterials can be employed as catalysts to react with toxic and noxious gases such as nitrogen oxide and carbon monoxide, in power generation equipment and automobile catalytic converters, to avoid environmental pollution caused when gasoline and coal are burnt.

High Energy Density Batteries

Traditional and rechargeable batteries are used in nearly all applications that necessitate electric power. These applications include laptop computers, automobiles, toys, electric vehicles, personal stereos, cordless phones, cellular phones, watches, and next-generation electric vehicles (NGEV) that reduce environmental pollution. The energy density (storage capacity) of these batteries is very low, necessitating frequent recharging. The life of traditional and rechargeable batteries is also low.

Nanocrystalline materials produced using sol-gel methods have a foam-like (aerogel) structure that can store significantly more energy than their traditional equivalents. Hence, they are highly suitable for separator plates in batteries. Moreover, nickel-metal hydride (Ni-MH) batteries made of nanocrystalline nickel and metal hydrides have been predicted to necessitate much lesser recharging and to last considerably longer because of their large grain boundary (surface) area and improved chemical, physical, and mechanical properties.

High-power Magnets

A magnet's strength is measured in terms of saturation magnetization and coercivity values. These values will increase when there is a decrease in the grain size and an increase in the specific surface area (surface area per unit volume of the grains) of the grains. It has been demonstrated that magnets made of nanocrystalline yttrium-samarium-cobalt grains have highly uncommon magnetic properties because of their extremely large surface area.

Common applications for these high-power rare-earth magnets include ultra-sensitive analytical instruments, quieter submarines, land-based power generators, automobile alternators, motors for ships, and magnetic resonance imaging (MRI) in medical diagnostics.

High-sensitivity Sensors

Sensors use their sensitivity to detect the variations in different parameters they are programmed to measure. The parameters include chemical activity, thermal conductivity, electrical resistivity, magnetic permeability, and capacitance. All of these parameters depend a lot on the microstructure (grain size) of the materials used in the sensors.

A variation in the sensor's environment is revealed by the sensor material's physical, chemical, or mechanical characteristics, which is leveraged for detection. For example, a carbon monoxide sensor made of zirconium oxide (zirconia) applies its chemical stability to identify whether carbon monoxide is present.

When carbon monoxide is present, the oxygen atoms in zirconium oxide react with the carbon in carbon monoxide to reduce zirconium oxide partially. This reaction activates a modification in the sensor's characteristics, such as capacitance and conductivity (or resistivity).

The rate and the degree of this reaction are significantly increased by a decrease in the grain size. Therefore, sensors made of nanocrystalline materials are highly sensitive to variations in their environment. Common applications for sensors made using nanocrystalline materials are ice detectors on aircraft wings, smoke detectors, automobile engine performance sensors, etc.

Automobiles with Greater Fuel Efficiency

Existing automobile engines waste substantial amounts of gasoline, thus adding to environmental pollution by burning the fuel incompletely. A traditional spark plug is not made to burn the gasoline totally and efficiently. This problem is amplified by faulty, or worn-out, spark plug electrodes.

Since nanomaterials are harder, stronger, and considerably more erosion-resistant and wear-resistant, they are currently being proposed for use as spark plugs. These electrodes extend the service life of the spark plugs and help burn fuel far more efficiently and fully. A totally new spark plug design known as the "railplug" is also in the prototype stage.

This railplug applies the technology derived from the "railgun"—a spin-off of the famous Star Wars defense program. However, these railplugs produce much stronger sparks (with an energy density of almost 1 kJ/mm²). Hence, traditional materials erode and corrode very quickly, and quite often are not of any practical use in automobiles.

By contrast, railplugs made of nanomaterials are much more long-lasting than even the traditional spark plugs. Furthermore, automobiles waste substantial amounts of energy by losing the thermal energy produced by the engine. This is particularly true with diesel engines. Hence, plans have been proposed to coat the engine cylinders (liners) with nanocrystalline ceramics, such as alumina and zirconia, so that they preserve heat in a more efficient manner, thus ensuring full and efficient fuel combustion.

Aerospace Components with Enhanced Performance Characteristics

Owing to the hazards involved in flying, aircraft manufacturers aim to make the aerospace components tougher, stronger, and last longer. One of the main properties needed in aircraft components is fatigue strength, which declines as the age of the component increases. By manufacturing the components using more robust materials, the aircraft's life can be significantly increased.

The fatigue strength increases with the decrease in the grain size of the material. Nanomaterials offer such a considerable reduction in the grain size over traditional materials that fatigue life is increased by an average of 200%–300%. Moreover, components made using nanomaterials are stronger and can work at higher temperatures, enabling aircraft to fly faster and more efficiently (using the same quantity of aviation fuel).

In spacecraft, higher-temperature strength of the material is critical as the components (for example, thrusters, rocket engines, and vectoring nozzles) work at much higher temperatures than aircraft and at greater speeds. Nanomaterials are ideal contenders for spacecraft applications, as well.

Better and Future Weapons Platforms

Traditional guns, such as 155 mm howitzers, cannons, and multiple-launch rocket systems (MLRS), use the chemical energy derived by burning a charge of chemicals (gun powder). The penetrator can be propelled at a maximum velocity of about 1.5–2.0 km/second.

Conversely, electromagnetic launchers (EML guns), or railguns, utilize electrical energy, as well as the concomitant magnetic field (energy), for propelling the penetrators/projectiles at velocities of up to 10 km/second. Such an increase in velocity causes greater kinetic energy for the same penetrator mass. The amount of energy is directly proportional to the damage imparted to the target. Therefore, the DoD (particularly the U.S. Army) has undertaken wide-ranging research on the railguns.

Since a railgun works on electrical energy, the rails have to be excellent conductors of electricity. Furthermore, they need to be so strong and inflexible that the railgun does not sag while firing and collapse due to its own weight. Copper is the apparent choice when it comes to high electrical conductivity.

However, railguns made of copper wear out considerably faster because of the erosion of the rails by the hypervelocity projectiles. Moreover, they lack high-temperature strength. The erosion and wear of copper rails call for very frequent barrel replacements.

To fulfill these necessities, a nanocrystalline composite material made of copper, tungsten, and titanium diboride is being assessed as a potential candidate. This nanocomposite exhibits the necessary electrical conductivity, satisfactory thermal conductivity, outstanding high strength, hardness, high rigidity, and wear/erosion resistance.

This results in erosion-resistant and wear-resistant railguns that last longer and can be fired more often than their traditional equivalents.

Longer-lasting Satellites

Satellites are being employed for both civilian and defense applications. These satellites make use of thruster rockets to stay in or alter their orbits because of various factors, such as the impact of gravitational forces applied by the earth. Hence, thrusters are needed to reposition the satellites.

To a great extent, the life of these satellites is governed by the amount of fuel they can take on board. In reality, repositioning thrusters waste over one-third of the fuel carried aboard by the satellites, caused by partial and inefficient burning of the fuel, such as hydrazine. The partial and inefficient combustion occurs due to rapid wearing out of onboard ignitors that stop performing effectively as a result.

Nanomaterials like nanocrystalline tungsten-titanium diboride-copper composite are promising options for improving the performance features and life of these ignitors.

Longer-lasting Medical Implants

In general, medical implants like heart valves and orthopedic implants are made of stainless steel and titanium alloys. These alloys are mainly used in humans as they are biocompatible, that is, they do not adversely react with human tissue. These materials are comparatively non-porous when used in orthopedic implants (artificial bones for hip, etc.).

If an implant must mimic a natural human bone in an effective way, the adjacent tissue must penetrate the implants, thus offering the implant the necessary strength. Since these materials are comparatively impermeable, human tissue does not penetrate the implants, thus minimizing their effectiveness.

Moreover, these metal alloys wear out fast, requiring frequent and very expensive surgeries. But nanocrystalline zirconia (zirconium oxide) ceramic is hard, corrosion-resistant (biological fluids are corrosive), wear-resistant, and biocompatible.

It is also possible to make nanoceramics porous as aerogels (aerogels can endure up to 100 times their weight) if they are produced using sol-gel methods. This would lead to much lesser implant replacements and thus a substantial reduction in surgical expenses. Nanocrystalline silicon carbide (SiC) is a potential material for artificial heart valves mainly because of its low weight, wear resistance, extreme hardness, high strength, corrosion resistance, and inertness (SiC does not react with biological fluids).

Ductile and Machinable Ceramics

As such, ceramics are extremely hard, brittle, and tough to machine. These properties of ceramics have dissuaded prospective users from leveraging their advantageous properties. However, these ceramics have been used more and more with reduced grain size. Zirconia, a hard, brittle ceramic, has even been made a superplastic, that is, it can be deformed to greater lengths (up to 300% of its initial length). However, these ceramics must have nanocrystalline grains to be superplastic.

Actually, nanocrystalline ceramics like silicon carbide (SiC) and silicon nitride (Si_3N_4) have been used in automotive applications such as ball bearings, high-strength springs, and valve lifters. This is because they have good machinability and formability, together with superior physical, mechanical, and chemical properties. They are also used as components in high-temperature furnaces.

It is possible to press and sinter nanocrystalline ceramics into different shapes at considerably lower temperatures. By contrast, it would be very hard, if not impracticable, to press and sinter traditional ceramics even at high temperatures.

Large Electrochromic Display Devices

An electrochromic device comprises materials wherein an optical absorption band can be added, or a current band can be modified by passing current through the materials, or by applying an electric field.

Nanocrystalline materials like tungstic oxide ($WO_3 \cdot xH_2O$) gel are used in huge electrochromic display devices. The reaction controlling electrochromism (a reversible coloration process that is influenced by an electric field) is the double-injection of ions (or protons, H^+) and electrons, which form tungsten bronze by combining with the nanocrystalline tungstic acid. These devices are mainly used in ticker boards and public billboards to send information.

Electrochromic devices are analogous to liquid-crystal displays (LCD) generally used in watches and calculators. But electrochromic devices display information by changing in color in response to an applied voltage. The color gets bleached upon reversing the polarity. The resolution, contrast, and brightness of these devices largely rely on the tungstic acid gel's grain size. Therefore, nanomaterials are being investigated for this purpose.

References

- Dendrimers: properties and applications. Acta Biochim Pol. 2001;9:199–208

- Nanoporous-material, materials-science: sciencedirect.com, Retrieved 20 January, 2019

- Carbon Materials and Nanotechnology. Wiley-VCH; Weinheim: 2010

- Nanotechnology in the Real World: Redeveloping the Nanomaterials Consumer Products Inventory. Journal of Nanotechnology, 6, 1769-1780

- Nanostructured Oxides in Chemistry: Characterization and Properties. Chem. Rev., 104, (9), 4063-4104, (2004)

Properties of Nanomaterials

Nanomaterials possess different physical, mechanical, electrical and optical properties. Structure, field emission, hardness, elastic modulus, adhesion, conductivity, scattering, absorption, etc. fall under the domain of these properties. This chapter has been carefully written to provide an easy understanding of the various properties of nanomaterials.

Physical Properties

Structures

Auger electron spectroscopy (AES) and X-ray photonelectron spectroscopy (XPS) can be utilized to characterize the surface impurities to a depth of 0.5~1 nm with a spatial resolution of 0.2 μm for AES and 0.2 mm for XPS, and a sensitivity of 0.3% for both analyzers. The TEM electron diffraction pattern and the powder X-ray diffraction pattern can clearly display the crystal composition and structure. The Scherer's equation can be exploited to estimate the average grain size R for a knowing X-ray wavelength λ at the diffraction angle θ from the equation as given by:

$$\text{FWHM} = 0.94\lambda/R \cos \theta$$

Where the FWHM is the full width at the half maximum of the characteristic spectrum in units of radians, R and λ are in nm. The impurity levels can be quickly examined by the electron probe micro-analysis (EPMA) from a scanning electron microscope (SEM). The chemical bonding modes that attached on the surface of the metal particles can be enunciated by the observation of the infrared absorption spectra. The metallic nanoparticles or the grounded powder of sol gel glass embedded with metal particles are mixed with transparent KBr or polyethylene glycol powders for observing at wavelengths of 400~500 cm−1 and 650~200 cm⁻¹, respectively, in a 1:10 volume ratio and then pressed into disks.

Carbon nanotubes with the inlet of ammonia gas, the growth on Cu-20 wt% Fe substrate reveals a thick layer of cotton-like substance. The outlook is a blue-purple, fluffy structure with a thickness more than 3 mm. The SEM image as shown in figure illustrates an extremely delicate filament of 500 nm in length. We can juxtapose the electron diffraction and morphology of the HR-TEM as shown in figure. The inset shows weak diffraction rings due to embedded polycrystalline metal particles and each spot composed of five parts arising from tubes of different sizes. It clearly indicates the chiral structure and the tube diameter of about 25 nm with longitudinal black-and-white strias of 0.29 nm in width. As counting from the point of white arrow, repetitive five-layers of tube-body-like carbon atoms follow two-layers of wall-like with a canted angle between wall and body being 45°. It is similar to a chiral tube as reported by Dresselhaus. There are 7~8 chiral tubes present throughout the whole cross-section. Apparently, the reports of Dresselhaus yield succinct recapitulation to the projection of single tube and single wall tube. However, there are exceptions in this study where many extremely delicate tubes accumulate in a bundle.

The Raman spectrum as shown in indicates that the intensity of G band is obviously higher than D band which also occurs a shoulder at 1620 cm⁻¹ implying a structure of graphite layer. Whereas the selected diffraction pattern as shown in the inset of figure reveals a hexagonal close packed (HCP) pattern of multi-tubes and a strong ring caused by carbon gel on copper net in accompany with two vague concentric-rings meaning that the amorphous carbon layers are still remnant in the sample. In this scenario single and multi-wall carbon tubes are accumulated in bundles embedded with amorphous carbon between tubes. This "carbon nano-polytubes" (CNPT) are expected to have the best hydrogen storing for CNTs for their pure physical process. A tedious and genius experiment of the HRTEM as illustrated in figure for the solid cross-section of a bundle reveals the well-arranged concentric strias. The inset indicates white-background black-centered little hexagon spots for the exaggeration of the arrowed point. The diameter of these spots is about 0.53~1.25 nm with angled, singled, or doubled tubes bunched together. At the center part of the cross-section appearing in gray, more tubes accumulated with closer wall-to-wall resulting from fewer embedded amorphous carbon than those at the periphery parts.

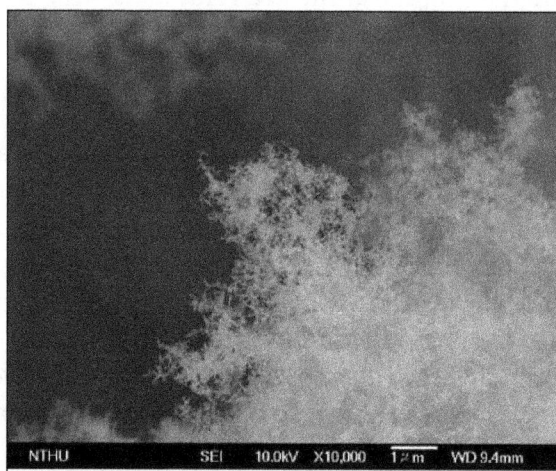

SEM of carbon polytubes (CNPTs) grown on Cu-20 wt% Fe bulk alloy substrate with H_2 and NH_4 as diluted gas.

Another laborious HRTEM picture as shown in figure illustrates a multi wrapped bundle of CNT for the sample deposited at substrate temperature of about 50° lower than that of figure. Since the column is inclined at about 15° to the electron beam, we can see simultaneously the column and the cross-section.

Carbon-related films are expected to have important applications in electronic devices owing to their inheriting with the strong electron field emission. Conventionally in many reports, the Fowler-Nordheim (F-N) equation was controvertibly implemented to delineate the results of field emission of diamond films and to calculate the work function. The F-N equation is derived for electron field emission from normal metal surfaces at high fields and low temperatures, and cannot delineate the temperature effect of the field emission for diamond films. There were some improved theoretical models for studying the field emission of diamond films. Further, some experimental results were reported illustrating the temperature dependence on the field emission of phosphorus-doped diamond films. But, there are few literatures studying the temperature dependent on field emission of boron-doped diamond films by theoretical models. We have developed a theoretical model for solving the field emission of semiconductor which conclusively can express the experimental works of the temperature dependence on field emission of boron doped diamond and diamond-like films.

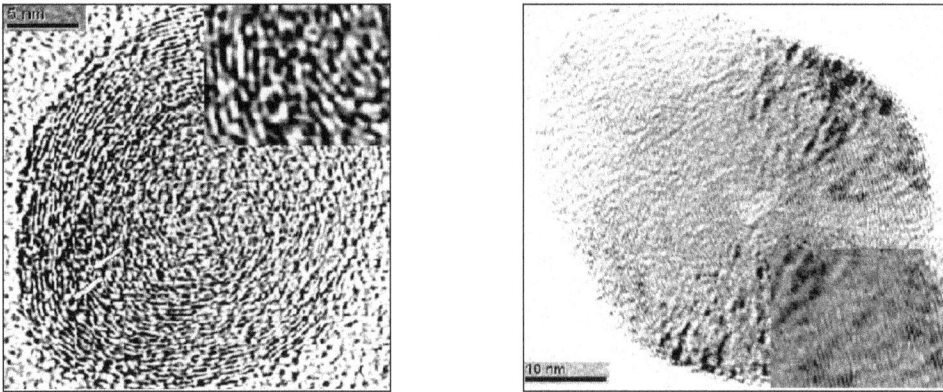

In figure: (a) The HRTEM for the solid cross-section of a bundle reveals the well-arranged concentric strias. The inset indicates whitebackground black-centered little hexagon spots for the exaggeration of the arrowed point. (b) HRTEM of a multi-circularly wrapped nanobundle (TWCNT). The insets are the exaggerated local points.

Field Emissiont

Firstly polycrystalline boron-doped diamond films were grown on p-type (100)-oriented Si substrate by an Astex 5400 microwave plasma enhanced chemical vapor deposition (MPECVD) system. To study the temperature effect on the field emission, the samples were mounted on the cooling station of a helium closed-cycle refrigerator. The diamond film was separated from the anode (indium-tin oxide coated glass), by a 50μm Teflon spacer. Typical measurements were made as the sample cooled down from room temperature to 20 °K. Aprogrammable current source provides a stepwise constant current with output voltages varying from 0 to 1100 V. On account of the low electron affinity, electrons emitted from p-type semiconductor can be presumed to comply with the emission current of metal-insulator semiconductor (MIS) Schottky barrier diode with total emission current density as giving by:

$$J_{total} = J_c + J_v + J_s$$

where J_c, J_v, and J_s are the emission current densities, which are proportional to the tunneling probabilities T_c, T_v, and T_s from the conduction band, the valence band and surface state, respectively. Using the WKB approximation. The transmission probability for electrons to tunnel through the potential barrier can be readily derived to yield the current density as given by $J_c = A^*T^2 T_c e^{-q(\phi_{Bn} - V_s)/KT}$, where A* is the Richardson constant, Tc is the electron tunneling coefficient, φ_{Bn} is the barrier height of n-type semiconductor, and Vs is the applied voltage developed across the semiconductor. Bandis et al. had alluded that field emission from surface states is unlikely implying the negligibility of J_s. A tedious manipulation of equation $FWHM = 0.94\lambda / R\cos\theta$ can yield a simple formalism as given by:

$$J_{total} = a_1 T^2 e^{-\theta/KT} e^{-b_1 \chi^{3/2}/F} + a_2 \frac{1}{\chi + E_g} F^2 e^{-b_2(\chi + E_g)^{3/2}/F}$$

where,

$$a_1 = \frac{qmk^2}{2\pi^2\hbar^3}, b_1 = \frac{4\sqrt{2m}}{3q\hbar}v(y),$$

$$a_2 = \frac{\pi q^3 m_p \hbar^3}{2mt^2(y)}, b_2 = \frac{4\sqrt{2m}}{3q\hbar}v(y)$$

v(y) is a tabulated function involving elliptic integrals, χ is the electron affinity, t(y) is another tabulated function, θ is the difference energy between surface conduction band minimum and Fermi level, and F is the external electric field.

We tacitly assumed that the emission current dropped due to the increase of surface barrier height as temperature decreases attributed to the result from the prediction of F-N equation. But the F-N equation contains only one parameter of the material, the effective surface barrier ϕ, is insufficient to describe the band diagram of diamond. In this study, we include two fundamental parameters, the electron affinity χ and surface band bending $E_{cs} - E_f$ which can clearly elucidate the band diagram as shown in figure for field emission process. It illustrates electrons tunnel through the surface barrier. In cases of the presence of density of states with the Fermi level higher than those of P-type bulk diamond films, the electrons flow cause band bending near the surface. If the Fermi level of surface states lowers down as temperature increases, less electrons flow from surface state to bulk p-diamond film. Consequently the bending down of surface band becomes shallow. The threshold turn-on voltage rises as temperatures decrease due to the enhancement of band bending down. The catastrophic bending down of surface conduction band at lower temperatures results in a lower electron emission. Similar result of B-doped diamond-like films is shown in figure. The electron affinity retains its original value at various temperatures. The work function of diamond-like films has a larger value than that of diamond films resulting in a lower electron emission current at the same applied bias.

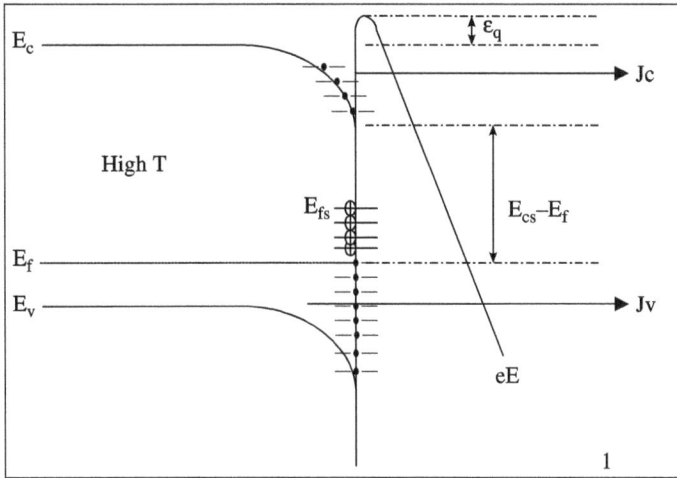

Schematic energy band diagrams for (a) high temperature; and (b) low temperature, respectively.

This work illustrates that the field emission current from the valence band has negligible dependence on temperatures and is negligibly small for p-type diamond and diamond-like films which dominates the emission current at low temperatures, while the field emission current from the conduction band depends on temperature largely. The conventional F-N equation is applicative to address the emission current at high field and low temperatures. Although the F-N equation can fit the emission current–voltage curve with various work functions at different ranges of temperatures, whilst the work function is an intrinsic invariant constants. We proposed that the behavior of the field emission characteristics can be well expressed by a similar theory of the thermionic emission current of MIS Schottky barrier diodes. The empirical fitting results show that the electron affinity is unaffected while the bending down of surface states increases as the temperature decrease.

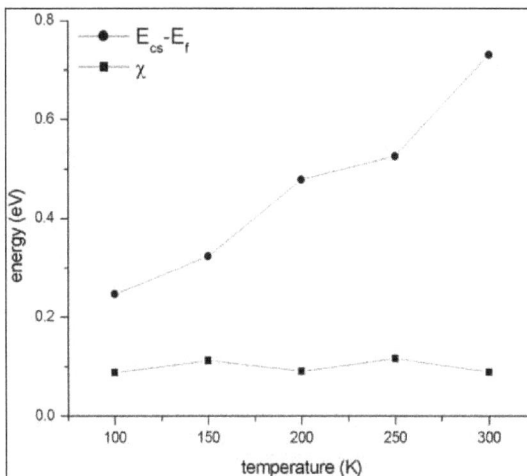

Theoretical fitting for B-doped diamond-like films with χ the electron affinity, $E_{cs} - E_f$ the energy difference between surface conduction level and Fermi level.

To study the temperature effect of field emission properties of CNTs, the emission current was measured in cycles of cooling down and warming up. The fitting results of the electric field dependence on current density at various temperatures are depicted in figure. The curves illustrate that the emission currents sensitively depend on temperatures. This behavior is quite different from the temperature dependence of the FE characteristic of well-aligned CNT films. The semiconductor thermionic equation that can successfully address the field emission of CNTs involves three fundament parameters of material properties i.e., the electron affinity χ, the energy gap $E_{g,}$ and the interface barrier height $E_{cs} - E_f$. Figure illustrates theoretical fitting parameters as deduced from the data of figure, in which β is 367. It implies that the energy difference between the conduction band and the Fermi level ($E_{cs} - E_f$) decreases as temperature decreases, while the electron affinity χ and the energy gap $E_g (\sim 2.5\ eV)$ is almost independent of temperatures. Thus we can elucidate the FE mechanism of random aligned R-CNT, which is similar to the results of well aligned A-CNT. At low temperatures, few electrons flow out from the semiconductor tip of the metallic multi-wall carbon nanotubes (MWCNTs) to surface localized states due to less ionization of localized state. The Fermi level of interface localized states is high resulting in lowering the interface barrier-height as the temperature decreases.

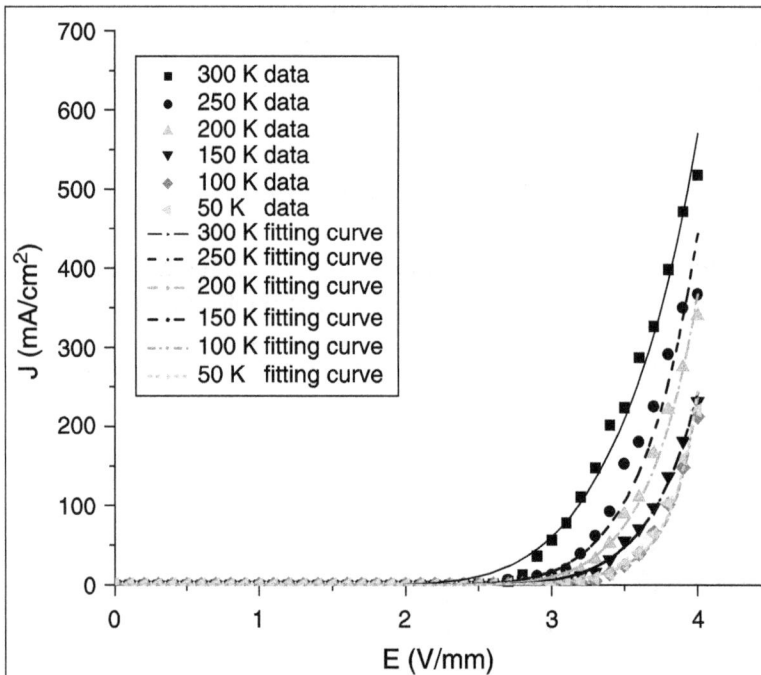

The temperature dependence of field emission
for randomly oriented MWCNT films.

As compared to the field emission properties of well - aligned CNTs as shown in figure, it is beyond our consensus that the threshold voltage of randomly oriented CNT films is smaller than that of well-aligned CNTs. The threshold field of randomly oriented CNT is about 27000 V/cm, while that of well-aligned CNT is about 33000 V/cm. The corresponding emission current density of R-CNT is $5.5 * 10^{-4}$ A/cm^2, while that for A-CNT is $1.0 * 10^{-5}$ A/cm^2. The enhancement factor of R-CNT ($\beta = 367$) is also larger than that of A-CNT that is also controvertible to physical consensus. The ideal value of enhancement factor β of CNT is the ratio of the average height h to the radius r of the tube that is crucially determined by the geometry of the body, and independent to the material. For randomly oriented CNT, the β is nearly equal to 1. Theoretically, the field emission current only contributes from the tube-end not from the tube-body. However, some authors reported that randomly distributed CNTs have impressive field emission capabilities.

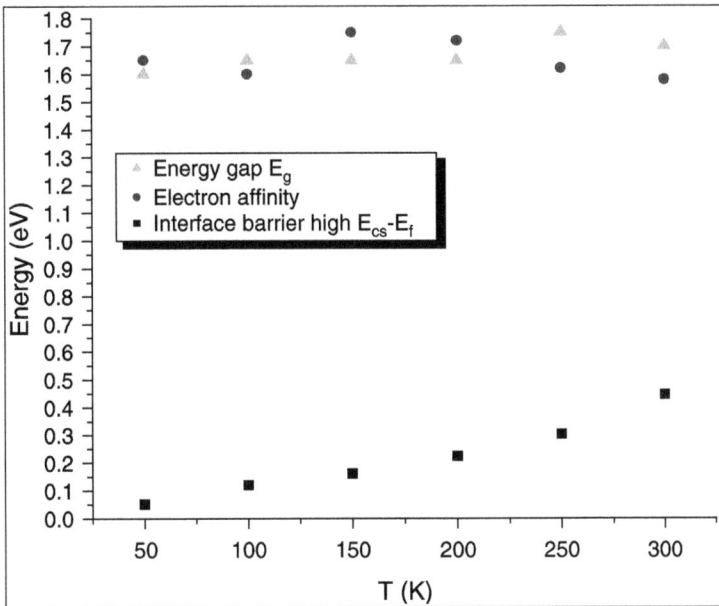

The theoretical fitting for randomly-oriented
MWCNT films.

The enhancement factor is attributed to the CNTs that are aligned perpendicular to the substrate. The experimental result is contradictory to the apparent consensus that the enhancement factor of R-CNTs will be less than that of A-CNTs. Though there are some CNTs that are perpendicular to the substrate, where the screening effect can be ignored. Zou tacitly assumed that the contribution of field emission efficiency from the local π electrons decreases from the tip to the body. However, some experiments and theoretical results appraised that defects distributed on the body play essential role in field emission. Envisaging the screen effect of the well aligned CNTs, the emission efficiency will be degraded. Furthermore, the fitted work function of CNT is about 1.65 eV, which probably results from the average of two terms attributing to amorphous carbon and defects. Amorphous carbon is the dominant part of the film with an energy gap of about 1~3 eV while defects localize the electron wave vectors of CNTs along tube axes and have an energy gap about 2.5 eV in the tube end.

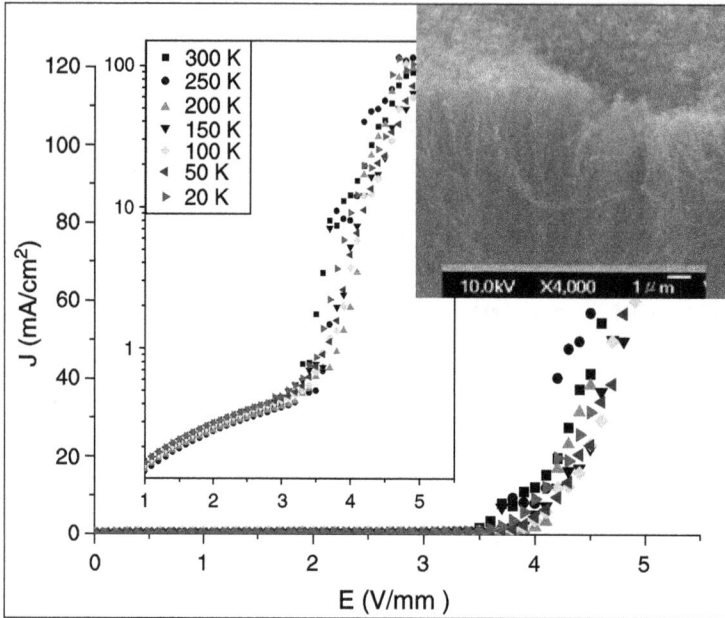

The temperature dependence of field emission for aligned MWNT films. The inset
is the SEM image of densely aligned CNT films.

The Raman spectra of CNT are similar to the graphite layered structure whilst with
the exact shift and line shapes depending on the real structure. Both single and multi-
wall CNTs reveal strong Stokes Raman peak near 1580 cm^{-1} relating to the E_{2g} active
modes. On accounting the defect structure of the multi-wall CNTs, the Raman D-line
near 1350 cm^{-1} shall be accompanied. The breathing modes of SWCNT ranging from
100~300 cm^{-1}, with real shifts closely relating to the inverse of tube diameter d are
usually implemented to mechanical and electrical properties, whilst the entangled tan-
gential stretching modes near 1589 cm^{-1} are used to determine the number of structures. Figure shows a typical Raman spectrum of MWNTs indicating two characteristic
peaks. The G-line peak at 1580 cm^{-1} corresponds to the high-frequency Raman-ac-
tive E_{2g} mode of graphite. The strong and broad D-line at 1350 cm^{-1} and a weak D
-band at around 1621 cm^{-1} was sophistically attributed to disorder-induced carbon features arising from finite particle size effect or lattice distortion.

Another prospect of the origins of D-line is attributed to defects in the curved graphite
sheets, tube ends, and finite size crystalline domains of the tubes. The films contain
many curled CNTs, figure demonstrates the popularly existed graphite nanoparticles
and amorphous carbon. The corresponding micro Raman spectra are shown in figure
for different species. The observed G-line (at 1580 cm^{-1}) illustrates that these two zones
popularly contain graphite sheet structures, while CNT nanoparticles or curled
structures reveal the characteristic peak at 1350 cm^{-1} resulting from the breakdown of
translational and point lattice symmetries at the edge plane. In one zone popularly
distributed with curve MWCNTs, the G-line shifts to a higher frequency from 1580 cm^{-1}
embodied in the significant existence of sp^2 short-range order in the tubes. The

appearing of stronger D-line intensity than the G-line is interpreted as due to the presence of finite size (nanometer order) of the crystalline domains or defects. As an evidence from the micro-Raman spectra, it concludes that the frequency shift of the G-line and the stronger D-line intensity in one sample is caused by defects in the tube which is also alluded by the SEM showing that the wall of curled tubes contain many defects like pentagons.

In other another zone distributed graphite grains the characteristic peaks are similar to disordered graphite. The decrease of the intensity ratio of D-line and G-line (ID/IG) is attributed to the accompanied large amounts of amorphous carbon particles. From above discussions, the presence of G line with a small bump at the high-energy side (1620 cm^{-1}) justifies the sample being MWCNTs as compared to the data dwelled in literatures.

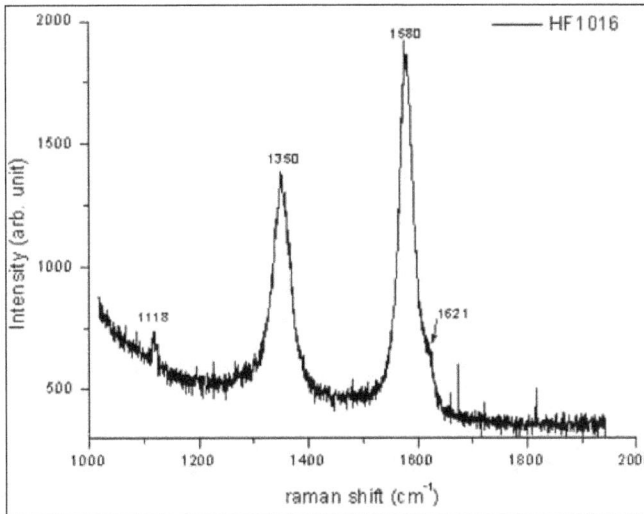

Atypical Raman spectrum of MWNTs.

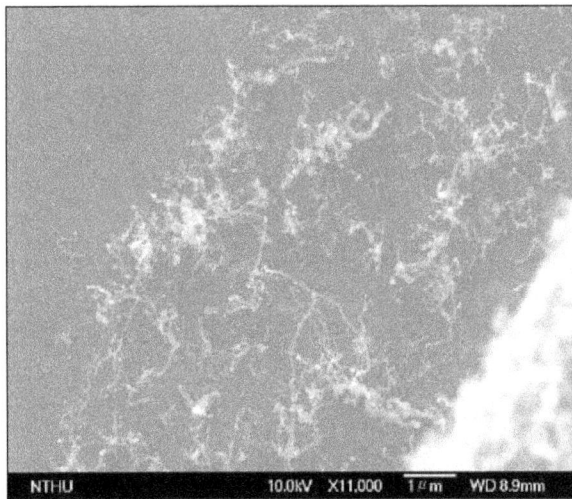

The SEM image for the sample at HF1113-NiFe-wb zone.

The SEM image for the sample at HF1113-NiFe-B zone.

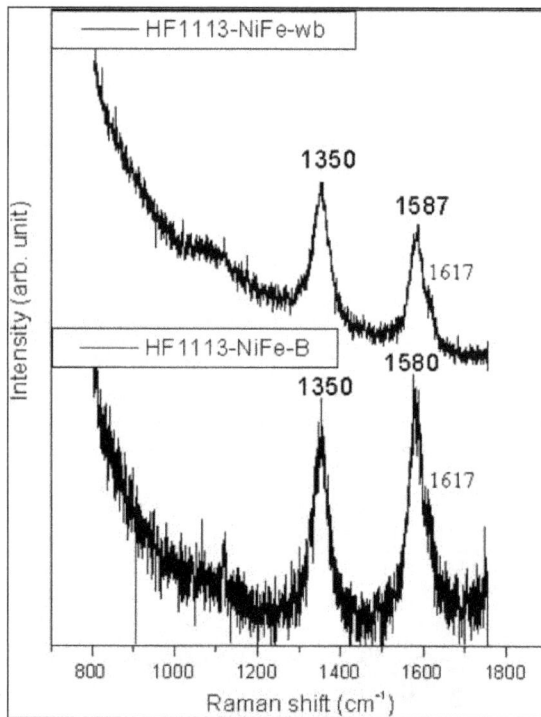

The micro Raman spectrum corresponding to above samples.

Magnetic Properties

Quantum Tunneling in Magnetic Nanoparticles

Magnetite Fe_3O_4 nanoparticles have been popularly found in animal cells for cruising. One of the fascinating properties of magnetic nanoparticles is the reduction from multidomains to a single domain as the particle size reduces to some limit values. Besides the vanishing of magneto hysteresis and the large reduction of coercive field for nanoparticles, the macroscopic quantum tunneling of the magnetic moment becomes possible. Aquantum phase transition can differ fundamentally from a classical thermal transition because of its non-analyticity in the ground state energy of the infinite lattice system. Unusual electronic and magnetic characteristics are prevalent at nonzero temperatures such as the metal-insulator transition in transitionmetal oxides , non Fermi-liquid behavior of highly correlated f-electron compounds, abnormal symmetry states of high-T_c superconducting cuprates, and novel bistability of semiconductor heterostructures. The investigation of the remarkable properties of these systems attracts great efforts of researchers in condensed matter physics. The physics underlying the quantum phase transitions described above is quite involved and in many cases, has not been completely understood so far. In the high-T_c superconductors, for example, the superconductivity gives a direct way to study the quantum order-disorder transition. In heavy-Fermion materials, the characterization of the magnetic instability at T = 0 is complicated due to the presence of charge carriers and substitutional disorder. In spin glasses , one can vary the strength of quantum fluctuations to tune the spin glass phase into the paramagnetic phase.

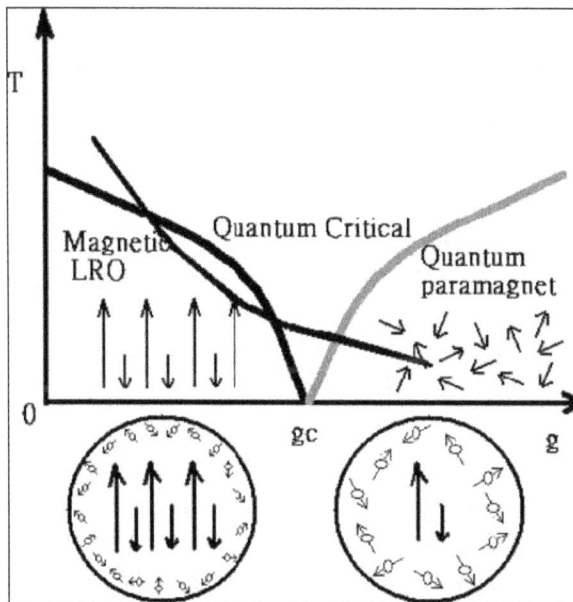

The phase diagram of quantum critical behavior and the anisotropic surface
spin glass induced superparamagnetic states.

A surface spin-glass layer is proven to be ubiquitous in magnetic nanoparticles at low temperatures. Alarger surface to volume ratio of the small nanoparticles implies a stronger surface anisotropic field to frustrate and disorder the inner spins, causing quantum tunneling at higher temperatures. The phase diagram showing the quantum critical point at T = 0 with a dimensionless coupling function $g = g_c$, in which the Hamiltonian $H(g) = H_0 + gH_1$ at the presence of transverse anisotropic field H_1, is sketched in figure where the anisotropic field-induced quantum tunneling due to the surface spin-glass layer is also speculatively plotted. There can be a level-crossing field where an excited state becomes the ground state at the critical field and creates a point of nonanalyticity of the ground state energy. The second-order phase transition for quantum driven phase transition usually occurs at the physically inaccessible T = 0 where it freezes into a fluctuationless ground state. The critical field emanating thermally driven phase transition occurring at T > 0 is smaller then the critical value g_c for T = 0 and decreases as the particle sizes reduce or the transverse field increases.

The low temperature magnetic viscosity of these systems shows a constant value below a finite temperature reflecting the independence of thermally over-barrier transitions and is the signature of quantum tunneling of the magnetization. Although there are some related evidences about our results, we provide an alternative theory and experimental tool to survey analytically. Electron spin resonance (ESR) spectrometry is exploited to study the magnetic states of single domain spinel ferrite nanoparticles. As the temperature decreases, the spectrum changes from super-paramagnetic resonance (SPR) to blocked SPR and arrives at quantum SPR as the temperature lowers down further. Ananoparticle system of a highly anisotropic magnet can be qualitatively specified by a simple quantum spin model, or the Heisenberg model with strong easy-plane anisotropy. Disordered spin-glass-like nanoparticles become quantum paramagnets under anisotropy-assisted quantum tunneling. We tacitly assumed that an alternative approach would lead to a better understanding of the fascinating interplay in the vicinity of the quantum critical point in magnetic nanoparticles.

A surface spin-glass layer is proven to be ubiquitous in magnetic nanoparticles at low temperatures. A larger surface to volume ratio of the small nanoparticles implies a stronger surface anisotropic field to frustrate and disorder the inner spins, causing quantum tunneling at higher temperatures. The phase diagram showing the quantum critical point at T = 0 with a dimensionless coupling function g = g_c, in which the Hamiltonian H(g) = H_0 + gH_1 at a presence of transverse anisotropic field H_1, is sketched in figure where the anisotropic field-induced quantum tunneling due to the surface spin-glass layer is also speculatively plotted. There can be a level-crossing field where an excited state becomes the ground state at the critical field and creates a point of nonanalyticity of the ground state energy. The second-order phase transition for quantum driven phase transition usually occurs at the physically inaccessible T = 0 where it freezes into a fluctuationless ground state. The critical field emanating thermally driven phase transition occurring at T > 0 is smaller then the critical value g_c for T = 0 and decreases as the particle sizes reduce or the transverse field increases.

The Hamiltonian of the Heisenberg model with strong easy-plane anisotropy with internal transverse fields σ_i^x i without applying an external field is given by:

$$H = -\sum_{i,j}^{N} J_{ij}\sigma_i\sigma_j - \Gamma\sum_i^N \left(\sigma_i^x\right)^2$$

where $\sigma's$ are the Pauli spin matrices, $J_{ij} > 0$ are the longitudinal exchange couplings and is the transverse anisotropy parameter for the spin–spin interaction causing quantum tunneling. Long-range force dominates the system for $J_{ij} \gg \Gamma$.

We can express $H = H_0 + H_1$, where H_0 and H_1 correspond, respectively, to the first and second terms in the right hand side of $H = -\sum_{i,j}^{N} J_{ij}\sigma_i\sigma_j - \Gamma\sum_{i}^{N}\left(\sigma_i^x\right)^2$ and commute with each other. The ground state of H_0 is long-range magnetically ordered and prefers ferromagnetism at low temperatures, while the ground state of H_1 favors the quantum paramagnetism. As the particle size decreases, the anisotropic field. increases up to a critical value, upon which a point of non-analyticity in the ground state energy is generated. The ground state of the total system varies from the magnetic long-range-order ground state H_0 to the paramagnetic ground state H_1. This means that the ground state energy is not continuous across the critical point at T = 0. But many experiments demonstrated that at some nonzero temperatures, though very low, an interplay between quantum and thermal fluctuations occurs.

In the case of applying an external transverse field, we consider the corresponding Hamiltonian in the same Heisenberg model as:

$$H = -\sum_{i,j}^{N} J_{ij}\sigma_i\sigma_j - \Gamma'\sum_{i}^{N}\sigma_i^x$$

The ground state of the first term prefers that the spins on neighboring ions are parallel to each other and become ferromagnetic for $J_{ij} \gg \Gamma$, whereas the second term allows quantum tunneling between the spin up $|\uparrow\rangle_j$ and spin down $|\downarrow\rangle_j$ states with amplitudes being proportional to the transverse field. Γ' Both the off-diagonal terms σ_i^x

i in. $H = -\sum_{i,j}^{N} J_{ij}\sigma_i\sigma_j - \Gamma\sum_{i}^{N}\left(\sigma_i^x\right)^2$ and $H = -\sum_{i,j}^{N} J_{ij}\sigma_i\sigma_j - \Gamma'\sum_{i}^{N}\sigma_i^x$ flip the orientation

of the spin on a site by quantum tunneling. There can be a level-crossing field where an excited state becomes the ground state at the critical field and creates a point of nonanalyticity of the ground state energy as a function of Γ'. The second-order quantum phase transition usually occurs at the physically inaccessible T = 0 where it freezes into a fluctuationless ground state. The transverse critical field emanating a quantum phase transition occurring at T > 0 is smaller than the critical value Γ_c for T = 0 which decreases as the particle sizes reduce.

At high temperatures, single domain magnetic nanoparticles are thermally free to orient their spin directions and exhibit superparamagnetic properties. The super paramagnetic state is blocked as temperature lowers down to enhance the exchange inter-actions between particles. This critical temperature increases with the particle volume V and the magnetic anisotropic constant Ka. Below the blocking temperature TB, depending on a typical time scale of measurements t, the slow down of thermal motion implies the magnetic nanoparticles to undergo a transition from superparamagnetic to blocked SPR which behaves like a ferromagnetic state for the total system. However, the zerofield-cooled magnetization measurement indicates that the super paramagnetic relaxation time is estimated to be $\tau_m \sim 10^2$ s. Since the time scale for observing the ESR spectra is much shorter than that for magnetization measurements, the blocking temperature T_B^E for ESR is much higher than that of the magnetization measurements T_B^m The ESR provides an excellent method to detect the quantum phase transition at temperatures higher than T = 0.

The temperature dependent EPR spectra for Fe_3O_4 nanoparticles obtained from 220 K to 4 K are specified by curves as shown in figure. The tiny spectrum centered at g ~ 4.3 is attributed to the isolated spin $^6S_{5/2}$ of the remnant Fe^{3+} ions when the second–order crystal field coefficient with axial symmetry vanishes while that with rhombic symmetry persists. The relatively narrow SPR line (~100 Gauss) fades and the broad blocked SPR resonance line (~1500 Gauss) manifests as the temperature decreases to about 35 K. The linewidth reveals abnormal broadening (~1500 Gauss) below the blocking temperature (63 K) and the broad line grows and becomes prevalent until the temperature reaches 22 K. The narrowing of the SPR linewidth at high temperatures is attributed to the thermal fluctuations of the magnetic nanoparticles while the broadening of the blocked SPR results from the line up of the magnetizations of all particles that enhances the anisotropic field at low temperatures. As temperatures decrease to 20 K, there is a renascence of an anomalous paramagnetic resonance with the amplitude growing and decaying until the temperature decreases to about 8 K. An anomalous paramagnetic resonance prevails behaving like a free exchange-coupled giant spin, as expressed to be a quantum superparamagnetic state. The anisotropic field $K\perp$ increases as temperature decreases resulting in a higher tunneling rate. The domain size of the quantum SP particle decreases attributing to prominent transfer of magnetic domains into surface spin glass state.

Considering that the strong surface anisotropic field would destroy the internal exchange force making the longrange ferromagnetic state to become paramagnetic can better elucidate this quantum paramagnetic state existing at low temperatures. Below 8 K, the amplitude of paramagnetic resonance decreases resulting from the commencement of maximizing the anisotropic field and the reducing of thermally assisted paramagnetic resonance.

Figure illustrates the occurrence of the amplitude peak at low temperatures and the linewidth variations for super paramagnetic, ferrimagnetic, and quantum paramagnetic resonance, which are about 100 G, 1500 G, and 10 G, respectively. The spin susceptibility which is proportional to the integration of the intensity is hidden in the very broad spectrum line at low temperatures. The line width of the paramagnetic resonance signal arising from quantum fluctuations is independent of temperatures. Two prominent critical points associated with the classically thermally driven from SPR to blocked SPR and the quantum tunneling from magnetic long-range order to quantum paramagnet are appraised. The sharply narrowing down of the linewidth for quantum SPR may be attributed to largely reducing of the domain size by quantum tunneling.

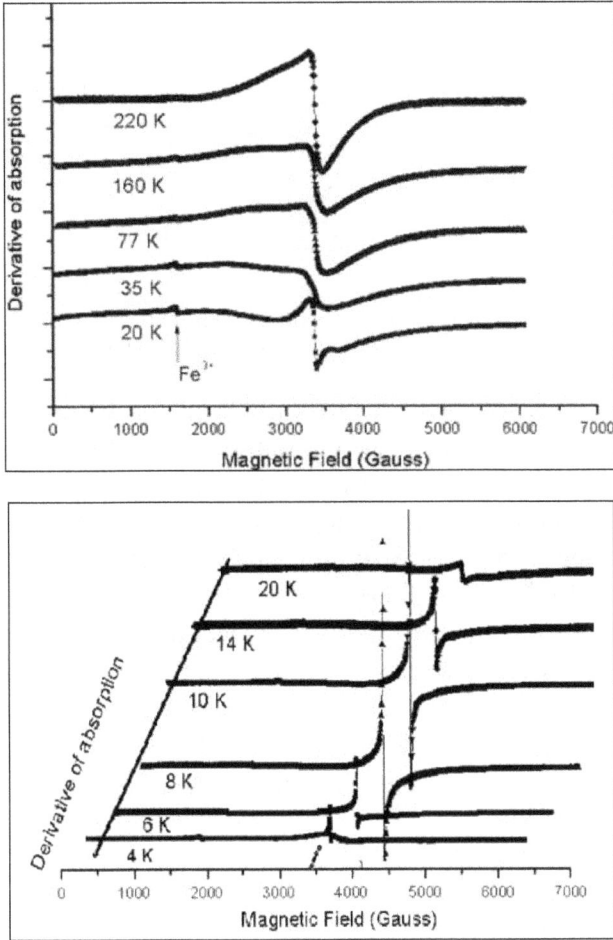

The ESR spectra for Fe_3O_4 ferrofluid measured at various temperatures between 220 K and 4 K. The spectra below 20 K are canted to express the same resonance position without turbidity of signals.

Finally we compare the ESR results to the magnetization measurements governed by apparently slower observing time scales. To justify the existence of SP, blocked SP and quantum SP states at various temperatures, we used the Coferrite nanopartiles which have a much larger anisotropic field. The magnetization for $CoFe_2O_4$ ferro fluid as a function of temperatures was measured by a MPMS2 superconducting quantum unit interference device (SQUID) as shown in figure. Above T_B the particles are superparamagntic where the field cooled (FC) and zero field cooled (ZFC) curves merge that elucidates the alignment of the anisotropic spins for the field cooled spectrum. Two remarkable transition points were denoted as T_B and T_c in the ZFC curve to represent the blocking state between 24 K and 11 K and the cross temperature at 11 K. Above T_B, the Co-ferrite nanoparticles exhibit superparamagnetism due to thermal fluctuations of the magnetic moments while blocked to the original ferrimagnetic order at temperatures below T_B.

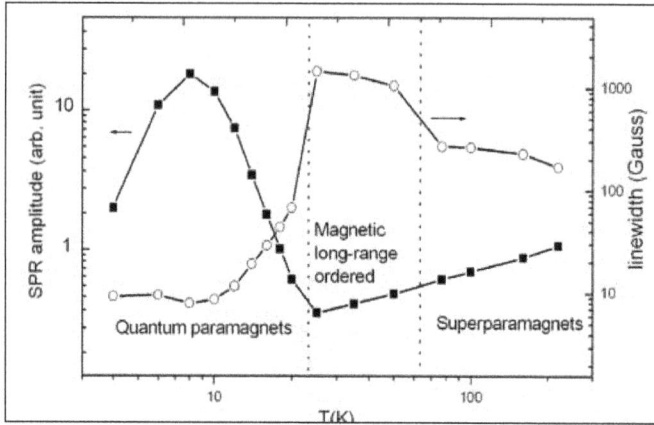

The ESR line widths and amplitudes at various temperatures for Fe_2O_3 ferrofluid are displayed to specify three magnetic states.

We have studied the ESR response of ferrofluid Fe_3O_4 samples as a function of temperature. The experimental data can be consistently explained in the framework of a qualitative model of the evolution of the nanoparticle magnetic system with decreasing temperature from the superparamagnetic, to blocked superparamagnetic and finally to the quantum tunneling regime. Size and anisotropy dependence of the transition temperatures agree with the Heisenberg model with strong easy-plane anisotropy. The critical temperatures of the quantum superparamagnetic resonance spectra are also proportional to the intensity of transverse magnetic field in accord to the Heisenberg model in the external transverse field. Plausibility of a quantum phase transition might occur as a consequence of the critical exponent $\gamma = 1.7 \sim 2.3$. More evidences for clues of this eventual QT should be provided such as measurements of heat capacity and ac susceptibility. The possibility of QPT in magnetic nanoparticles is vital in theoretical and experimental point of view.

Domain Walls in Thin Magnetic Films

For thin magnetic films, the mostly interested problems are magnetic domain formation and the related magnetoresistance. An magnetic force microscopy MFM with a high image resolution is required for both scientific explorations and technical applications associated with the study of highly dense thin-magnetic-film recording media, whose density is limited by domain walls and ripple structures that cause noises in the feedback signal. Magnetization perpendicular to the surface has been found in the ultra-thin films of sandwich structures such as Fe/Ag , $(Co/Pt)_N$, $(Co/Au)_N$, and Cu/Ni/Cu(001) where N is the number of repetition layers. In a magnetic field parallel to the film plane, the magnetic domain size initially decreases and then increases as the film thickness increases. The in-plane magnetization for films with thicknesses greater than five monomolecular layers (ML) reveals a reduction of the magneto static energy. In MFM images, the distribution of magnetization parallel or anti-parallel inside the walls appears dark or bright lines, respectively. Various domain structures, including Bloch walls, Neel walls, Bloch lines , cross-ties, and 360° walls have been observed in Co, Fe, and permalloy films. The formation mechanism for the 360° domain walls has not yet been clearly understood. The 360 domain walls are generally believed to form normally near the defects where the interaction between a wall and an inclusion is strong. In spite of many reports on MFM images of magnetic thin films, an investigation of a high resolution (with sizes) of domain structure is still desirable.

A demarcation occurs when the Ni film thickness is around 50 nm, at which film thickness, Neel walls are usually observed. Striped Bloch walls are dominated for thicker films. Both the domain area of the Neel wall and the spaces between the striped lines in the Bloch walls are proportional to the square root of the film thickness. The periodic spaces between the striped domains increases with the applied field perpendicular to the surface. The spaces become zigzag for thick films. The formation of magnetic domain walls crucially depends on the anisotropy energy, the magneto static energy and the mechanical stress of magnetic thin films. Magnetic energies comprise of:

- The domain-wall energy E_w due to the exchange energy between nearest neighbors characterized by the exchange coupling constant J (erg/cm).

- The magneto crystalline anisotropy energy E_a expressing the interaction of the magnetic moment with the crystal field characterized by the constant K_v (erg/cm³).

- The magneto static energy E_m arising from the interaction of the magnetic moments with discontinuous magnetization across the bulk and the surface.

- The surface magneto crystalline anisotropy energy E_{ks} resulting from a correction of symmetry broken near the surface characterized by a constant K_s (erg/cm²).

- The magneto-restrictive energy E_r arising from mechanical stress or defect-induced force on the film resulting in an introduction of effective anisotropy into the system characterized by K_m (erg/cm³).

A competition of above energies implies various domain walls including Bloch walls, Neel walls, asymmetric Bloch walls, Bloch lines, cap switches, and 360° domain walls. For stripe domains, the magnetic potential follow the Laplace function within the periodic stripe spaces, and the static magneto-energy is expressed as :

$$\epsilon_m = \frac{I_s^2 d}{\pi^2 \mu_0} \sum_{n=0}^{\infty} \frac{1}{n^2 d} \int_0^d \sin n\left(\frac{\pi}{d}\right) x \, dx = \frac{2I_s^2 d}{\pi^3 \mu_0} \sum_{n:odd}^{\infty} \frac{1}{n^3}$$
$$= 5.40 \times 10^4 \, I_s^2 d \, J / m^2$$

where I_s is the saturation magnetization, d is the period of stripe lines, and μ_0 is the permeability. The domain wall energy for thin stripe domains can be approximated as:

$$\epsilon_w = \frac{\gamma \ell}{d}$$

where γ is the surface energy per unit area of the domain and ℓ is the film thickness. Neglecting the anisotropic energy for stripe domains, the minimizing of the free energy implies the most probable width of the domain walls as:

$$d = 3.04 \times 10^{-3} \frac{\sqrt{\gamma \ell}}{I_S} \text{ in MKS unit of } d \propto \sqrt{\ell}$$

In thick films inherited with Bloch walls, considering the finite width for spins that rotate from one direction to the opposite direction, the wall energy E_w including the exchange energy E_{ex} and anisotropic energy within the Bloch walls existed in thick films is:

$$E_w = E_{ex} + E_a = \frac{\pi^2 JS^2}{Da} + \frac{K_1}{2} D$$

where S is the spin, D is wall thickness, a is the lattice constant, and K1 is the anisotropic field. A minimize of Ew with respect to D implies an equilibrium value of:

$$E_w = \pi (AK_1)^{1/2}, \text{ and } D = \pi \left(\frac{A}{K_1} \right)^{1/2}$$

where $A = (2JS^2)/a$. With the value of K1, we can evaluate the wall thickness D.

For magnetic thin films of thickness larger than 5 monolayers (ML), the magnetization prefers to lie on the plane to reduce the magneto static energy exhibiting the Bloch wall with the total magnetic energy as:

$$E_B = E_m + E_w = A \left(\frac{\pi}{D} \right)^2 D + \frac{K_1}{2} D + \frac{2\pi D^2}{\ell + D} M_e^2$$

where M_e is the effective magnetization perpendicular to the domain plane. The magnetization for Bloch walls occurring in thick films is normal to the surface while that for Neel walls occurring in thin films is parallel to the film surface. The Neel walls prevail to minimize the free energy at sufficiently low thickness to wall-width ratio. The total magnetic energy for Neel walls is:

$$E_N = E_m + E_w = A \left(\frac{\pi}{D} \right)^2 D + \frac{K_1}{2} D + \frac{\pi D \ell}{D + T} M_S^2$$

Asymmetric Bloch wall contains a Bloch core in the film center surrounded by a Neel surface cap. The Bloch line behaves such that the wall contrast changes abruptly from bright to dark within a distance of 1 nm. No micromagnetic theory of Bloch lines has been established. A cap switch undergoes a change of the sense of the rotation of wall surface magnetism ensuing a change in Kerr contrast. A 360° domain means that the magnetization rotates 360° within a wall.

The structure of the magnetic domain changes in complying with the applied external field to balance the increment of the work due to domain wall displacement. The stripe walls in a sufficiently high field become zigzag resulting in increasing the negative wall energy to balance the static energy. The radius of curvature r of the domain wall relates to the magnetic pressure on the wall as:

$$\frac{\gamma}{r} = 2I_s H \cos \theta$$

Where γ is surface energy density on the wall, and θ is the cant angle between the external field H and the normal of the film. The wall changes from stripe to zigzag for $r = \ell / 2$ suggesting the estimated critical value of H.

Magnetic domains are of various kinds, and competitively contribute to the anisotropy energies. The transition region between the domains, called the domain wall, is not continuous across a single atomic plane. Profiles of a domain wall can be defined according to the sign of the magneto static interaction between the local surface position and the tip.

The AFM topography and the corresponding MFM images for Ni films deposited at room temperature is shown in figure. Atypical circular domain wall was performed in figure, which had been carried out with several different scan speeds, scan positions, directions and tip magnetizations. An alternative Bloch line was observed in the same plate, which also shows ripples suffered from strong tip-sample interaction as expressed by the micromagnetic calculation.

The relationship between the topography and local magnetic properties can be established by a combination of the high surface resolutions of DFM and MFM. The line profile of the wall suppression by local particles as shown in figure adduces that the surface topography affects the wall formation. The wall terminated by an inclusion as a white dot and indicated by "a". Domain configurations in general would be disturbed by the presence of particles and domain walls that will be kept away from an inclusion. The arrow "b" indicates a little shift of the Bloch wall with a cap switch due to the presence of a particle. The plausible reason may be that it dissipates less anisotropy energies for the wall to walk in thin films. Another explanation is that the inclusion prevents the occurrence of the wall-cross. The tilting of the wall is due to the local inclined anisotropy. The arrow "c" illustrates that the spin orientation change by 180 degrees under a lateral distance of only 3 nm. It also reported in references showing a lateral distance only of 1~2 nm. The basic assumption of micromagnetic theory, i.e., a small canting angle between adjacent Heisenberg spins, is no longer valid. The formation of this alternative Bloch line seems to be originated from the rotational nature of magnetization due to stress, oblique anisotropy, impurity, vacant space, and irregular ingredients.

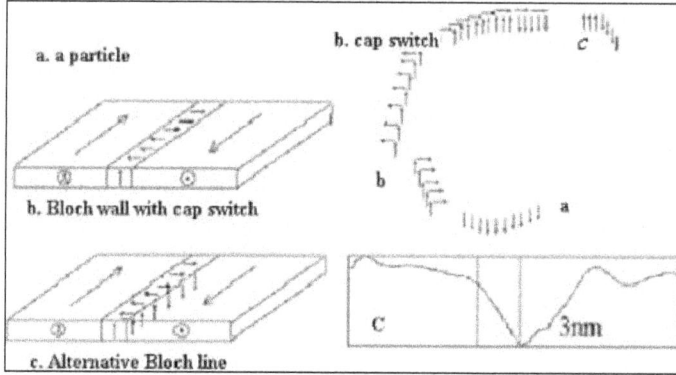

The corresponding line profile and spin orientations.

The domain size increases with the film thickness according to the formula $d \propto \sqrt{\ell}$ as proposed by Kittel as expressed in $J_{total} = a_1 T^2 e^{-\theta/KT} e^{-b_1 \chi^{3/2}/F} + a_2 \dfrac{1}{\chi + E_g} F^2 e^{-b_2(\chi+E_g)^{3/2}/F}$

where d is the domain size, and ℓ is the film thickness as developed in figure. In deriving this formula, Kittel considered only the magneto static energy, the exchange energy, and the magneto crystalline anisotropy energy. This simplified theory can not express the deviation of the results of the experiment from the theoretical curve for the domain size of about 291 nm ± 25 nm at film thickness of 75 nm. Neglecting the anisotropy caused by the existence of defects and inclusions in thicker films may be responsible for the deviation from the ideal square root law.

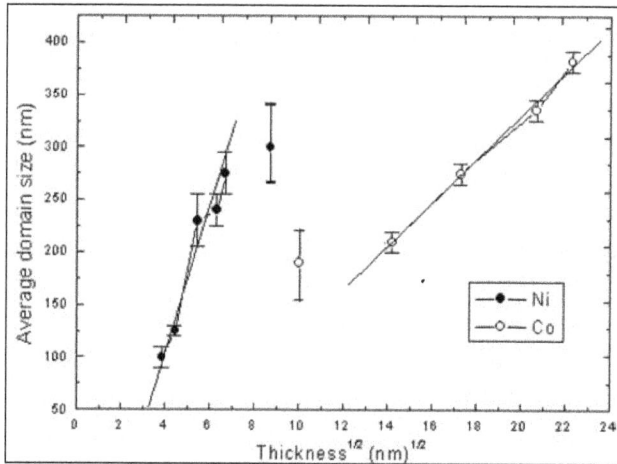

A plot of domain sizes as functions of the square-root
of nickel and cobalt film thickness.

Magnetoresistance in Thin Films

Researches in domain wall resistance have grown dramatically in recent owing to the great advances in the fabrication of magnetic memory devices. Specifically, large negative magnetoresistance (MR) observed at room temperature in cobalt films behaving with strip-domain walls was hotly investigated and reported in terms of giant domain-wall (DW) scattering that contributes to the resistivity. Measurements of resistivity for currents conducting parallel (CIW) and perpendicular to DW's play the essential role of MR studies. Two excellent reviews concerning the domain-wall scattering of magneto resistance were reported. Berger proposed that on account of the shorter wavelength of conduction electrons in comparing to the domain wall-width, the electronic spin follows the local magnetization adiabatically and gradually as it traverses across the wall. Whereas Cabrera and Falicov treated the problem of domain-wall-induced electrical resistivity in iron analytically by examining the difference in reflection coefficients at a domain wall for up and down spin electrons. The domain wall essentially presents a potential barrier where the barrier heights are different for the two-spin channels owing to the existence of exchange field.

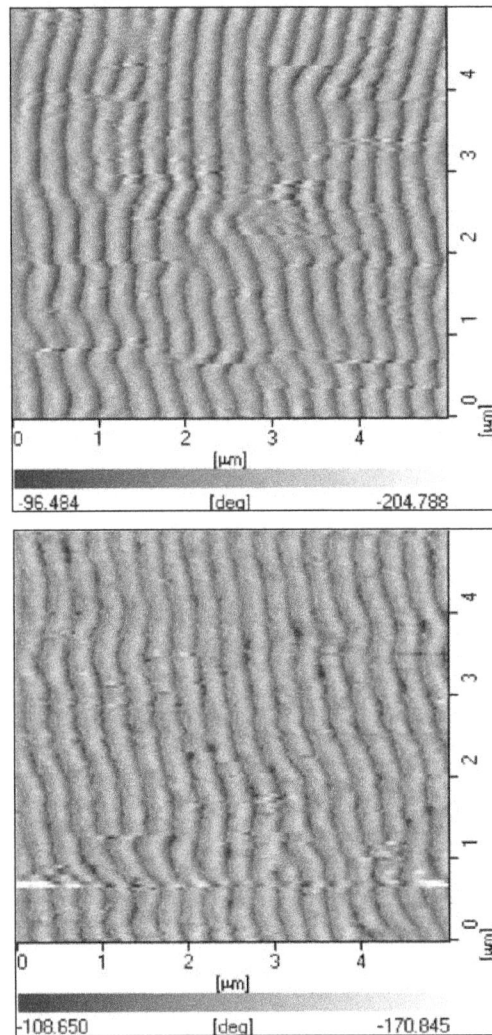

In figure: represented the MFM images with a $5\mu m \times 5\mu m$ size of magnetic domains for a 100 nm-thick Ni film (a) virgin, (b) with a magnetic field parallel to the strip line, and (c) with field perpendicular to the strip line, respectively.

The magnetoresistance (MR) measurement yields detailed information concerning small magnetization changes. Recently, several groups studied the width dependence of the magnetization reversal process in narrow ferromagnetic wires by measuring the MR, and reported that the coercive force and the switching field are proportional to the diamagnetic field along the wire axis. Figure represented the MFM images with a $5\,\mu m \times 5\,\mu m$ size of magnetic domains for a 100 nm-thick Ni film under various directions of magnetic fields. The dark and bright contrasts can be identified with the up and down magnetic domains showing (a) straight line distributed along the y-direction (parallel to the strip-line) for H = 0, (b) domain-width increased for applying H = 15 T along the y-direction, (c) the strip-domain oriented to the x-direction (transverse to the strip-line) for applying a H = 15 T perpendicular to the y-direction, and (d) the stripe domains changed to a labyrinth shape for the applied field along the z-direction (perpendicular to the surface), respectively. The MFM images for the 800 nm-thick Co film are shown in figure with domain structures as given by (a) straight lines distributed along the y-direction, and (b) the stripe domain became a bubbly shape under a bias magnetic field of 1.5 T in the y-direction, respectively.

The temperature dependence on the resistivity for the Ni film with a thickness of 250 nm measured at 20 to 100 K with a DC current of 1 mAis shown in figure. The upper curve a shows the resistivity with the beam current transverse to the domain walls (CPW) without applying external magnetic fields. The conductance is 15% smaller than when the current is parallel to the domain walls (CIW) as sketched in curve b. The CPW value is even 32% smaller than that under an applied magnetic field of H = 1.5 T as dictated in curve c. Here the transverse current in the legend refers to the current perpendicular to the domain walls and the longitudinal current refers to the current parallel to the domain walls, respectively.

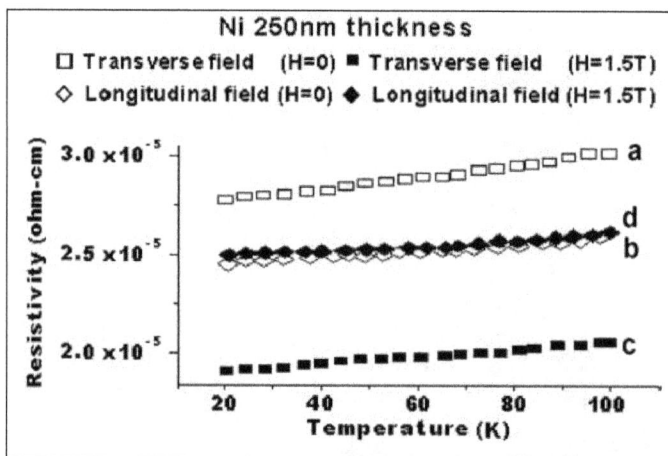

The R–T curves of the 250 nm-thick Ni film with different domain-wall structures and current flow directions. The magnetoresistivities reduce by 15% when the current is parallel instead of transverse to the domain wall and reduces by 32% in the transverse current when the width of stripe domain is increased under a magnetic field of 1.5 T.

The parametric effect arisen from the reflection of incoming electron waves by the ferromagnetically ordered domains as they entered the twisted spin structure of a wall, and the diamagnetic effect due to the zigzag character of the orbital motion of electrons when they pass between the up and down spin regions of the domain. This diamagnetic effect is the source of a negative MR. The resistivities of current parallel to the domain wall (CIW) and perpendicular to the domain wall (CPW) are given theoretically on account of boundary scattering as given by:

$$\rho_{CIW} = \rho_0 \left[1 + \frac{\xi^2}{5} \frac{\left(\rho_0^\uparrow - \rho_0^\downarrow \right)^2}{\rho_0^\uparrow \rho_0^\downarrow} \right]$$

$$\rho_{CPW} = \rho_0 \left[1 + \frac{\xi^2}{5} \frac{\left(\rho_0^\uparrow - \rho_0^\downarrow \right)^2}{\rho_0^\uparrow \rho_0^\downarrow} \left(3 + \frac{10\sqrt{\rho_0^\uparrow \rho_0^\downarrow}}{\rho_0^\uparrow + \rho_0^\downarrow} \right) \right]$$

The magneto resistance ratio R due to walls, which is defined as:

$$R_{CIW} = \frac{\rho_{CIW} - \rho_0}{\rho_0} = \frac{\xi^2}{5} \frac{\left(\rho_0^\uparrow - \rho_0^\downarrow \right)}{\rho_0^\uparrow + \rho_0^\downarrow}$$

$$\frac{R_{CPW}}{R_{CIW}} = 3 + \frac{10\sqrt{\rho_0^\uparrow \rho_0^\downarrow}}{\rho_0^\uparrow + \rho_0^\downarrow}$$

where ρ_0^s is the resistivity for spin states s of the ferromagnetic material, $\rho_0^{-1} = 1/\rho_0^\uparrow + 1/\rho_0^\downarrow$ is the conductivity of the ferromagnet without the appearance of domain walls, $\pi \hbar^2 k_F / 4mdJ$, and d is the domain wall width.

To estimate the MR due to wall scattering, Levy chose the commonly accepted values of Fermi wave vector kF = 1 Å⁻¹, the exchange splitting energy J = 0.5 eV, and $\rho_0^\uparrow / \rho_0^\downarrow = 5 - 20$ for typical ferromagnetic materials of Co, Fe and Ni at room temperature.

The MR_{CPW} = −32% of the 250 nm-thick Ni film measured at temperature 100 K is shown in figure for the current transverse to the domain wall under zero magnetic filed, when a saturation magnetic field H = 1.5 T is applied. Physically, several magnetization dependent scattering processes influence the electrical transport. This fact can be summarized in a general formula expressing the components of the electric field generated by a current density flowing through a homogeneous ferromagnet (providing the Matthiessen's rule is valid) by neglecting the extraordinary Hall effect and the possible deviation of the current lines while crossing different domains, such as that induced by the Hall effect. The electric field is given by:

$$\vec{E} = \rho(B)\vec{J} + \rho_{AMR}\left(\vec{\alpha} \cdot \vec{J} \right)\vec{\alpha} + \rho_0 \vec{B} \times \vec{J} + \rho_{wall}\vec{J}$$

With $\vec{M}$ the magnetization, $\vec{\alpha}$ the unitary vector along the electric-field direction, and $\vec{B}$ the internal magnetic induction vector. The first term represents the usual longitudinal resistance contribution, which varies like B^2 at low temperature (Lorentz contribution) and decreases almost linearly with B at higher temperatures (magnon damping). The second term is the anisotropic magneto resistance (AMR) along the magnetization direction. Its projection perpendicular to the current lines is called the planar Hall effect. The third term is the standard Hall effects composed of the ordinary effect proportional to B and the last contribution is related to the resistance due to spin scattering in domain walls. All of these items contribute to the negative GMR by 32%.

For the 800 nm-thick Co film as shown in figure, the $MR_{CPW} = 40\%$. From the MFM domain configuration, we can expect that the domains becoming bubbly shape increase the scattering probability and induce the positive GMR by 40%. The conductivities increase as the magnetic field increases when the magnetic field is applied along the electron transport direction at room temperature and 4 K as shown, respectively, in figures. The conductivity increases steadily with the increase of magnetic field at high temperatures on account of the reduction of the radius of the spiral motion of electrons that escape from surface scattering implying elongation of the mean free path. The MR measured at 4 K decreases rapidly at the beginning, which is presumed to be arising from the largely enhancement of electron scattering with the boundaries of Co wire at the presence of a small magnetic field at low temperatures.

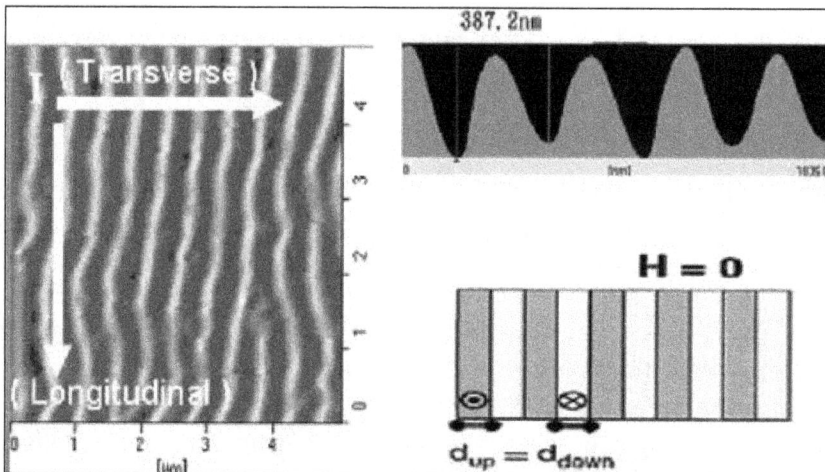

The MFM domain configuration of above Ni film for (a) H = 0, (b) along the y-direction (⇑) under a bias magnetic field of 1.5 T, respectively. The stripe domain-width increases and fewer domain walls are remnant.

Figures show the M-R measurement when the magnetic field was applied along the y-direction at room temperature and at 4 K. The conductivity does not change obviously as the magnetic field increases. The Sondheimer oscillation appears for magnetic field being nearly normal to the surface due to periodic striking the surface for electrons traveling in circular motion on a plane canting to the surface. Figure shows the MR measurement when the magnetic field was applied along the z-direction at room temperature. The conductivity decreases slightly as the magnetic field increases in comparing with the result for the field being applied along the transverse direction. Apparently the MR curve as portrayed in figure is not always smooth but with bumps with the aperiodic oscillation much slower and irregular than the Aharomov-Bohm oscillation. At this time we are unable to address this phenomenon clearly, which is tacitly assumed this fact to be arisen from the quantum size effect that results from the phase coherence in the electron waves scattered by different defects, and the distribution of the phase relationships by the magnetic field and the domain-wall trap.

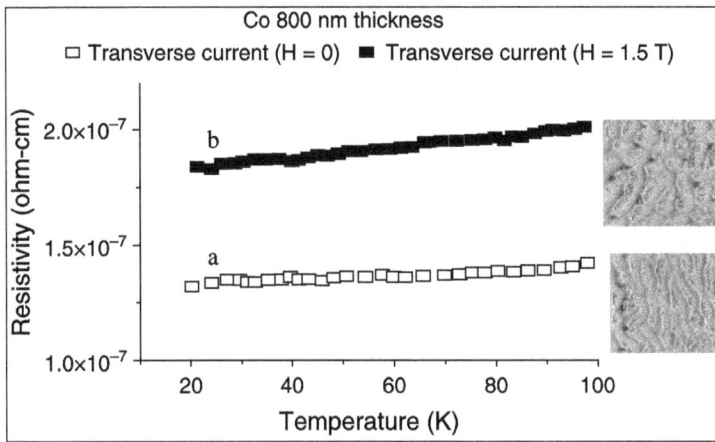

The R–T curve of the 800 nm-thick Co film with different domain wall structures under a transverse current. The resistivity increases 40% when the domain wall becomes bubbly shape from a straight line under a bias magnetic field of 1.5 T.

Concerning the reduction of the radius of the electron spiral motion under magneto size effect, the ratio of the conductivities for thin and thick films as the magnetic field is applied longitudinal to the current are:

$$\frac{\sigma_f}{\sigma_0} = 1 - \frac{3}{16k}\left[2 - \frac{k_r^2}{4k^2 + k_r^2}\left(1 + e^{-2\pi k/k_r}\right)\right]$$

$$\frac{\sigma_f(B) - \sigma_f(B=0)}{\sigma_f(B=0)}$$

$$\frac{1 - \frac{3}{16k}\left[2 - \frac{k_r(B)^2}{4k^2 + k_r(B)^2}\left(1 + e^{-2\pi k/k_r(B)}\right)\right]}{1 - \frac{3}{16k}\left[2 - \frac{k_r(B=0)^2}{4k^2 + k_r(B=0)^2}\left(1 + e^{-2\pi k/k_r(B=0)}\right)\right]} - 1$$

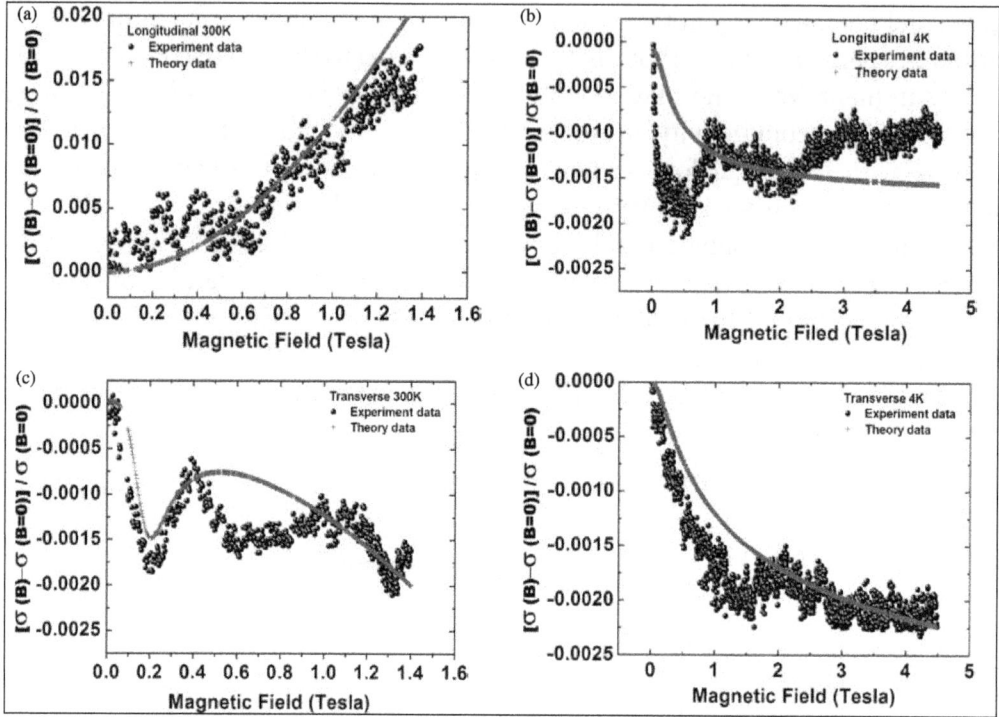

The experimental and theoretical values of the ratio of, $\sigma_f(B) - \sigma_f(B=0)$ for (a) the magnetic field being applied along the transport x-direction at 300 K, (b) at 4 K, (c) the magnetic field being along the y-direction at 300 K, and (d) at 4 K, respectively.

Where $k = d / \lambda_0$, and $k_r = d / D, D$ is the radius of electron spiral motion, and λ_0 is the electron transport mean free path. We suppose $\lambda_0 \cong 10\,\text{nm}$ and 100 nm, at room temperature 300 K and 4 K, respectively. Samples studied in this work have the Co wire thickness of 15 nm, which implies the reduction of k from 1.5 to 0.15 as temperature decrease from 300 K to 4 K resulting from the increase of mean free path at low temperatures.

The solid lines in figure show the experimental and theoretical data derived from equation:

$$\frac{1 - \dfrac{3}{16k}\left[2 - \dfrac{k_r(B)^2}{4k^2 + k_r(B)^2}\left(1 + e^{-2\pi k / k_r(B)}\right)\right]}{1 - \dfrac{3}{16k}\left[2 - \dfrac{k_r(B=0)^2}{4k^2 + k_r(B=0)^2}\left(1 + e^{-2\pi k / k_r(B=0)}\right)\right]} - 1$$

At temperature 300 K and 4 K, respectively. We can see that the experimental values are closed to the theoretical curves. The implemented theoretical data of the radius for the electron spiral motion is derived from the Faraday's law expressing as:

$$D = (mv) / (qB) = m / (qB) \cdot q / m \cdot \left(\Gamma(s + 5/2)\right) / \left(\Gamma(5/2)\right) \cdot \tau_0 \cdot E$$

where s is a characteristic exponent, and τ_0 is a relaxation time constant.

For the case of applying the transverse magnetic fields, the conductivity ratio is:

$$\frac{\sigma_f}{\sigma_0} = \frac{A^2 + \xi^2 B^2}{A}$$

The conductivity deviation ratio is:

$$\frac{\sigma_f(B) - \sigma_f(B=0)}{\sigma_f(B=0)} = \frac{\dfrac{A(B)^2 + \xi(B)^2 B(B)^2}{A(B)}}{\dfrac{A(B=0)^2 + \xi(B=0)^2 B(B=0)^2}{A(B=0)}} - 1$$

where,

$$A = \frac{3}{2}\left\{ -\frac{1}{2}\mu + \mu^2 + \frac{\mu}{2}\left(1 - \mu^2 + \xi^2\mu^2\right)\mathrm{In}\left[\frac{\left(1-\mu^{-1}\right)^2 + \xi}{1+\xi^2}\right]\right.$$

$$\left. -2\xi\mu^3 \tan^{-1}\left\{\frac{\xi}{\mu}\cdot\frac{1}{\xi^2 + 1 + \mu^{-1}}\right\}\right\}$$

$$B = \frac{3}{2}\left\{ -\mu^2 + \mu^3\,\mathrm{In}\left[\frac{\left(1-\mu^{-1}\right)^2 + \xi}{1+\xi^2}\right]\right.$$

$$\left. +\frac{\mu}{\xi}\left(1 - \mu^2 + \xi^2\mu^2\right)\tan^{-1}\left\{\frac{\xi}{\mu}\cdot\frac{1}{\xi^2 + 1 + \mu^{-1}}\right\}\right\}$$

$$\xi = \lambda_0 / D \quad \mu = k\left[\mathrm{In}(1/p)\right]^{-1}$$

With the parameters giving by $p = 0.13, \mu = 0.75$ and $p = 0.004, \mu = 0.018$ at temperature 300 K and 4 K, respectively, we can calculate the conductivity deviation ratio. The theoretical curves derived from equation,

$$\frac{\sigma_f(B) - \sigma_f(B=0)}{\sigma_f(B=0)} = \frac{\dfrac{A(B)^2 + \xi(B)^2 B(B)^2}{A(B)}}{\dfrac{A(B=0)^2 + \xi(B=0)^2 B(B=0)^2}{A(B=0)}} - 1$$

which agree satisfactorily with the theory except the bending curve that occurs near at $B = 0.6\mathrm{T}$ at temperature 300 K. This smallness of the specularity parameters p is attributed to the strong surface scattering for the ion-sputtered films. The largely reducing of the p values at low temperatures results from the surface diffusion of the catastrophic molecular desorption.

Magnet Micro-strips and Ferromagnetic Resonance for Magnetic Films

The large demand in communication and video applications intrigues us an impetus to develop the design of monolithic microwave micro-strip circuits. Microwave techniques allow high sensitivity measurement of the dependence of the conductivity of thin magnetic films on temperature and magnetic field. Apermissive investigation on the resonance frequency tunable by magnetic or electric field for filters is desired. Measurement of the magnetic field dependence of the resonant frequency shift of a microwave micro-strip has been performed by Tsutsumi. There have been a lot of studies on the measurement of the complex permeability over a broad frequency band. The dynamic susceptibility deduced from the ferromagnetic resonance spectra in magnetic films with a non-uniform magnetic configuration is reported.

The response of the magnetic moment M under an effective field H_{eff} is described by the Landau-Lifshitz equations of motion as:

$$\frac{dM}{dt} = -\gamma\left(M \times H_{eff}\right) - \frac{4\pi\mu_0\lambda}{M^2} M \times \left(M \times H_{eff}\right)$$

where γ is the gyromagnetic ratio, and λ is the damping factor in units of S^{-1}. In the static case the total magnetic moment has to be parallel to the total effective field, the magnetic anisotropy energy density E_a is related to the effective field by:

$$H_{eff} = -\frac{\partial E_a}{\partial M}$$

where $E_a = K_{u1}\left(\alpha_1^2\alpha_2^2 + \alpha_2^2\alpha_3^2 + \alpha_3^2\alpha_1^2\right) + K_{u2}\alpha_1^2\alpha_2^2\alpha_3^2 + K_{u2}\alpha_1^2\alpha_2^2\alpha_3^2 + ...$, and $\alpha_1, \alpha_2, \alpha_3$ are the direction cosines of the saturation magnetization with respect to the , and crystallographic axes respectively, and K_{u1} and K_{u2} are the second and forth order terms of the perpendicular uniaxial anisotropic energy. For cubic anisotropy with the easy axis along the direction, the in-plane effective field is approximated as $h_{eff} \approx 4K_{u1}/3M_s$ where M_s is the saturation magnetization. Consider a specimen of a cubic ferromagnetic crystal with ellipsoid anisotropic magnetization, the Lorentz field $(4\pi/3)M$ and the exchange field λM do not contribute to the torque because their vector product with M vanishes identically. In applying a field parallel to the film (the xz plane), we can derive the ferromagnetic resonance (FMR) frequency from equation $\frac{dM}{dt} = -\gamma\left(M \times H_{eff}\right) - \frac{4\pi\mu_0\lambda}{M^2} M \times \left(M \times H_{eff}\right)$ as:

$$\omega = \gamma\sqrt{\left(H_0 + H_a\right)\left(H_0 + H_a + 4\pi M\right)}$$

where H_0 is the external field. Here we have implied the demagnetizing factor for an infinite plane of thin film to be $N_x = N_z = 0, N_y = 4\pi$ for the external magnetic field along the xz plane.

A typical micro-strip transmission line is shown in figure where the geometrical parameters are also described there. For the micro-strip with small insertion loss, a matched 50 ohm transmission line should be considered. In this design we follow the well-known quasi-TEM formulae derived from Wheeler's and other workers. The substrate is a sapphire plate with (0001) orientation and a thickness of $500\,\mu m$. The calculated line width w is 430μm, and the effective dielectric constant $\varepsilon_{\mathrm{eff}}$ is 7.27. The frequency dependence of the effective dielectric constant $\varepsilon_{\mathrm{eff}}(f)$ reported by Kobayashi will be incorporated to obtain a more accurate result.

The surface resistance R_s of metallic films is related to the attenuation constant α_c of a stripling, which has been given by Pucel. The other losses of the micro-strip resonator are the dielectric loss and the radiation loss. Sapphire is a good insulating dielectric material with loss tangent usually being less than 0.0001. The radiation loss increases with the square of frequency and gap discontinuity of the micro-strip.

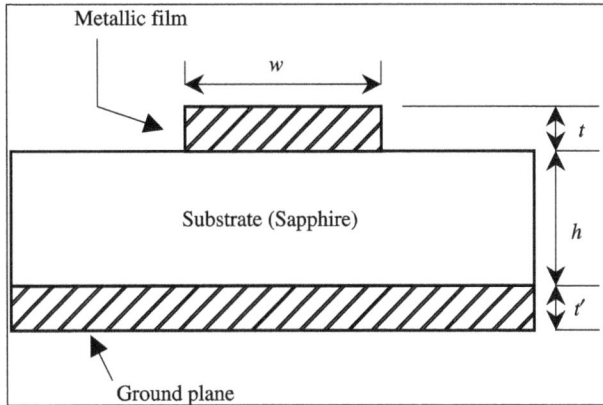

The front view of a micro-strip transmission line. The line width is w, thickness is t, substrate thickness is h, and thickness of the ground plane is t'.

A T-junction micro-strip, which has a lower radiation loss than other kinds of mirostrips is depicted in figure. Here, the open-ended stub represents a quarter-wave resonator. The resonance condition is:

$$f_{\mathrm{res}} = \frac{nc}{4(L+\Delta L)\sqrt{\varepsilon_{\mathrm{eff}}(f)}}, n = 1,3,5,7\ldots$$

where ΔL is the effective length of a open end effect. The measured unloaded quality factor, Q_0, is related to α by:

$$Q_0 = \frac{\beta}{2\alpha}$$

where $\beta = \dfrac{2\pi}{\lambda_g}$ and $\alpha = \dfrac{8.686}{Q_0\lambda_g} = \dfrac{8.686 f_{\mathrm{res}}\sqrt{\varepsilon_{\mathrm{eff}}(f)}}{cQ_0}(\mathrm{dB/m})$

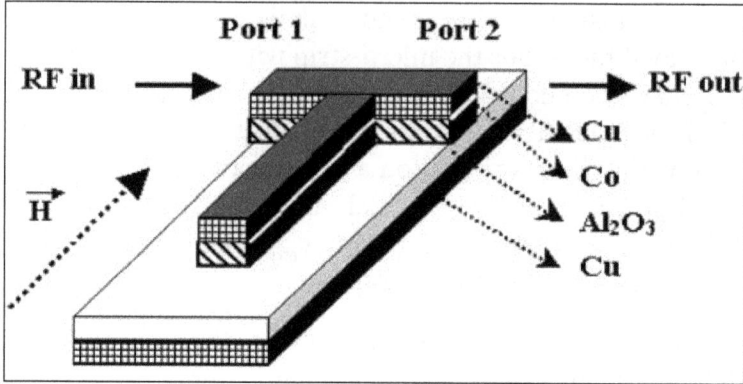

Schematical diagram of a T-junction. The length of the stub is L,
and the arrow indicates the direction of the applied magnetic field.

The input impedance of the quarter-wave stub resonator can be characterized as:

$$Z_{in} = R_0\left(1 + j2Q_0\Delta\right)$$

where Δ is defined by:

$$\Delta \equiv \frac{\omega - \omega_0}{\omega_0}$$

At resonance, the input impedance will be reduced to the bare dc resistance R_0. The absolute value of S_{21} thus is given by,

$$|S_{21}| = \frac{2R_0/(2R_0+Z_c)\sqrt{1+(2Q_0\Delta)^2}}{\sqrt{1+(2R_0/(2R_0+Z_c))^2(2Q_0\Delta)^2}}$$

where Z_c is the charateristic impedance of the micro-strip. The upper and the lower frequencies deviated from the resonant frequency f_0 define the bandwidth $\Delta f = f_2 - f_1$. The resonance peak of S_{21} versus frequency is readily applicable for the determination of the quality factor.

A plot of the resonant frequencies of the co-existed FMR and micro-strip structure modes with respect to the applied fields reveals a nearly straight line making a clue of equation $\omega = \gamma\sqrt{(H_0+H_a)(H_0+4\pi M)}$. The Q factor decreases firstly with the applied fields and then increases at 257 mT as shown, respectively, in figures. This field implies that the FMR frequency coincides with the S_{21} resonance frequency that leads the largest microwave field to be dissipated in the side arm.

From the values of the Q factor, we can evaluate the loss due to FMR for resolving the change of attenuating constants at several S_{21} resonance frequencies when sweeping the magnetic fields. The fields at the peak loss correspond to the occurrence of the FMR is the closest to that of the transmission resonance. The sheet resistance can be calculated readily from the transmission loss and is depicted in figure. The conductivity of the nickel film can be accessed accordingly from the multiplication of the film thickness and the sheet resistance $\Omega/$. The magneto resistances also have resonant peaks occurring at the FMRs where the attenuation of microwave field is a maximum.

The value of the magnetization M depending on the demagnetization factor that varies with different orientations of the external magnetic field. The demagnetization factors are $N_x = 4\pi, N_y = N_z = 0$, for the field parallel to the surface. The resonant frequency $f = \omega/2\pi$ can be plotted almost linearly proportional to the field as shown in figure. In this case, the simulated anisotropic field $H_a \approx 533$ Gauss, while the $M \approx 149$ Gauss. On the other hand for the external field perpendicular to the surface, we have $N_z = 4\pi$, and $N_x = N_y = 0$.

The calculated FMR frequency at different magnetic field for cobalt films is plotted in figure by the simulated electronic circuits. The linear plot can be fitted with equation $\omega = \gamma\sqrt{(H_0 + H_a)(H_0 + 4\pi M)}$ to solve the anisotropic field H_a and the saturation magnetization M_s. The same plot for nickel film with silver film over coated on nickel films to reduce the radiation lose is shown in figure. The low field region has a better linearity than that plot without considering the FMR at zero applied fields. Table dictates the saturation magnetization and anisotropic field for bulk and thin film of Co, and Ni in which the bold-faced data are measured in this work. This experiment results for thin films are in congruent with the reported bulk values.

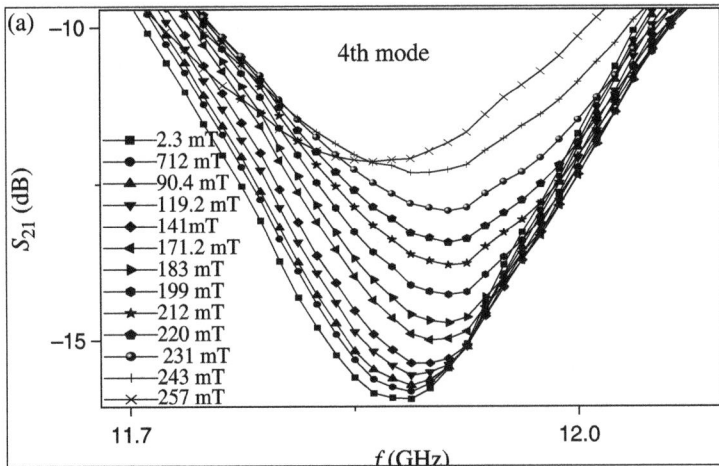

(a)

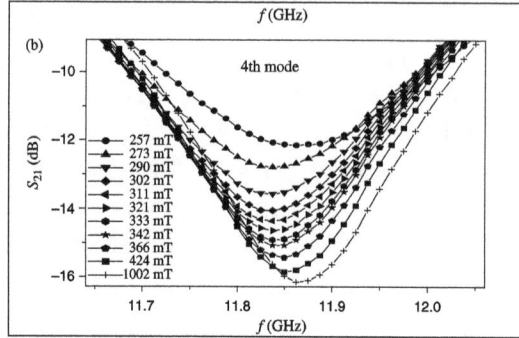

The spectra of the 4th structure mode peak under applied magnetic field for (a) peak rises up under increasing magnetic field, and (b) peak falls down under increasing the magnetic field.

In this scenario, we found the co-existence of the ferromagnetic resonance and the structure resonance of a T-microwave micro-strip transmission line under proper values of magnetic fields. The conductivity, the anisotropic field, and the magnetization factor of the magnetic film can be simultaneously determined from the same spectrum. The physical properties of metallic magnetic films can be determined preceding to lay out the spintronic devices. This work advocates a fertilized method to study the magnetic properties of thin films with a simple microwave network analyzer without implementing an involved microwave cavity for the conventional FMR.

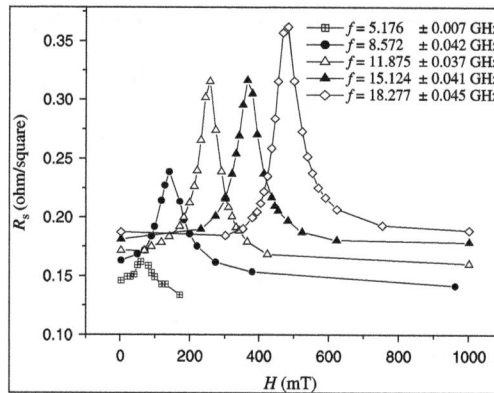

Surface resistance versus applied magnetic field at different structure resonate modes.

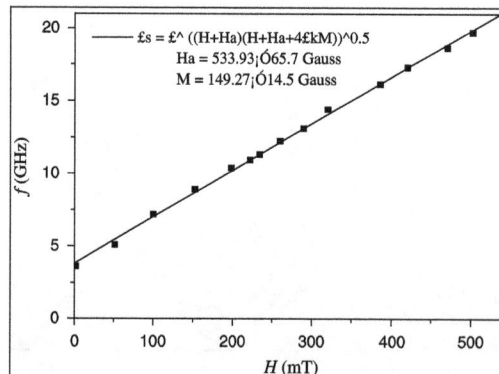

The FMR peak and fitting curve of a meander micro-strip line.

Mechanical Properties

Nanoparticles show different mechanical properties relative to microparticles and bulk materials, providing more effective options for the surface modification of many devices in the mechanical strength, or to improve the quality of nanomanufacturing/nanofabrication, etc. To be more specific, on the one hand, the mechanical effects of nanoparticles can affect the tribological properties of lubricants with nanoparticles as well as reinforce composite coatings. In a lubricated contact, the comparison in the hardness between nanoparticles and the contacting surface determines whether particles are deformed or indented into the surface when the contact pressure is sufficiently large. This information could reveal how the particles behave in the contact. On the other hand, nanoparticles are usually used as abrasives in the nanopolish of ultra-smooth surfaces by chemical mechanical polishing (CMP), which is the most effective planarization tool in the manufacture of an integrated circuit (IC), till now. Good controls over the mechanical properties of particles and their interactions with the polished surface etc are important for improving the surface quality and enhancing material removal. Successful applications in these fields usually need a deep understanding of the basics of the nanoparticles' mechanical properties, such as hardness and elastic modulus, interfacial adhesion and friction, movement law, as well as their size-dependent effects. In order to acquire more of this information, different testing methods have been developed, e.g., nanoindentation with atomic force microscopy (AFM), in situ compression by a force probing holder based on the observation with transmission electron microscopy (TEM). However, the obtained results are still inadequate and some are controversial. For instance, there is still no definite conclusion as to whether the elastic modulus of nanoparticles measured with AFM is affected by the particle size and the indentation depth. Furthermore, the contact mechanics, especially the frictional and mechanical behaviors related to nanoparticles, have not been fully understood. The applicability of classic theories, e.g., the Hertzian theory, for describing the contact behaviors in the case of particle sizes down to the nanoscale.

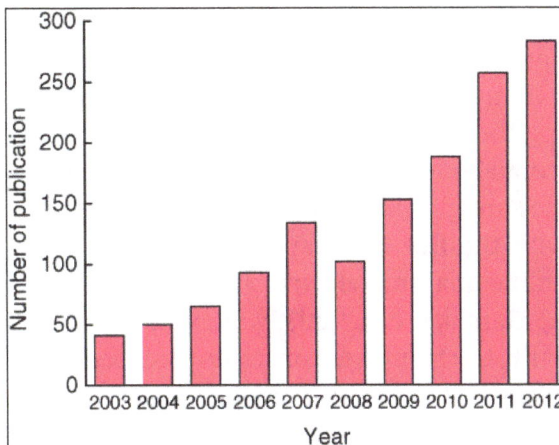

The number of publications related with the topic on the mechanical properties of nanoparticles in the past decade.

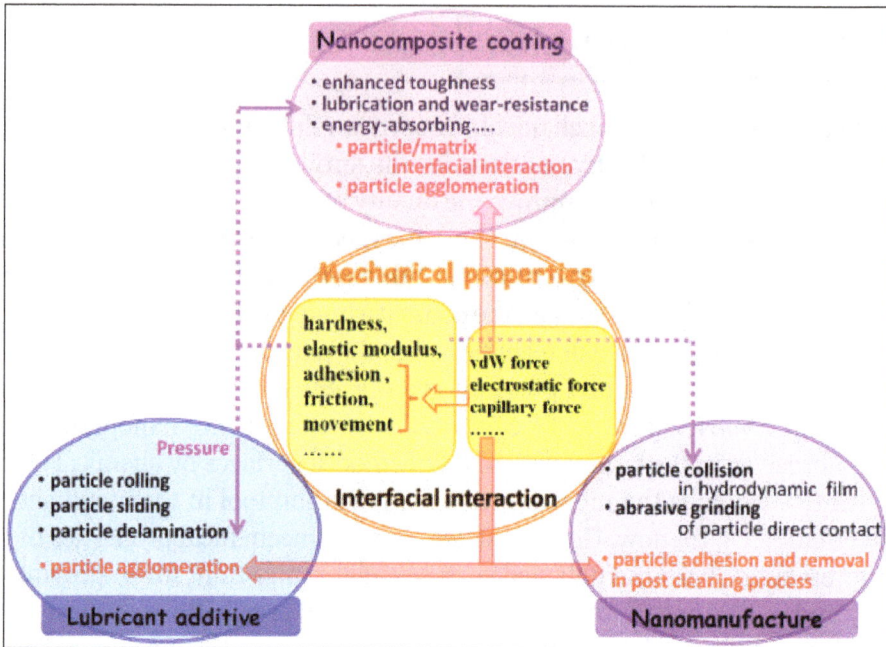

Van der Waals (vdW) Forces

VdW forces are the weak interaction between all molecules and particles, which play important roles in the particles' mechanical properties. This kind of force includes three parts: one is the orientation force (the Keesom force), resulting from the interaction between the permanent dipole moment of polar molecules. The second is the induction force (the Debye force), which comes from the interaction between the permanent dipole moment of the polar molecule and the induced dipole moment. The third is the dispersion force (the London force), which exists in a wide variety of polar and nonpolar molecules, coming from the induced instantaneous dipole polarization. VdW energies are usually from several to dozens of thousands of Joules per mole, one or two orders of magnitude smaller than the chemical bond energy. The vdW forces are long-range forces and can be effective in a large range of distances, varying from long distances greater than 10 nm down to atomic scale distance (about 0.2 nm). The methods for calculating the vdW interaction forces or energies between small molecules or large macroscopic bodies have been well established. Several of the common vdW forces and energies are given in table. The vdW forces of objects with any shape can be transformed with the Derjaguin approximation to those between two planes per unit area. Based on the quantum electrodynamics theory, Lifshitz deduced the expression for calculating Hamaker constants, which can be used to solve problems with media involved. Typically, the Hamaker constants for interactions in a medium are an order of magnitude lower than those in a vacuum. The vdW force is always attractive between identical materials, but it may be repulsive between dissimilar materials in a third medium (usually liquid).

Table: Several of the common vdW energies and forces.

Types	vdW energies	vdW forces
Molecular-plane	$W = -\dfrac{\pi C_{vdw}\rho}{6D^3}$	$F = -\dfrac{\pi C_{vdw}\rho}{2D^4}$
Sphere–sphere	$W = -\dfrac{A}{6D}\dfrac{R_1 R_2}{R_1 + R_2}$	$F = -\dfrac{A}{6D^2}\dfrac{R_1 R_2}{R_1 + R_2}$
Sphere–plane	$W = -\dfrac{AR}{6D}$	$F = -\dfrac{AR}{6D^2}$
Plane–plane	$W = -\dfrac{A}{12\pi D^2}$	$F = -\dfrac{A}{6\pi D^3}$

C_{vdw} is a coefficient related to the atomic pair potential, R is the sphere radius, R_1 and R_2 are the radii of two spheres, respectively, D is the distance between two surfaces, $A = \pi^2 C_{vdw}\rho_1\rho_2$ is the Hamaker constant and ρ is the atomic density.

Electrostatic Force and Electrical Double Layer Force

For particles suspended in water or any liquid with a high dielectric constant, they are usually charged and can be prevented from coalescing due to the repulsive electrostatic force. The charging of a surface in a liquid has three main sources: (1) the ionization or dissociation of surface groups; (2) the adsorption or binding of ions from the solution onto a previously uncharged surface; (3) when two dissimilar surfaces are very close, charges can hop across from one surface to the other. The surface charges are balanced by an oppositely charged ion layer in the solution at some distance away from the surface, forming the EDL. The idea of the EDL was first formally proposed by Helmholtz, who derived the charge distribution in the solution based on the simple molecular capacitor model. In reality, the thermal motion of ions in the solution introduces a certain degree of chaos causing the ions to be spread out in the region of the charged surface, forming a 'diffuse' double layer. In that case, the analysis of the electronic environment near the surface is more complex and requires more detailed analyses. Gouy, Chapman and Stern put forward more accurate models for analysing the surface and electrolyte interfaces, making great contributions to the development of EDL theories. Gouy and Chapman independently developed theories of a so called 'diffuse double layer', in which the change in the concentration of the counter ions near a charged surface follows the Boltzmann distribution. The Gouy–Chapman theory provides a better approximation of the real system than the Helmholtz theory, but it still has limited quantitative applications. It assumes that ions behave as point charges and that there is no physical limit for the ions in their approach to the surface. Then, the Gouy–Chapman diffuse double layer was modified by Stern so that ions have a finite size and cannot approach the surface closer than a few nanometres: the first layer of ions in the Gouy–Chapman diffuse double layer are not at the surface, but at some distance away from the surface. As a result, the potential and concentration of the diffuse part of the layer is low enough to justify treating the ions as point charges. Stern also assumed that some ions are probably adsorbed by the surface in a plane; this layer is known as the 'Stern layer'. Within this layer, thermal diffusion is not strong enough to overcome the electrostatic forces. In the diffusive outer layer, the ions are far enough from the solid surface and are subjected to weak electrostatic forces from the surface only, hence they remain mobile.

A double layer is formed to neutralize the charged surface, which in turn causes an electrokinetic potential between the surface and any point in the mass of the suspending liquid. This voltage difference is of the order of millivolts and is referred to as the surface potential. The magnitude of the surface potential is influenced by the surface charge and the thickness of the double layer. Starting from the surface, the potential drops off roughly linearly in the Stern layer and then exponentially through the diffuse layer, approaching zero at the imaginary boundary of the double layer. The potential curve is useful because it can suggest the electrical force strength between particles and the critical distance within which this force comes into play. A charged particle's mobility is related to the dielectric constant and the viscosity of the suspending liquid, as well as the zeta potential, which is a potential at the boundary between the moving particle and the liquid. The boundary is called the slip plane and usually defined as the point where the Stern layer and the diffuse layer meet. The common EDL model is shown in figure. The EDL interaction energy and the force between the bodies of different geometries can be referred to.

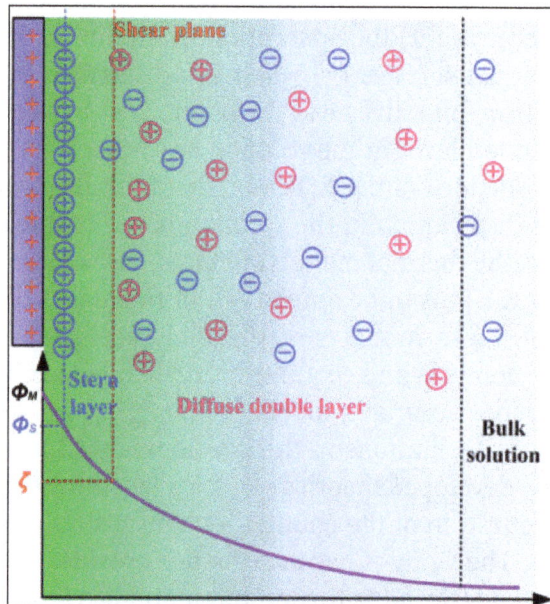

Schematic model of EDL.

Capillary Force

Capillary force is mainly due to the formation of liquid menisci (also termed the meniscus force), the significance of which was realized by Haines and Fisher. Capillary force can be classified into two types: normal capillary force and lateral capillary force. Capillary forces should be considered in the studies on powders, soils and granular materials, the adhesion between particles or particles to surfaces and the stiction in micro/nano-electromechanical systems. It is also relevant to nanoparticle assembling or living cells self-assemble technologies.

The normal capillary force arises from the Laplace pressure within the curved meniscus formed by liquid condensation or vapour bridges around two adhering solid surfaces. It can be attractive or repulsive depending on whether the capillary bridge is concave or convex. Two equations are important to understand the capillary forces, i.e. the Young–Laplace equation and the Kelvin equation. The Young–Laplace equation relates the curvature of a liquid interface to the pressure difference, while the Kelvin equation describes capillary condensation, which is the physical basis for many adhesion phenomena. Capillary condensation is the condensation of vapour into capillaries or fine pores even for vapour pressures below the saturation vapour pressure. The Kelvin equation relates the actual vapour pressure to the surface curvature of the condensed liquid. The normal capillary force is owing to two actions: one is the pressure difference across the curved interface and the other is the action of the surface tension force exerted around the annulus of the meniscus. Butt and Kappl gave the usual derivations and expressions for capillary forces between different geometries.

The origin of the lateral capillary forces is the overlap of the perturbations in the shape of a liquid surface due to the presence of attached particles. The larger the interfacial deformation created by the particles, the stronger the capillary interaction between them. The lateral capillary forces are effective in controlling small colloidal particles and protein macromolecules confined in liquid films to form fine microstructures.

Solvation, Structural and Hydration Forces

Apart from vdW forces and EDL forces, some other forces, i.e. solvation, structural or hydration forces, come into play when two surfaces or particles approach very close (separation less than a few nanometres) in the liquid. These forces can be monotonically repulsive, monotonically attractive or oscillatory and they can be much stronger than either the vdW forces or EDL forces at small separations. Solvation, structural or hydration forces (in water) arise between two particles or surfaces if the solvent or water molecules become ordered by the surfaces. When the ordering occurs, an exponentially decaying oscillatory force with a periodicity equal to the size of the confined liquid molecules, micelles or nanoparticles appears. Solvation forces depend not only on the properties of the liquid medium but also on the surface physicochemical properties, such as hydrophilicity, roughness, crystalline state, homogeneity, rigidity and surface micro-texture. These factors affect the structure of the confined liquids between two surfaces, which in turn affects the solvation force. The hydration force is a strong short-range repulsive force between the polar surfaces separated by a thin polar liquid layer (thickness <3 nm); the force magnitude decays exponentially with the liquid layer thickness. A well known interpretation of hydration force is that the solvent molecules are bound strongly and are restructured by polar surfaces. An ordered-solvent layer was formed at the surface-solution interface, which exponentially decays away from the surface; the overlap of the ordered-solvent layers near the two mutually approaching surfaces creates a force. The hydration force could determine the behaviors of many diverse systems, e.g., the colloidal dispersion stability, the swelling of clays and the interactions of biological membranes.

DLVO Theory

The DLVO (Derjaguin–Landau–Verwey–Overbeek) theory was introduced by Derjaguin and Landau in 1941 and Verwey and Overbeek in 1948 for describing the stability of colloidal dispersions. The theory combines the effects of the vdW attraction and the electrostatic repulsion. It can explain many phenomena quantitatively in colloidal science, e.g., the adsorption and the aggregation of nanoparticles in aqueous systems, and describe the force between charged surfaces interacting through a liquid medium. figure shows the schematic plot of the DLVO interaction potential energy E of model nanoparticles (diameter: 100 nm and surface potential: 20–40 mV) which are dispersed in aqueous salt solutions.

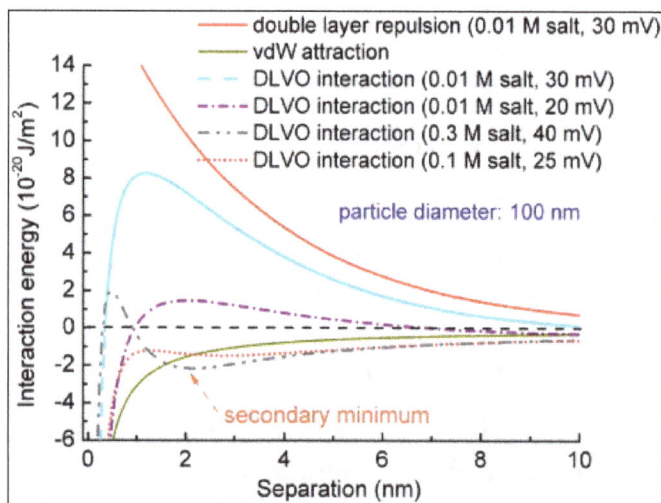

Schematic plots of the DLVO interaction potential energy E (the Hamaker constant A is 1.5×10^{-20} J).

It can be seen that a strong long-range repulsion with a high energy barrier is present for highly charged surfaces in dilute electrolyte (i.e. long Debye length). When the surface charges are reduced or the concentration of the electrolyte solutions are increased, a small secondary minimum in the potential energy curve appears. Colloid particles may undergo a reversible flocculation due to the secondary minimum because of its weak energy barrier, resulting in slow particle aggregation for the surface with a low charge density. Below a certain surface charge or above a certain electrolyte concentration (known as the critical coagulation concentration), the energy barrier falls below the zero axis and particles then coagulate rapidly. Consequently, the colloid system becomes unstable.

Although the DLVO theory is the basis for understanding colloid stability and has a considerable amount of experimental support, it is inadequate for the colloid properties in the aggregated state. This is because short-range interactions are dominant in this state and the specific properties of ions should be taken into account rather than regarded as point particles. Most deviations of experimentally measured forces from those expected from the DLVO theory are due to the existence of a Stern-layer or non-DLVO forces, e.g., ion-correlation, solvation, hydrophobic and steric forces.

Contact, Adhesion and Deformation Theories of Nanoparticles

In traditional contact theories for two objects in contact with each other under external forces, for instance, the simplest case of two interacting elastic spheres deduced by Hertz in 1882, surface forces were not included. In these models, the displacement and the contact area are equal to zero when no external force is applied. However, as the size of the object is decreased to the nanoscale, the surface forces play a major role in their adhesion, contact and deformation behaviors. Modern theories of the adhesion mechanics of two contacting solid surfaces are based on the Johnson–Kendall–Roberts (JKR) theory or the Derjaguin–Muller–Toporov (DMT) theory. The JKR theory is applicable to easily deformable, large bodies with high surface energies. Strong, short-range adhesion forces dominate the surface interaction; the effect of adhesion is included within the contact zone. In contrast, the DMT theory better describes very small and hard bodies with low surface energies. In this case, the adhesion is caused by the presence of weak, long-range attractive forces outside the contact zone. Tabor introduced a nondimensional physical parameter, often referred to as Tabor's parameter, to quantify the limits of JKR, DMT and the cases between them. The intermediate regime between the JKR and the DMT theories has also been described by Maugis using the Dugdale model; a 'transition parameter' roughly equivalent to Tabor's parameter was defined. A summary of the different conventions used for defining the 'transition parameter' was given by Greenwood. Carpick et al provided a simple analytic equation to determine the value of the 'transition parameter'; it could closely approximate Maugis' solution. The expansion of the JKR theory by Maugis and Pollock leads to the additional description of plastic deformation. Table summarizes the relations between the contact radius, deformation and the adhesion force for two spheres contacting each other according to the three mostly used theories.

Table: Relations between the contact radius a, the contact radius a_0 due to adhesion force without an external load, the deformation δ and the adhesion force for two spheres contacting each other according to the Hertz, JKR and DMT theories.

	Hertz	JKR	DMT
a	$\left(\dfrac{R^*P}{E^*}\right)^{1/3}$	$\left\{\dfrac{R^*}{E^*}[P+3\pi R^*\gamma+(6\pi R^*\gamma P+(3\pi R^*\gamma)^2)^{1/2}]\right\}^{1/3}$	$\left[\dfrac{R^*}{E^*}(P+2\pi R^*\gamma)\right]^{1/3}$
δ	$\dfrac{a^2}{R^*}$	$\dfrac{a^2}{R^*}$	$\dfrac{a^2}{R^*}-\left(\dfrac{8\pi a\gamma}{3E^*}\right)^{1/2}$

a_0	0	$\left(\dfrac{6\pi R^{*2}\gamma}{E^*}\right)^{1/3}$	$\left(\dfrac{2\pi R^{*2}\gamma}{E^*}\right)^{1/3}$
P_{ad}	0	$2\pi R^*\gamma$	$\dfrac{3\pi R^*\gamma}{2}$

R* is the reduced radius defined as $1/R^* = (1/R_1) + (1/R_2)$, γ is the adhesion work per unit area. P is the external force and E* is the reduced Young's modulus defined as $(1/E^*) = \dfrac{3}{4}[((1-v_1^2)/E_1)+((1-v_2^2)/E_2)]$, E_1, E_2, and v_1, v_2 are Young's moduli and Poisson's ratios of the two spheres, respectively.

Although the Hertz, JKR and DMT theories have been widely used to study the mechanical properties of nanoparticles, whether or not the continuum mechanics can be used to describe a particle at the nanometre scale is still in discussion. The molecular dynamics (MD) simulation method provides an opportunity to understand the atomistic processes in the contact region. Luan and Robbins researched the contact between two nanocylinders by MD simulations and found that the atomic-scale surface roughness produced by discrete atoms led to dramatic deviations from the continuum theory. Contact areas and stresses may be changed by a factor of two, whereas friction and lateral contact stiffness by an order of magnitude. Also Miesbauer et al analysed the contact between two NaCl nanocrystals with MD simulations. It was found that the Hertzian theory was a suitable description of the studied system when the system size was larger than 50 Å; the discrepancy became more obvious as the particle was even smaller. Cheng and Robbin investigated the nanoscale contact with MD simulations to test the adaptability of continuum contact mechanics at the nanoscale; the results suggested that the continuum contact models could be applied to the case where the forces averaged over the areas containing many atoms. Nonetheless, the continuum theory, because of its concise expression, is still widely applied in the mechanical analysis at the nanoscale, such as designing micro/nano-devices, creating nanostructured materials with optimized mechanical properties and understanding the molecular origins of friction and adhesion.

Main Techniques for Studying Nanoparticles

The research methods frequently used in studying the mechanical properties of nanoparticles will be briefly introduced as follows:

AFM Techniques

AFM is a powerful technique that can be used to obtain both high-resolution images on many kinds of solid surfaces and the vertical force as well as lateral force between a sharp tip and the surface. The schematic diagram of the basic working principle of AFM is shown in figure, including a cantilever with a sharp tip on its end, piezotube scanner, scanning and feedback systems, a four quadrant photoelectric detector and the computer. Briefly, the sharp tip scans over the sample and the deflection of the cantilever is quantified through a laser beam reflected off the backside of the cantilever and received by the photoelectric detector. If a constant force is kept between the tip and sample during scanning, the topographic image of the sample surface can be obtained by plotting the height of a sample stage on the piezoscanner, which is controlled by a feedback system. Alternatively, the interaction force between the tip and sample can be obtained with the cantilever's vertical deflection using the force-versus-distance curves, briefly called force curves, together with Hooke's law. These curves can provide valuable information on some of the important properties of nanoparticles, such as hardness, elastic modulus and the adhesion between nanoparticle and substrate. The lateral force is closely related to the torsional deflection of the cantilever; an accurate value can be obtained after careful calibration of the cantilever's torsional coefficient.

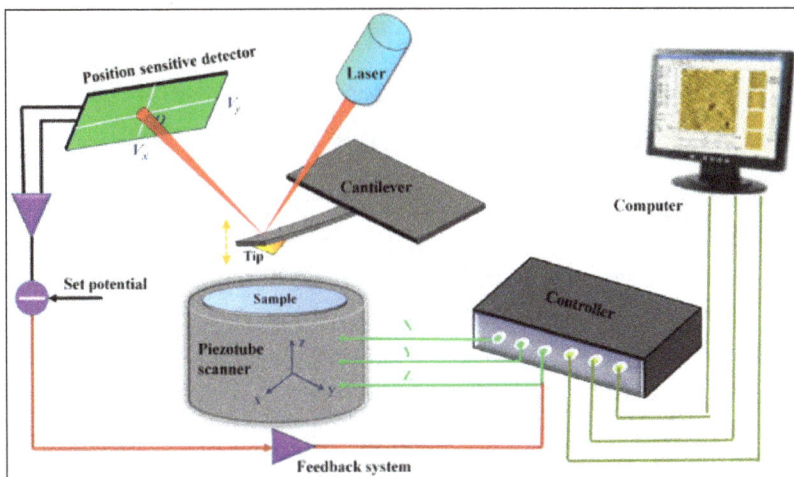

Schematic diagram of the basic working principle of AFM.

Particle Tracking Velocimetry (PTV)

PTV is an image-based velocimetry method of measuring the velocity field and tracking individual particles in fluidic systems. Fluorescent particles are usually used as tracers within a defined area where those particles are illuminated; then pictures of these particles are taken. The motion trajectories of the particles can be reconstructed by locating them in those pictures and the velocities of the particles can be calculated correspondingly. Based on these, deep insight into some of the complex and low-velocity flows in a region can be acquired. It is a technique that is slightly different from particle image velocimetry (PIV) where the particles' displacements within a segment of an image are averaged. Currently, there are mainly two different PTV methods, i.e. two-dimensional particle tracking velocimetry (2D-PTV) and three-dimensional particle tracking velocimetry (3D-PTV). The defined area is a thin light sheet for 2D-PTV while it is an illuminated volume for 3D-PTV, which is usually based on a multiple-camera-system.

In Situ TEM

TEM could provide images with a significantly higher resolution than a light micro-scope by using electrons as 'light source' which have a much lower wavelength. The ba-sic principle is that a beam of electrons passes through a very thin sample and, after in-teracting with the atoms in the sample, some unscattered electrons reach a fluorescent screen to form an image. The image is shown in varied darkness indicating the material density in different parts of the specimen. The image is magnified and can be studied directly from the screen or recorded with a camera for post-analysis. In situ TEM offers the capability of real-time observation of the responses of the microstructural evolu-tion of nanostructures to external active stimuli and their relationship with properties. Active stimuli applied to the sample examined in the microscope during simultaneous imaging include mechanical, thermal and electrical ones, etc.

MD Simulation

Computational simulations are usually considered as very useful complementary tools to experimental studies on the mechanical properties of nanoparticles. Among many different kinds of computation methods, MD simulation is an important aspect which could model the time evolution of the physical motions of interacting atoms or mol-ecules. It is a computation method that is based on statistical mechanics; statistical ensemble averages are normally hypothesized to be equal to the time averages of the system. Mostly, in MD simulation, Newton's equations of motion for the atoms or mol-ecules in a system are numerically solved to get their positions and velocities and finally to describe the thermodynamic behaviors of the system. The interactions and potential energy between atoms or molecules are defined by a molecular mechanics force field.

Basic Mechanical Properties of Nanoparticles

Hardness and Elastic Modulus of Nanoparticles

Understanding some basic mechanical properties of nanoparticles, such as the hard-ness and the elastic modulus, will aid a lot in the proper design of particles in specific applications, as well as evaluating their roles and action mechanisms. the measure-ment of the mechanical properties of microparticles has been developed for decades. The microindentation technique was used by Steinitz in 1943 to test the hardness of microparticles with indented areas of larger than 100 μm^2 and a minimum indenter size of 20 μm^2 . About ten years ago, nanoindentation was employed by Shorey et al to measure the elastic properties of particles (average size: 5 μm) used for magnetorhe-ological finishing. Their methods were aimed at measuring the film of particles rather than individual particles. The deformation behaviors of polystyrene microspheres (di-ameter: 20 μm) by using AFM against a mica surface was firstly investigated by Biggs and Spinks in 1998. Since then, protocols of calculating the mechanical characteris-tics (e.g., the elastic modulus) of nanoparticles have developed rapidly, primarily by measuring the particles' deformation with AFM. Typically, quantitative computation of the elastic modulus of nanoparticles requires the measurement of indentation h by converting AFM force-displacement curves into force-indentation curves instead of measuring the contact area radius. The latter is hard to obtain directly. The external load P applied through the cantilever (its spring constant denoted as k) to the tip can be described with the Hooke law,

$$P = k \cdot \delta_c$$

where δ_c is the cantilever deformation. The indentation depth h of the tip into the sample surface is:

$$h = z - \delta_c$$

where z is the piezo displacement. The relative displacements and deformations of the particle-AFM tip system in the indentation process are shown in figure. Since there is often system thermal drift, the deflection offset, δco should be considered. In this case, equation $P = k \cdot \delta_c$ can be rewritten as:

$$P = k \cdot (\delta_c - \delta_{c0}).$$

Also, if the position for the tip initially touching the sample surface is considered, resulting in another height offset z0, equation $h = z - \delta_c$ becomes:

$$h = (z - z_0) - (\delta_c - \delta c_0).$$

In this way, the force–indentation curves can be obtained for the calculation of the particles' elastic modulus by evaluating the slope of the loading region on the curves with contact theories. Mostly based on the previous method, the elastic modulus of a variety of nanoparticles have been measured by compressing or bending particles primarily with AFM the hardness of some nanoparticles is also given. The nanoparticles' hardness and elastic modulus often deviate from their bulk materials' and some show obvious size-dependent behaviors. Typical related results and the underlying mechanisms can be divided as the following three categories.

- In the case of spherical polymer nanoparticles, there are yet no uniform size-dependent behaviors of the mechanical properties. For instance, the compressive moduli of the polystyrene nanoparticles (diameter: 200 nm) were found to be slightly less than those of the corresponding bulk materials due to the presence of hydrated ionic functional groups. In contrast, the work conducted by Paik et al showed that the elastic modulus of polypropylene (PP) nanoparticles was higher than that of the bulk material. It was thought that the glass transition temperature (T_g), the crystalline phase and crystallinity etc could affect the deformation of the polymer chain inside and thereby result in the change of the particle's elastic modulus.

- For crystalline metal nanoparticles, dislocations inside the particles have been demonstrated as one of the factors contributing to the change in the mechanical behavior of nanoparticles, which is in contrast to the traditional view that no dislocation is present in crystalline nanoparticles. The experimental work done by Ramos et al indicated that the hardness and elastic modulus of six-fold symmetry gold nanoparticles were higher than the bulk phase due to the formation of stacking faults and dislocations in specific crystallographic directions. Mordehai and Nix et al performed nanoindentation and compression tests combined with theoretical simulation to reveal the deformation behaviors of single-crystal gold nanoparticles on sapphire substrates. The particle strength under indentation increased with the lateral dimension of the particle due to the competition between the generation of dislocations beneath the indenter and their drainage from the particle. Under compression with a flat diamond punch, the compressive stress of the particle increased with the decrease of the particle size since the nucleated dislocations resulted in the stress gradient along the slip planes. In situ TEM nanoindentation experiments showed the direct evidence of the presence of dislocations in metal nanoparticles during deformation but they disappeared during the unloading process, as shown in figure. Wang et al recently demonstrated a new kind of stacking fault related with dislocations in gold nanocrystals, which could nucleate, migrate and annihilate under mechanical loading with in situ TEM and MD simulation. For silicon nanoparticles, similar behaviors were observed by Gerberich et al that their hardness (particle diameter: 40 nm) was four times greater than the value of bulk silicon. They proposed that the dislocations or line defects inside the particle are the main factors resisting high pressures. Furthermore, atomistic simulation conducted by Zhang et al confirmed that the superhard silicon nanoparticles resulted from the nucleation and movement of dislocations. Apart from dislocations or defects, the changes of the lattice strain and the bond energies of nanoparticles to the compressive stress were proposed as another cause for the strengthening and weakening of the mechanical properties of nanoparticles. Furthermore, first-principles electronic-structure calculations made by Cherian et al suggested the size dependence of the bulk moduli of several semiconductor nanoclusters correlated with the strong interaction with the passivant.

- For nanowires or nanotubes, it has been typically found by Jing et al and Cuenot et al that the elastic moduli of silver and lead nanowires decreased with the increasing radial diameter. They proposed that the increase in the modulus was attributed to the effects of the surface stress, the oxidation layer and the surface roughness, or the surface tension effect. MD simulations conducted by Yang et al showed the bulk modulus of Ni/Ni_3Al nanowires increased but the surface energy decreased with the increasing wire perimeter size. However, only the fracture properties rather than the elastic behavior of ZnO nanowires were affected by the surface effects due to the presence of surface cracks and defects. Worth mentioning is the fact that measuring the mechanical properties of individual nanoparticles is very complex; many influencing factors could affect the finally measured results. These factors include the uniform dispersion of nanoparticles on an ideally hard substrate, the precise locating of particles and the proper application of loads onto the particles, as well as the measurement of the minimum particle deformation, etc. In addition, many uncertainties during measuring and calculating the mechanical properties of nanoparticles with AFM, e.g., uncertainties associated with the instrument calibration and the calculation models, should be considered.

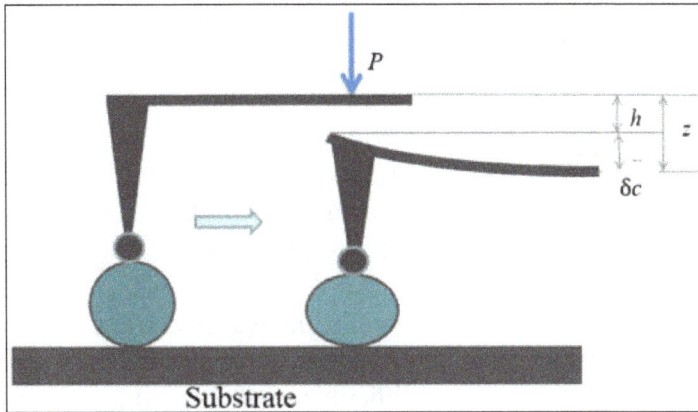

In figure: relative displacements and deformations of the particle-AFM tip system during the indentation process. Left: the AFM tip just touches the particle without deformation of the particle. Right: the particles deformation occurs due to the applied force by the AFM tip.

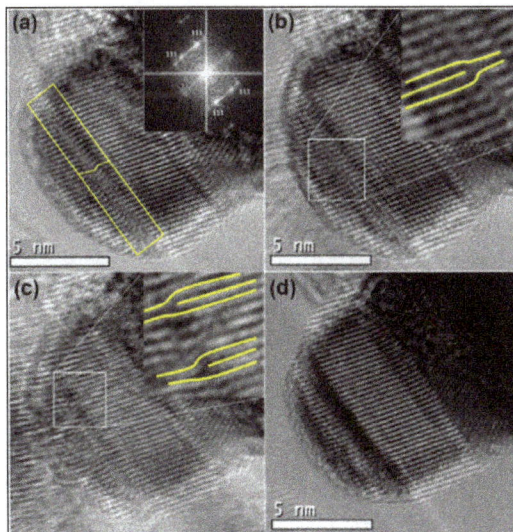

In figure: high-resolution TEM images of a silver nanoparticle before and after compression: (a) before compression (twin highlighted); (b) at the initial stage of compression (an edge dislocation highlighted); (c) at a stage of further compression (two additional dislocations shown in the inset); (d) after the removal of the compression (no dislocation observed).

Adhesion and Friction of Nanoparticles

The adhesion and the friction of nanoparticles play important roles in nanofabrication, lubrication, the design of micro/nano devices, colloidal stabilization and drug delivery. In this case, characterizing the adhesion and friction behaviors of nanoparticles has attracted significant research interest over the past decade. So far, AFM has been proved to be a powerful tool to measure the adhesion and friction between a nanoparticle and a solid surface. The AFM tip itself can also be thought of as a nanoparticle; then the adhesion force as well as the friction force can be easily obtained by the cantilever's deflection. However, the use of AFM is practically limited by the tip material and its geometric shape. By attaching the particle to the force sensor in the microscope, the force between a surface and a colloid particle was directly measured with AFM by Ducker et al in 1991. Since the properties of the attached particle, such as the size, the shape and the material were controllable, the uncertainties in the force measurement caused by the irregular shape of the AFM tip etc could be avoided. Hence, the colloidal probe technique is more effective for studying the adhesion and friction of micro/nanoparticles. Nevertheless, it is actually very difficult to attach a single nanoparticle with the size of less than 1 μm on the AFM force sensor; the colloid probes in most references have sizes larger than 1 μm. A chemical method was used by Vakarelski et al to place individual gold nanoparticles (20–40 nm) on the tip of an AFM cantilever to measure the adhesion force between nanoparticles and mica. Ceria nanoparticles (50 nm in diameter) were attached on the AFM tip with epoxy glue by Ong and Sokolov to measure the adhesion force between nanoparticles and a flat silica surface. Other various methods include measuring the adhesion force of the tip against a film of nanoparticles and manufacturing a tip with a certain curvature by thermal oxidation, etc.

Besides the direct adhesion measurement by the vertical deflection of the AFM cantilever, nanoparticle movement manipulation by the cantilever's torsional deflection was firstly used to push C_{60} islands grown on a NaCl surface in 1994. Since then, this method has been increasingly popular to characterize the intriguing nanoadhesion/friction behaviors of nanoparticles. For instance, the frictional anisotropies for molybdenum oxide (MoO) nanoparticles were investigated by Sheehan and Lieber. The maximum sliding friction force between polymer latex spheres (radius between 50 and 100 nm) and a highly oriented pyrolytic graphite (HOPG) surface was obtained by Ritter et al. More recently, the interfacial friction between antimony (Sb) nanoparticles and a HOPG surface was successfully measured through pushing nanoparticles with the AFM tip by Dietzel. In addition, the adhesion forces between nanoparticles with different sizes and the surface were measured by Guo et al.

In the most general case, the adhesion force is a combination of electrostatic force, vdW force, meniscus or capillary force, solvation force and structure force, etc. The adhesive contact between elastic surfaces is usually described by single-asperity theories such as JKR, DMT or M-D (Maugis-Dugdale) theories, as mentioned previously. The adhesion force of micro/nanoparticles has been extensively studied and most of the equations for the continuum contact theories can be applied extremely well, even at the submicron scale. A linear dependence of the adhesion force on the reduced radius was found by Heim et al for the adhesion between silica spheres, proving that the DMT theory was also valid for the particle with dimensions below 1 μm. The simulation of the adhesion between a nickel AFM tip and a gold surface by Landman et al showed good agreements with the JKR theory for both the mean positions of atoms and the stress distribution. Individual nanoparticles with varying size from about 50 to 500 nm were manipulated on a silicon surface using AFM by Guo et al. The results showed that the friction forces between the particles and the substrate were proportional to the two third power of the radius, which was in agreement with the Hertzian theory, as shown in figure. The situations where the continuum contact theories are no longer applicable involve changing surface energy with time, viscoelastic materials and rough surfaces. All of these factors could give rise to hysteresis and time-dependent effects.

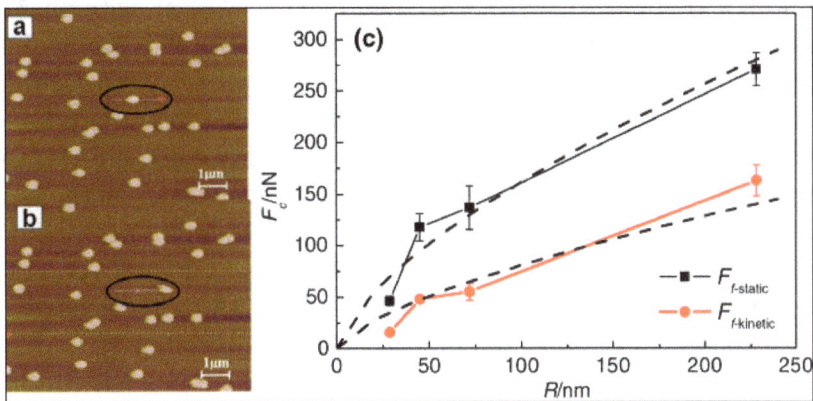

AFM images of a nanoparticle on the substrate (a) before and (b) after manipulation; (c) the dependence of the friction force of polystyrene particles on the silicon surface on the particle radius (R).

Under ambient conditions, the capillary force (meniscus force) was demonstrated to make the largest contribution to the adhesive force. The capillary force between a plate and a sphere was calculated by O'Brien and Hermann, proving the meniscus dimension was of 1 nm. The adhesion between particles in aqueous media was found to be mainly influenced by the electrostatic force, solvent force and structure force. For small particles of nanoscale size, more subtle effects beyond continuum theories have been observed. Specifically, the surface molecular structure, the distribution of terminal groups on the particles' surfaces and the surface energy variation due to particle deformation could influence and even dominate the adhesion behaviors.

In Amontons and Coulomb's friction theories for describing macroscopic dry sliding friction, the friction is proportional to the normal force, but independent of the contact area as well as the sliding velocity. However, the tribological properties at the micro/

nano scale cannot be explained with these empirical theories. Ever since 1987 when the frictional forces were detected with AFM by Mate et al for the first time, the friction at the micro/nano scale has been observed by many researchers to deviate considerably from the predictions based on established macroscopic laws. The nanoscopic friction is proportional to the true contact area, which is not necessarily proportional to the loading force. Furthermore, the friction in a nanoscale contact increases logarithmically with the sliding velocity, being in sharp contrast to empirical theories. The friction between the AFM tip and the substrate has been measured as a function of many parameters, such as the externally applied load, the sliding velocity, the tip radius and shape, the relative orientation between the scan direction and the substrate lattice, the temperature and the chemical nature of the sample. The method using the AFM tip to control the lateral manipulation of nanoparticles provides a powerful tool to measure the interfacial friction of nanoparticles with arbitrary materials and sizes. Polymer latex spheres (50–100 nm in radius) were manufactured by Ritter et al on a HOPG surface; the threshold force needed to overcome the static friction of a single latex sphere was found to depend on the sphere size, being in accordance with the JKR and DMT theories. Similarly, Sb nanoparticles on a HOPG surface were pushed by Dietzel et al with an AFM tip and two coexisting frictional states were observed: some particles showed finite friction and increased linearly with the interfacial areas, while other particles experienced a state of frictionless sliding. The transition from static to kinetic friction was also investigated in another of their work and a hysteretic character in the force domain was found. Polystyrene nanospheres with radii varying from about 30 to 200 nm on the polished nanosmooth silicon surface were manipulated by Guo et al with the contact mode of AFM; the typical results are shown in figure. The results indicated that the ratios between the kinetic friction $F_{f\text{-kinetic}}$ and the static friction force $F_{f\text{-kinetic}}$ were in the range of 0.3–0.6. Moreover, the ratio did not change whether the particles were located in different areas of the surface, the tip normal force was varied or even the surface was modified.

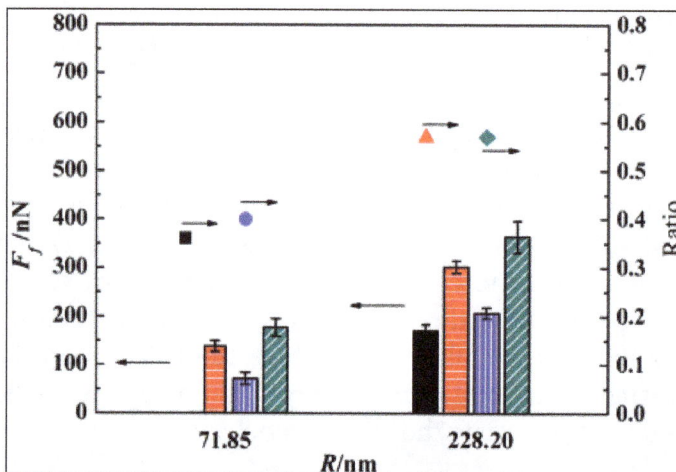

Static and kinetic frictions and their ratios for particles with radii of 71.85 nm and 228.2 nm on a non-hydroxylated surface and a hydroxylated surface, respectively.

The normal load is 348 nN. The columns with solid fill and horizontal stripes represent $F_{f\text{-kinetic}}$ and $F_{f\text{-kinetic}}$ on the non-hydroxylated surface, respectively. The columns with vertical and oblique stripes repesent those on the hydroxylated surface, respectively. The square and the circle represent the ratio of $F_{f\text{-kinetic}}$ and $F_{f\text{-kinetic}}$ for particles with R of 71.85 nm on the non-hydroxylated and the hydroxylated surface, respectively. The triangle and the rhombus represent those for particles with (R = 228.20 nm) on the non-hydroxylated surface and the hydroxylated surface, respectively.

Gold particles with a mean diameter of 25 nm were manufactured by Mougin et al on silicon substrates; it was found that the adhesion of the particles to the substrate was strongly reduced by the presence of hydrophobic interfaces. The friction and wear of spherical gold nanoparticles under dry conditions and submerged in water were studied by Maharaj and Bhushan; the results indicated that the addition of gold nanoparticles reduced friction and wear. Sitti and Hashimoto proposed an AFM-based force-controlled pushing system for the manipulation and assembly of nanoparticles. Interaction forces among the AFM probe tip, the nanoparticle and the substrate, including the vdW force, capillary force, electrostatic force, repulsive contact force and frictional force were analysed; several modes of particle motion including sliding, rolling and rotation were observed.

Movement of Nanoparticles

Various forces such as gravitational (buoyancy) forces, surface forces, viscous flow forces and the forces due to Brownian motion result in the movement of nanoparticles in the media in different ways. However, the experiments for the direct observation of nanoparticles' movement are limited primarily due to the small particle size preventing the application of the most commonly used imaging techniques. Fortunately, the rapid development of measurement technology provides opportunities for tracking individual nanoparticles or even single molecules. Up to now, several methods have been used for making high-resolution measurements of the motion of single nanoparticles. Among these methods, two groups can be classified: one is to passively track the particle motion without applying significant external stimuli and the other is to measure the particles' motions under external mechanical forces. To be more specific, studies based on two typical methods will be emphasized in the following parts.

The first method is particle tracking with the fluorescence technique. A system for observing nanoparticles was developed by Xu et al using a high-resolution fluorescence microscope and fluorescent core-shell SiO_2 nanoparticles of 50–60 nm in diameter were used as the seed nanoparticles. By using this system, the velocity profile of nanoparticles in a channel flow, the Marangoni flow in evaporating water droplets and nanoparticle–wall collision behaviors were investigated. The Marangoni flow in a droplet manifested with fluorescent nanoparticles revealed a stagnation point where the directions of the surface flow, the surface tension gradient and the surface temperature gradient changed, as shown in figure. The nanoparticle–wall collision experiments showed the nanoparticles adsorbed on the solid surface after collision in liquid were much easier to be removed than those deposited on dry surfaces. The reason for this observation was that the particles might be adsorbed at the secondary minimum of the particle–wall interaction when the collision occurred in water, rather than at the primary minimum for the particles deposited on dry surfaces, as described in the DLVO theory mentioned earlier. Another system for in situ observing nanoparticles' movement with the fluorescence technique in confined geometries where external loads and rotations could be applied was developed by Lei et al . With this system, it has been found that the velocities of free particles were much larger (20 times) than the rotating speed, providing evidence that nanoparticle impacting was also one of the main surface material removal factors during the surface planarization process.

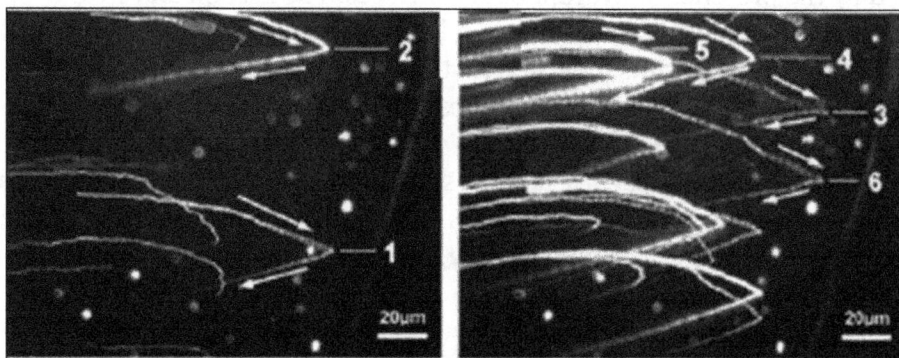

The particle trajectories in a water droplet during the evaporation process.

The second method is the TEM observations, which could give more delicate details of the particle movement and provide deeper understanding of the roles of particles in specific applications. The movement behaviors of a single MoS_2 nanoparticle in a dynamic contact were directly observed with in situ TEM by Lahouij et al; the results showed that either a rolling or a sliding process of the fullerenes could be possible during shearing. The motion of inorganic nanoparticles during fluid evaporation was observed using a TEM by Zheng et al. The observation of the self-assembled process of nanoparticles in a liquid medium with the particle size comparable to the molecular dimension of the liquid was made using an environmental TEM by Dai.

The movement of nanoparticles is very complicated due to the influence of many factors, e.g., complex forces, medium and environment. In this instance, the studies on the single nanoparticle's motion in the past were mostly qualitative in nature; more precise measurement methods or instruments with a combination of functions are needed for quantitative analyses in future works.

Applications Relevant to the Mechanical Properties of Nanoparticles

Nanoparticles in Lubrication

The mechanical properties of nanoparticles play a major role in influencing the tribological properties of lubricated systems with nanoparticles. The effects of the mechanical properties of nanoparticles as lubricant additives on the tribological properties differ in various materials. From a general point of view, the combined effects of rolling, sliding and the formation of a third body layer and tribofilms are the main reasons for the increased lubricating behavior after adding nanoparticles, as briefly described in the following parts.

- The rolling mode of nanoparticles in the lubricated contact area could provide very low friction and wear; however, the occurrence of this effect is strongly dependent upon some properties, e.g., the shape, the size and the concentration of the nanoparticles in the lubricant. Spherically shaped and mechanically stable nanoparticles without significant agglomeration are favourable for their rolling in the contact area between tribopairs. As far as the intrinsic mechanical properties of nanoparticles are concerned, whether the initial spherical shape of the nanoparticles in the contact area can be preserved or not have a close relationship with their hardness/elastic properties, which are also affected by the nanoparticle size.

- The sliding mode of nanoparticles could also result in low friction and wear. Sliding friction usually occurs when the particle is not very spherical in shape and has low adhesion to the tribopair surfaces. Besides, particle agglomeration in the contact area is another factor that could lead to sliding friction during the shearing of tribopairs. In this case, the nanoparticles play a role as a spacer in minimizing the direct contact between the asperities of two shearing surfaces. Externally applied pressure on the nanoparticles, the rigidity of the tribopair surfaces and the interaction forces between particles are very relevant to the above two modes of particle movement in the lubricated contact area. A smaller applied load and harder tribopair surfaces readily lead to rolling friction of nanoparticles in the contact area, because these would give less of a probability for the particles to mechanically deform or indent into the surface. Moreover, particle agglomeration is greatly determined by the interaction force between particles, thereby inhibiting rolling while promoting the sliding of nanoparticles in the contact area Another important aspect of the nanoparticle in the lubricant under a low applied pressure is that the viscosity of the lubricant could be enhanced and thereby the oil film formation properties in the lubricated contact could be improved, as shown in figures. It can also be noted from this figure that when the applied pressure increased further, the sliding effect of nanoparticles could give rise to the surface polishing effect.

- When the applied pressure is sufficiently large, nanoparticles become mechanically unstable and delamination of nanoparticles could happen. For instance, studies suggested that when the applied pressure was ~1 GPa and the tribopair operated in the boundary lubrication regime, exfoliation of inorganic fullerene-like (IF) nanoparticles as the lubricant additive would dominate. In this case, material layers of the broken particles could form as the third body and adhere on the tribopair surfaces separating the counterpart. These layers likely align themselves parallel to the tribopair surfaces due to adhesion and shear. It occurs more often for metal dichalcogenide and graphite nanoparticles, which have anisotropic layered structures with weak vdW forces as the bonding interaction between layers. In addition, valleys between asperities could be filled out by nanoparticles; then the tribopair surface could be partly smoothened out to reduce friction and wear. It is worth pointing out that nanoparticles as lubricant additives do not always give rise to favourable tribological properties. Increases in friction and wear, as well as lubricant starvation, were observed due to the abrasive effect of hard nanoparticles under large pressures and heavy aggregations of oils with high particle concentrations in the inlet of the contact area.

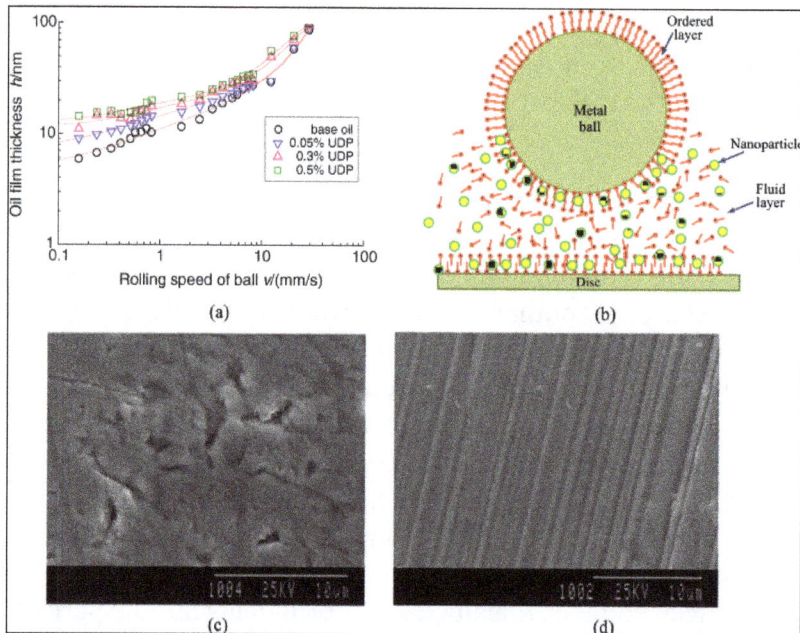

Typical results of lubrication properties of oils (polyester (PE)) with added diamond nanoparticles: (a) film thickness against ball rolling speed for PE with nanoparticles of different concentrations (applied pressure: 174 MPa); (b) physical model of nanoparticles as additive; (c) SEM image of the rubbing surface under PE lubrication (rubbing time: 30 min; applied pressure: 220 MPa); (d) SEM image of the rubbing surface under PE lubrication with nanoparticles (rubbing time: 30 min; applied pressure: 220 MPa).

Nanoparticles in Nanomanufacturing

CMP is an indispensable planarization tool in nanomanufacturing ICs. Abrasive and corrosive slurry is used to physically grind and chemically remove microscopic topographic features on a wafer to obtain a flat surface. In this process, abrasive nanoparticles in the slurry are a very important contributor to obtain controlled material removal without sacrificing planarity. They usually either embed in the polishing pad or remain immersed in the slurry, as schematically shown in figure. Among many factors that could affect the material removal rate and surface quality in CMP, the mechanical interaction between the nanoparticles and the wafer surface plays a critical role. For the material removal process, two models have been proposed to understand the mechanical behaviors of abrasive nanoparticles in CMP, i.e. the hydrodynamic model and the solid contact model.

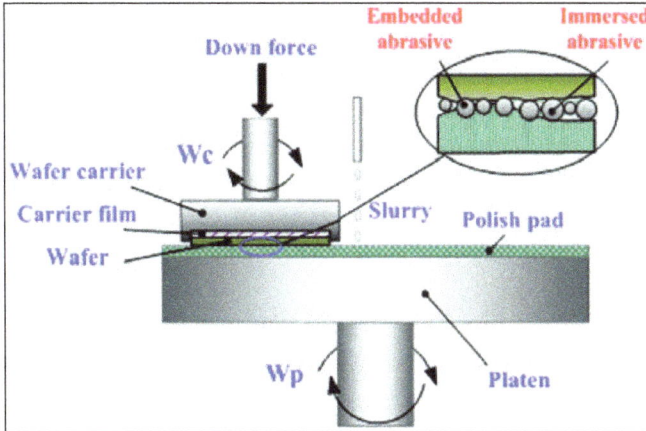

Schematic illustration of the CMP tool and the contact among wafer, pad asperity and abrasive particles during the CMP process.

- In the hydrodynamic model, the wafer and the polishing pad are separated by a thin liquid film; the material removal is primarily due to the collision of abrasive nanoparticles onto the wafer, or the fluidic shearing. The effects of the particle size, the incidence speed and angles etc on the collision between nanoparticles and the wafer surface have been investigated. Xu et al designed an experiment based on a fluorescent microscope system where fluorescent nanoparticles adsorbed on a glass surface were exposed to the vertical impact of a liquid with 15 wt% abrasive nanoparticles. The results suggested that the collision between the abrasives and the wafer surface had a negligible effect on the material removal at a liquid impact speed of 3 m s^{-1}. When the impacting speed was increased and the nanoparticle incidence angle was changed, damage to the wafer surface could occur. For instance, there were many pits and scratches on the surface on the wafer surface under a speed of 50 m s^{-1} and an incidence angle of 45°; heavy and heterogeneous deformation in the surface layer was observed with the high-resolution TEM, as shown in figure. MD simulation studies on the collision process of a nanoparticle onto a silicon or silica surface suggested the damage could be increasingly reduced with the increasing incident angle. Moreover, the critical velocity for the pileup formation on the silicon surface is affected by the incidence angle rather than the particle size.

- In the solid contact model, part of the polishing pad is in direct contact with the wafer surface. The particles embedded in the pad slide against the substrate surface, in a similar way to fixed abrasive grinding (referred to as fixed particles). The particles immersed in the slurry between the pad and the wafer can be referred to as free particles. Lei et al used a fluorescence based experimental system to track the movement of individual particles between the polishing pad and the solid surface. The results confirmed that some particles were fixed on the polishing pad and rotated with the pad, while the others moved freely in the slurry flow. Paul et al proposed that the ratio of the number of fixed particles and that of free particles was of great importance to the material removal mode. In regard to this model, it has been widely accepted that the material removal is due to the two-body abrasion between the polishing pad and the wafer surface, as well as that between nanoparticles and the wafer surface. Nevertheless, increasing evidence shows that the rolling of free nanoparticles in the slurry is not notably inferior to abrasive sliding for the material removal and surface finish on the atomic scale in the CMP process; a typical MD simulation result is shown in figure.

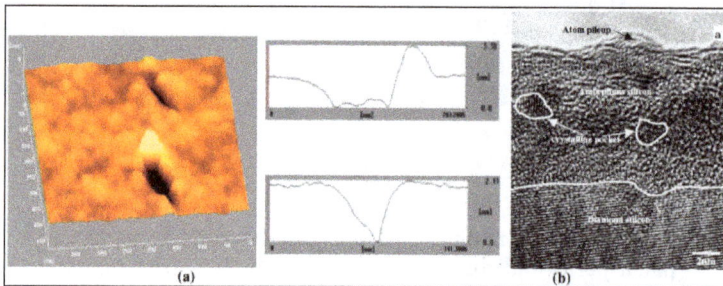

(a) AFM image of the surface after a 10 min exposure; (b) cross-section high-resolution TEM images of the specimen subsurface after exposure.

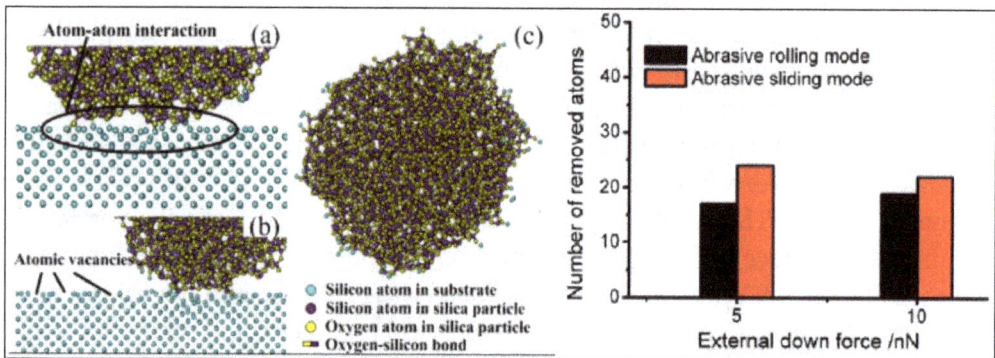

The rolling process of a silica particle under an external down force of 5 nN and a lateral driving force of 10 nN (left); The number of removed atoms against various external down forces with abrasive rolling and abrasive sliding (right).

Another important aspect is understanding the adhesion and the removal of nanoparticles on the wafer surface; this is a relevant problem in the post-CMP cleaning process. Interfacial forces, such as the vdW force, electrostatic force and capillary force in the vicinity of the nanoparticle and the wafer surface dominate the adhesion process. Many experimental factors could influence the adhesion strength between the particle and the wafer surface, as the following list demonstrates.

- The adhesion could increase with the contact time, since the contact area and then the interfacial forces increase as time progresses.

- Large atmospheric humidity could accelerate the adhesion formation.

- The size effect of nanoparticles on the adhesion strength has been a research focus and some contradictory results have been obtained. Heim et al found that the relationship between microparticle/wafer surface adhesion and the particle radius agreed with the prediction of contact theories. On the contrary, the results obtained by Thoreson et al suggested that no size effect of the particle/wafer surface adhesion could be observed. This trend was also confirmed by the experiments conducted by Lei et al, in which a series of heat-treated AFM probes with various curvature radii were employed to measure the particle/wafer surface adhesion. This result might be as a result of the reduction of the real contact area caused by asperities on the tip surface.

- In addition, the influence of capillary force should be also considered, since small particles could aggregate to form larger ones.

After nanoparticles are adsorbed onto the wafer surface due to the action of interfacial forces, they could be embedded in the surface by pad pressure if valleys or asperities are present on the surface. In this case, the number of residue nanoparticles on the wafer surface after the CMP process could reduce when the wafer surface hardness increases. These residue nanoparticles should be removed during the post-CMP cleaning process to avoid their unfavourable effects on the follow-up processe. Applying external mechanical stimuli, megasonic cleaning as well as some wet chemical effects are optional ways to overcome the adhesion force and remove physisorbed (in some cases chemisorbed) or partially embedded nanoparticles, as schematically shown in figure.

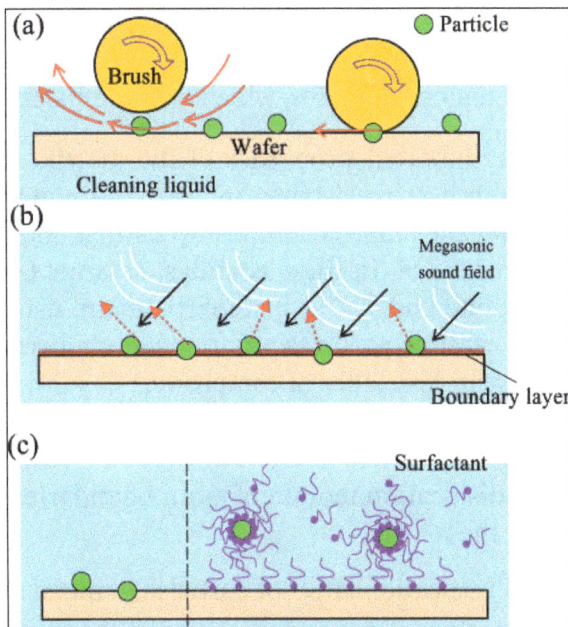

Basic schematic models of typical post-CMP cleaning of nanoparticles: (a) brush–particle mechanical interaction; (b) megasonic cleaning; (c) chemical additive (surfactant).

The mechanical removal forces are very complex, mainly including the contact elastic force, the hydrodynamic drag force and the friction between a brush and nanoparticles, etc. A schematic description of all the forces for a nanoparticle on the surface in the cleaning process was put forward by Huang et al. Furthermore, contact models and lubrication hydrodynamic theories were employed to analyse the fluid flow field and calculate the hydrodynamic drag force, as well as the surface roughness and the characteristics of the brush nodules were considered.

The mechanical brush scrubbing method is very efficient for removing residue particles; however, it becomes less effective when the particle size is very small, e.g., a nanoparticle. In this case, megasonic cleaning and chemically-activated removal could be adopted. Megasonic cleaning is to utilize a sound field with a frequency of typically 0.8–2 MHz to excite controlled cavitation, which is gentler and on a much smaller scale than that produced under ultrasonic cleaning. Increases in the megasonic frequency, the cleaning period and the solution temperature etc could improve the cleaning effects. Nevertheless, this cleaning method has some problems, e.g., the instability of the sound field for cleaning large size wafers and low cleaning efficiency. The basic idea of the chemically-activated removal, i.e. using a chemical additive, is to weaken the bonds between particles or to change the charges on the wafer surface and the particles (alter the solution pH or add surfactants) for controlling the electrostatic repulsion between the particles and the wafer surface.

Nanoparticles in Coatings

Incorporating different kinds of nanoparticles within a metal or polymer matrix to produce nanocomposites can deliver improved properties, such as enhanced mechanical properties, self-lubrication, wear-resistance and energy-absorbing capabilities. There are two main categories of nanocomposites, which are summarized as follows.

- Due to some of the inherent properties of the matrix, for instance the high strength and modulus, wear resistance and high thermal and electrical conductivity, metal or metal alloy matrix composite coatings show distinct advantages over polymeric composites. In these coatings, ceramic (Al_2O_3, TiO_2, SiC) and carbon-based (graphite and CNTs) nanoparticles are usually added. There are three reasons why ceramic particles are used as reinforcement to enhance the hardness and the wear-resistance of composites:

  ○ The high hardness and strength of particles.

  ○ Migration and dislocation motion of grain boundaries can be prevented by the particles in the matrix.

  ○ Heterogeneous nucleation effect of particles in metal or metal alloys.

The addition of graphite nanoparticles or CNTs in a metal matrix could, on the one hand, reduce the porosity of a pure metal coating, then the coating would be much denser and compact with fewer cracks. On the other hand, the crystalline size in the coating could be refined due to the presence of nanoparticles. In addition, the structural and chemical stability of CNTs, which have a higher stiffness and strength compared with the metal matrix are another important factor contributing to the strengthening effect.

- Modification of the physical properties of a polymer matrix can be achieved by adding inorganic or organic nanoparticles, causing some new characteristics of polymers to be obtained. For example, polymer composite coatings containing some inorganic nanoparticles can provide resistance to the initiation and propagation of cracks, fill cavities and initiate crack bridging, deflection and bowing. Basically, the mechanical properties of polymer based nanocomposites can be affected by many factors, among which the interface between the nanoparticle and the polymer matrix plays a dominant role due to the large specific surface area of the particle. Hence, a good design of a nanocomposite, by taking the complex interplay between matrix, interface and nanoparticles into consideration, could tailor the composite material system with desirable physical properties. Several underlying mechanisms responsible for the interface reinforcement are:

  ∘ The interaction between nanoparticles and the polymer matrix could result in the formation of special microstructures (for instance, a finer scale lamellar structure), correspondingly the improved mechanical properties could prevent rapid crack propagation in the coating.

  ∘ Nanoparticles could enhance their interaction with the matrix through chemical bonds (for example, increase the cross-linking densities in the coatings) or increase the physical interactions between macromolecular chains of the matrix. In this manner, effective pathways could be provided for nanoparticles to complement the poor mechanical and tribological performances of some polymer matrices, e.g., their poor resistance to surface abrasion and wear.

Uniform dispersion of nanoparticles in the matrix is very crucial in obtaining improved mechanical properties (e.g., strength and ductility) of nanocomposite coatings, since the maximum filling content of nanoparticles with a large surface area is limited. When the content of nanoparticles in the nanocomposite coatings exceeds a critical value, particle agglomeration would happen, resulting in deterioration of the mechanical properties (for instance, aggravated microcracks on the coating surface), decrease in Young's modulus and increase in the wear rate. In order to achieve good particle dispersion, appropriate preparation and processing methods are needed. Specifically, powder metallurgy and vapour phase processing, ultrasonic assisted melting and disintegrated melt deposition, mechanical alloying and friction stir processing, as well as layer-by-layer deposition have been used. In addition, the interface between the nanoparticle and the matrix can be modified precisely on the molecular/atomic level with techniques, such as atomic layer deposition (ALD) or molecular assembly to obtain some interesting structures, e.g., core/shell hybrid nanoparticles.

Electrical Properties

The properties like conductivity or resistivity are come under category of electrical properties. These properties are observed to change at nanoscale level like optical properties. The examples of the change in electrical properties in nanomaterials are:

- Conductivity of a bulk or large material does not depend upon dimensions like diameter or area of cross section and twist in the conducting wire etc. However it is found that in case of carbon nanotubes conductivity changes with change in area of cross section.

- It is also observed that conductivity also changes when some shear force (in simple terms twist) is given to nanotube.

- Conductivity of a multiwalled carbon nanotube is different than that of single nanotube of same dimensions.

- The carbon nanotubes can act as conductor or semiconductor in behavior but we all know that large carbon (graphite) is good conductor of electricity.

These are the important electrical properties of nanomaterials with their examples.

Optical Properties

Nanomaterials exhibit a variety of unusual and interesting optical properties that can differ significantly from the properties exhibited by the same bulk material. By carefully controlling the size, shape and surface functionality of nanoparticles a wide range of optical effects can generated with many useful applications. An optical response in a nanomaterial can be created through several different mechanisms, depending on the nanomaterial size, composition and arrangement, and each method may provide certain benefits depending on the target application.

Scattering, Absorption and Extinction

When light is incident on a nanoparticle, it can be scattered or absorbed (the sum of scattering and absorption is referred to as extinction). Nanoparticles are in the size regime where the fraction of light that is scattered or absorbed can vary greatly depending on the particle diameter. At diameters less than 20 nm, nearly all of the extinction is due to absorption. At sizes above 100 nm, the extinction is mostly due to scattering. By designing a particle with a larger or smaller diameter, the optimal amount of scattering and absorption can be achieved. Another byproduct of this relationship between size and absorption/scattering is that aggregation can increase the effective size of a nanoparticle resulting in an increase in scattering. This is why 20 nm diameter silica particles are clear in solution but re-suspensions of dried 20 nm silica particles (aggregated) will be a milky white color.

Plasmonic Nanomaterials

Nanoscale structures made of metals such as gold, silver and aluminum can support surface plasmon modes where the free electrons in the material naturally resonate at a frequency that depends on the composition, size and shape of the particle. When the wavelength of incident light matches the oscillation frequency, the particles can strongly absorb or scatter the light resulting in a strongly colored particle. By tuning the size and shape, the peak resonance wavelength can be shifted across the visible and into the infrared region of the spectrum allowing for a wide range of color tunability.

In addition to metal nanoparticles, there have been recent examples of metal oxide nanoparticles doped with other metal atoms that show strong and tunable plasmon resonances. By changing the particle size, dopant and dopant concentration, plasmon resonances in the near-infrared (NIR) and short-wavelength infrared (SWIR) have been fabricated.

Quantum Dots

Nanoparticles made of semiconductor materials, often referred to as quantum dots (QD), absorb and emit light at certain wavelengths that depend strongly on particle size and shape due to quantum confinement effects. By changing the size and composition of the quantum dots, their emission wavelengths can be tuned from the UV through the visible to the near infrared regions of the spectrum. For example, by tuning the size of CdSe QDs from 2 nm to 8 nm in diameter, the emission wavelength can be shifted across the visible spectrum, with the smaller particles emitting in the blue and the larger particles emitting red light.

As fluorescent materials, quantum dots offer a number of advantages over organic fluorescent dyes: in addition to the ability to easily tune optical properties by varying particle size, the QDs are less prone to photobleaching under high intensity illumination, offer comparable or larger quantum yields than organic dyes, and can be excited much further away from their emission peak, giving them a large effective Stokes shift and allowing more flexibility with imaging or choosing excitation sources to avoid auto-fluorescence in biological samples.

QDs are used in a variety of applications including photodetectors, solar cells, light emitting diodes (LEDs), televisions, and for medical imaging. An example is a metal cored quantum dot composite particle that is used in lateral flow diagnostic devices.

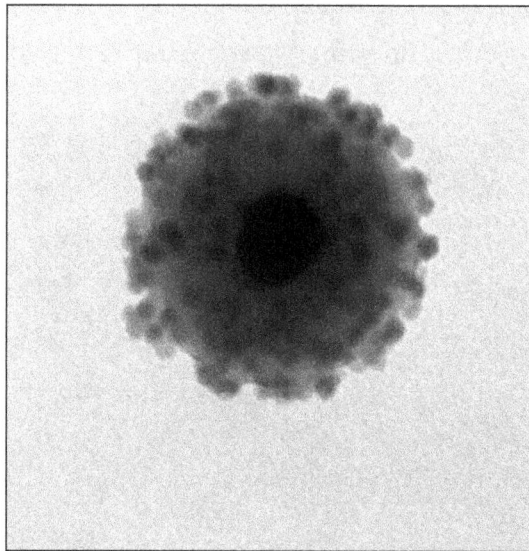

Photonic Crystals

Another method of generating color is by organizing nanoparticles into ordered structures, with structural elements similar in size to the wavelength of light. These structures can selectively reflect certain portions of the spectrum, producing films with optical responses that can be tuned by selecting the size of the component particles.

This type of "structural color" has many examples in nature, such as the iridescent blue appearance of a morpho butterfly wing, which is due to nanoscale periodic structures that reflect only blue wavelengths of light. Structural color is gives rise to the multi-colored iridescence of opal gemstones, which are made from highly ordered assemblies of silica particles.

Highly monodisperse nanoparticles can be used for the assembly of colloidal crystals where dried particles will organize into a periodic assembly that can diffract light. Structural color assemblies can also be created dynamically in solution using magnetically-responsive particles, such as the particles shown below which align into ordered structures in the presence of a magnetic field.

References

- Electrical-properties-of-nanomaterials: winnerscience.com, Retrieved 28 April, 2019

- Introduction-to-nanoparticle-optical-properties: nanocomposix.com, Retrieved 06 July, 2019

- Nanoparticles: From Theory to Application (Weinheim: Wiley-VCH)

- Nanoparticles at fluid interfaces J. Phys.: Condens. Matter 19 413101

Nanoparticles: A Comprehensive Study

Particles between the size of 1 and 100 nanometers are referred to as nanoparticles. Some of the processes associated with them are nanoparticle production, endocytosis and exocytosis of nanoparticles, etc. Common types of nanoparticles are magnetic nanoparticles, shell nanoparticles, superparamagnetic iron oxide nanoparticles, etc. All these related concepts of nanoparticles have been carefully analyzed in this chapter.

Nanoparticle is the ultrafine unit with dimensions measured in nanometres (nm; 1 nm = 10^{-9} metre) Nanoparticles exist in the natural world and are also created as a result of human activities. Because of their submicroscopic size, they have unique material characteristics, and manufactured nanoparticles may find practical applications in a variety of areas, including medicine, engineering, catalysis, and environmental remediation.

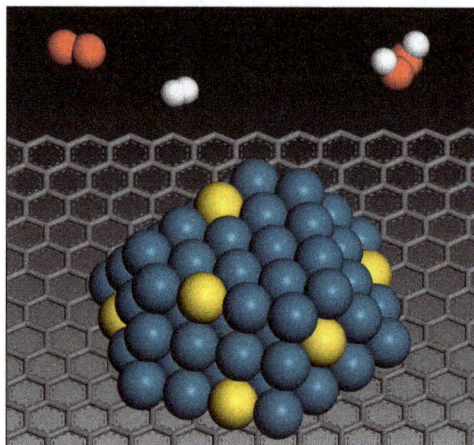

Image showing nanoparticles of an alloy of gold (yellow) and palladium (blue) on an acid-treated carbon support (gray). These particles were employed as catalysts for the formation of hydrogen peroxide from hydrogen (white) and oxygen (red).

Properties of Nanoparticles

In 2008 the International Organization for Standardization (ISO) defined a nanoparticle as a discrete nano-object where all three Cartesian dimensions are less than 100 nm. The ISO standard similarly defined two-dimensional nano-objects (i.e., nanodiscs and nanoplates) and one-dimensional nano-objects (i.e., nanofibres and nanotubes).

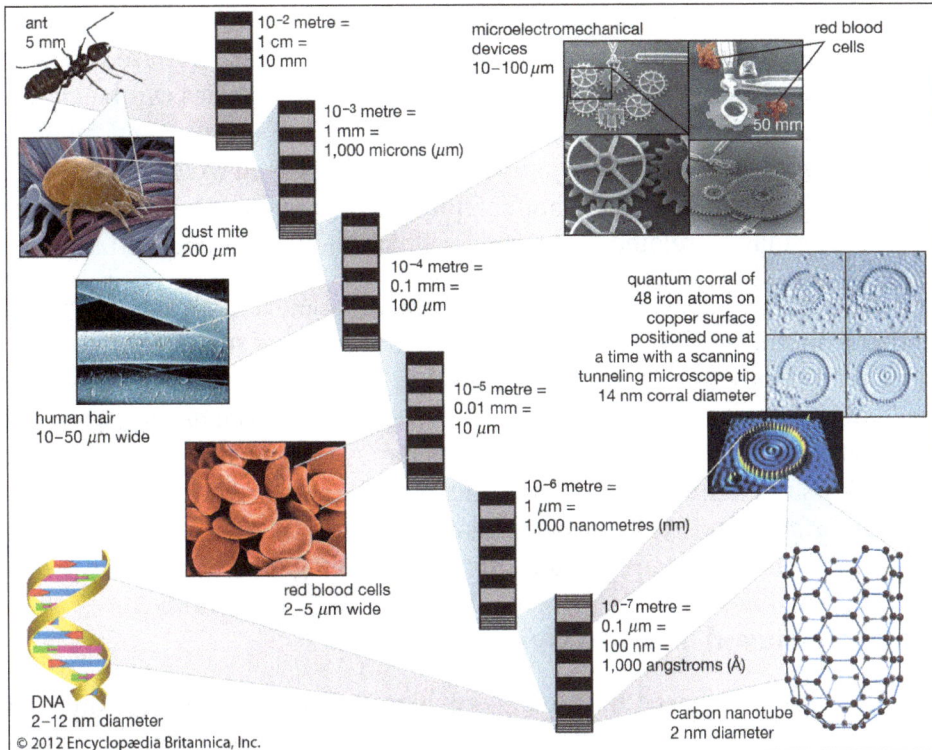

Examples from biological and mechanical realms illustrate various "orders of magnitude" (powers of 10), from 10^{-2} metre down to 10^{-7} metre.

But in 2011 the Commission of the European Union endorsed a more-technical but wider-ranging definition:

> A natural, incidental or manufactured material containing particles, in an unbound state or as an aggregate or as an agglomerate and where, for 50% or more of the particles in the number size distribution, one or more external dimensions is in the size range 1 nm–100 nm.

Under that definition a nano-object needs only one of its characteristic dimensions to be in the range 1–100 nm to be classed as a nanoparticle, even if its other dimensions are outside that range. (The lower limit of 1 nm is used because atomic bond lengths are reached at 0.1 nm.)

That size range—from 1 to 100 nm—overlaps considerably with that previously assigned to the field of colloid science—from 1 to 1,000 nm—which is sometimes alternatively called the mesoscale. Thus, it is not uncommon to find literature that refers to nanoparticles and colloidal particles in equal terms. The difference is essentially semantic for particles below 100 nm in size.

Nanoparticles can be classified into any of various types, according to their size, shape, and material properties. Some classifications distinguish between organic and inorganic nanoparticles; the first group includes dendrimers, liposomes, and polymeric nanoparticles, while the latter includes fullerenes, quantum dots, and gold nanoparticles. Other classifications divide nanoparticles according to whether they are carbon-based, ceramic, semiconducting, or polymeric. In addition, nanoparticles can be classified as hard (e.g., titania (titanium dioxide), silica (silica dioxide) particles, and fullerenes) or as soft (e.g., liposomes, vesicles, and nanodroplets). The way in which nanoparticles are classified typically depends on their application, such as in diagnosis or therapy versus basic research, or may be related to the way in which they were produced.

There are three major physical properties of nanoparticles, and all are interrelated: (1) they are highly mobile in the free state (e.g., in the absence of some other additional influence, a 10-nm-diameter nanosphere of silica has a sedimentation rate under gravity of 0.01 mm/day in water); (2) they have enormous specific surface areas (e.g., a standard teaspoon, or about 6 ml, of 10-nm-diameter silica nanospheres has more surface area than a dozen doubles-sized tennis courts; 20 percent of all the atoms in each nanosphere will be located at the surface); and (3) they may exhibit what are known as quantum effects. Thus, nanoparticles have a vast range of compositions, depending on the use or the product.

Nanoparticle-based Technologies

In general, nanoparticle-based technologies centre on opportunities for improving the efficiency, sustainability, and speed of already-existing processes. That is possible because, relative to the materials used traditionally for industrial processes (e.g., industrial catalysis), nanoparticle-based technologies use less material, a large proportion of which is already in a more "reactive" state. Other opportunities for nanoparticle-based technologies include the use of nanoscale zero-valent iron (NZVI) particles as a field-deployable means of remediating organochlorine compounds, such as polychlorinated biphenyls (PCBs), in the environment. NZVI particles are able to permeate into rock layers in the ground and thus can neutralize the reactivity of organochlorines in deep aquifers. Other applications of nanoparticles are those that stem from manipulating or arranging matter at the nanoscale to provide better coatings, composites, or additives and those that exploit the particles' quantum effects (e.g., quantum dots for imaging, nanowires for molecular electronics, and technologies for spintronics and molecular magnets).

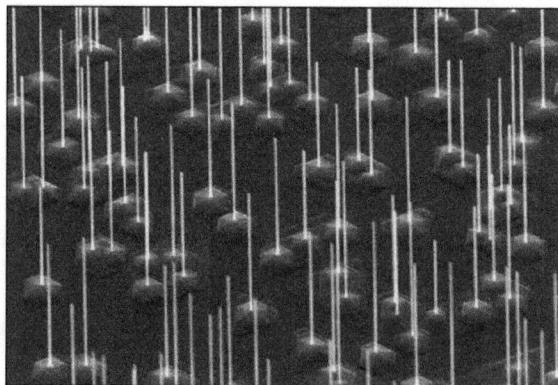

Nanowires as seen by a field-emission microscope.

Nanoparticle Applications in Materials

Many properties unique to nanoparticles are related specifically to the particles' size. It is therefore natural that efforts have been made to capture some of those properties by incorporating nanoparticles into composite materials. An example of how the unique properties of nanoparticles have been put to use in a nanocomposite material is the modern rubber tire, which typically is a composite of a rubber (an elastomer) and an inorganic filler (a reinforcing particle), such as carbon black or silica nanoparticles.

For most nanocomposite materials, the process of incorporating nanoparticles is not straightforward. Nanoparticles are notoriously prone to agglomeration, resulting in the formation of large clumps that are difficult to redisperse. In addition, nanoparticles do not always retain their unique size-related properties when they are incorporated into a composite material.

Despite the difficulties with manufacture, the use of nanomaterials grew markedly in the early 21st century, with especially rapid growth in the use of nanocomposites. Nanocomposites were employed in the development and design of new materials, serving, for example, as the building blocks for new dielectric (insulating) and magnetic materials. The following sections describe some of the many applications of nanoparticles and nanocomposites in materials.

Polymers

Similar to the way in which carbon and silica nanoparticles have been used as fillers in rubber to improve the mechanical properties of tires, such particles and others, including nanoclays, have been incorporated into polymers to improve their strength and impact resistance. In the early 21st century, increasing use of non-petroleum-based polymers that were derived from natural sources drove the development of "all-natural" nanocomposite polymers. Such materials incorporate a biopolymer derived from an alginate (a carbohydrate found in the cell wall of brown algae), cellulose, or starch; the biopolymer is used in conjunction with a natural nanoclay or a filler derived from the shells of crustaceans. The materials are biodegradable and do not leave behind potentially harmful or non-natural residues.

Food Packaging

Nanoparticles have been increasingly incorporated into food packaging to control the ambient atmosphere around food, keeping it fresh and safe from microbial contamination. Such composites use nanoflakes of clays and claylike particles, which slow down the ingress of moisture and reduce gas transport across the packaging film. It is also possible to incorporate nanoparticles with apparent antimicrobial effects (e.g., nanocopper or nanosilver) into such packaging. Nanoparticles that exhibit antimicrobial activity had also been incorporated into paints and coatings, making those products particularly useful for surfaces in hospitals and other medical facilities and in areas of food preparation.

Flame Retardants

Nanoparticles were explored for their potential to replace additives based on flammable organic halogens and phosphorus in plastics and textiles. Studies had suggested that, in the event of a serious fire, products with nanoclays and hydroxide nanoparticles were associated with fewer emissions of harmful fumes than products containing certain other types of additives.

Batteries and Supercapacitors

The ability to engineer nanocomposite materials to have very high internal surface areas for storing electrical charge in the form of small ions or electrons has made them especially valuable for use in batteries and supercapacitors. Indeed, nanocomposite materials have been synthesized for various applications involving electrodes. Composite materials based on carbon nanotubes and layered-type materials, such as graphene, were also researched extensively, making their first appearances in commercial devices in the early 2000s.

Nanoceramics

A long-term objective in materials science had been to transform ceramics that are brittle and prone to cracking into tougher, more resilient materials. By the early 21st century, researchers had achieved that goal by incorporating an effective blend of nanoparticles into ceramics materials. Other new ceramics materials that were under development included all-ceramic or polymer-ceramic blends, which combined the unique functional (e.g., electrical, magnetic, or mechanical) properties of a nanocomposite material with the properties of ceramics materials.

Light Control

In the 1990s the development of blue light-emitting diodes (LEDs), which had the potential to produce white light at significantly reduced costs, inspired a revolution in lighting. Blue LEDs brought about a need for composite materials that could be used to coat the diodes to convert blue light into other wavelengths (such as red, yellow, or green) in order to achieve white light. One way of obtaining the desired light is by leveraging the size or quantum effect of small semiconducting particles. The application of such particles facilitated the development of nanocomposite polymers for greenhouse enclosures; the polymers optimize plant growth by effectively converting wavelengths of full-spectrum sunlight into the red and blue wavelengths used in photosynthesis. Light conversion in the above cases is achieved with submicron particles of inorganic phosphor materials incorporated into the polymer.

Nanoparticle Production

Nanomaterials and nanoparticles are used in a broad spectrum of applications. Today they are contained in many products and used in various technologies. Most nano-products produced on an industrial scale are nanoparticles, although they also arise as byproducts in the manufacture of other materials. Most applications require a precisely defined, narrow range of particle sizes (monodispersity).

Specific synthesis processes are employed to produce the various nanoparticles, coatings, dispersions or composites.

Defined production and reaction conditions are crucial in obtaining such size-dependent particle features. Particle size, chemical composition, crystallinity and shape can be controlled by temperature, pH-value, concentration, chemical composition, surface modifications and process control.

Two basic strategies are used to produce nanoparticles: 'top-down' and 'bottom-up'. The term 'top-down' refers here to the mechanical crushing of source material using a milling process. In the 'bottom-up' strategy, structures are built up by chemical processes. The selection of the respective process depends on the chemical composition and the desired features specified for the nanoparticles.

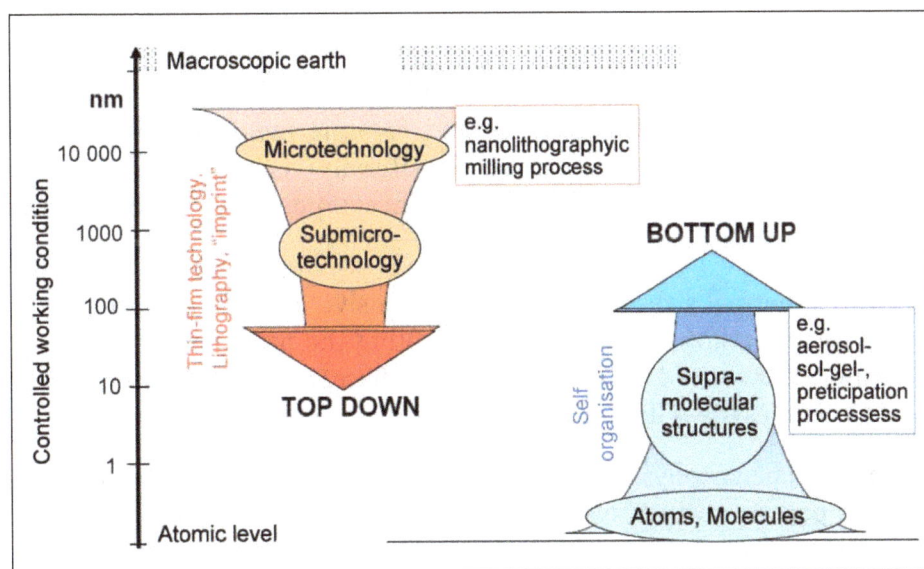

Methods of nanoparticle production: top-down and bottom-up.

Top-down or Mechanical-physical Production Processes

'Top-down' refers to mechanical-physical particle production processes based on principles of microsystem technology. The traditional mechanical-physical crushing methods for producing nanoparticles involve various milling techniques.

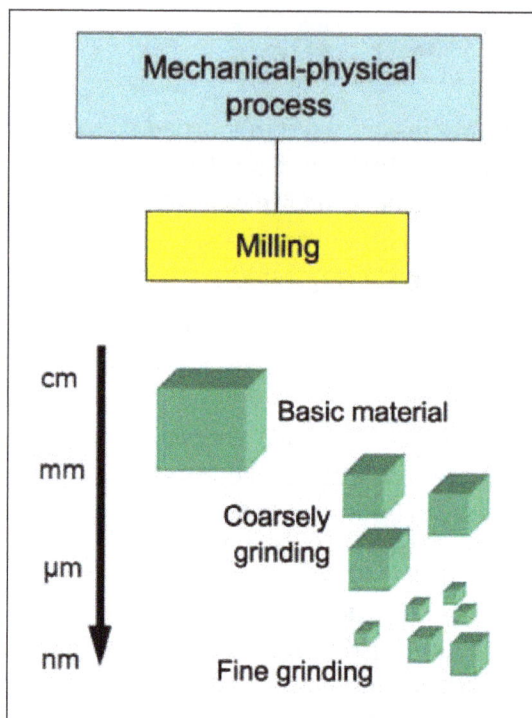

Overview of mechanical-physical nanoparticle production processes.

Milling Processes

The mechanical production approach uses milling to crush microparticles. This approach is applied in producing metallic and ceramic nanomaterials. For metallic nanoparticles, for example, traditional source materials (such as metal oxides) are pulverized using high-energy ball mills. Such mills are equipped with grinding media composed of wolfram carbide or steel.

Milling involves thermal stress and is energy intensive. Lengthier processing can potentially abrade the grinding media, contaminating the particles. Purely mechanical milling can be accompanied by reactive milling: here, a chemical or chemo-physical reaction accompanies the milling process.

Compared to the chemo-physical production processes, using mills to crush particles yields product powders with a relatively broad particle-size ranges. This method does not allow full control of particle shape.

Bottom-up or Chemo-physical Production Processes

Bottom-up methods are based on physicochemical principles of molecular or atomic self-organization. This approach produces selected, more complex structures from atoms or molecules, better controlling sizes, shapes and size ranges. It includes aerosol processes, precipitation reactions and solgel processes.

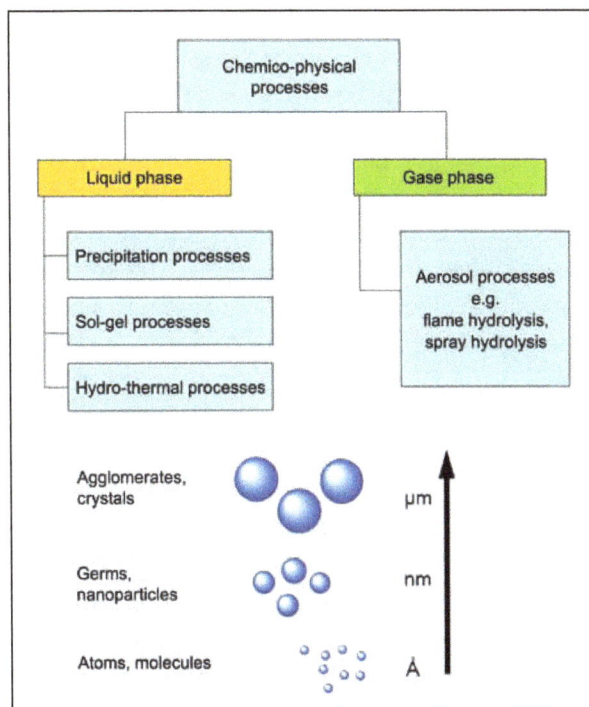

Chemo-physical processes in nanoparticle production.

Gas Phase Processes (Aerosol Processes)

Gas phase processes are among the most common industrial-scale technologies for producing nanomaterials in powder or film form.

Nanoparticles are created from the gas phase by producing a vapor of the product material using chemical or physical means. The production of the initial nanoparticles, which can be in a liquid or solid state, takes place via homogeneous nucleation.

Depending on the process, further particle growth involves condensation (transition from gaseous into liquid aggregate state), chemical reaction(s) on the particle surface and coagulation processes (adhesion of two or more particles), as well as coalescence processes (particle fusion). Examples include processes in flame-, plasma-, laser- and hot wall reactors, yielding products such as fullerenes and carbon nanotubes:

- In flame reactors, nanoparticles are formed by the decomposition of source molecules in the flame at relatively high temperatures (about 1200 to 2200 °C). Flame reactors are used today for the industrial-scale production of soot, pigment-titanium dioxide and silicon dioxide particles.

- In plasma reactors, plasma (ionized gas) provides the energy for the vaporization and for initializing the decomposition reactions.

- In laser reactors, lasers selectively heat the gaseous source material, utilizing its absorption wavelength, and decompose it to the desired product.

- In hot wall reactors, vaporization and condensation are applied. The source material is vaporized in an inert gas under low pressures (1 mbar). This removes the enriched gas phase from the hot zone. The particles created by the rapid cooling are collected on filters. Technically, hot wall reactors are used for example in producing nanoscale nickel- and iron powders.

- The chemical gas phase deposition process is used to directly deposit nanoparticles from the gas phase onto surfaces. Here, the source material is vaporized in a vacuum and condensed on a heated surface by a chemical reaction, i.e. deposited from the gas phase into the solid final product.

Droplet Formation Containing Particles

Particles can also be produced from droplets using centrifugal forces, compressed air, sonic waves, ultrasound, vibrations, electrostatics and other methods. The droplets are transformed into a powder either through direct pyrolysis (thermal cleavage of chemical compounds) or via direct reactions with another gas. In spray pyrolysis, droplets of the source material are transported through a high-temperature field (flame, oven), which rapidly vaporizes the readily volatile components or leads to decomposition reactions. The formed particles are collected on filters.

Liquid Phase Processes

The wet-chemical synthesis of nanomaterials typically takes place at lower temperatures than gas phase synthesis. The most important liquid phase processes in nanomaterial production are precipitation, sol-gel processes and hydrothermal syntheses.

Precipitation Processes

The precipitation of solids from a metal ioncontaining solution is one of the most frequently employed production processes for nanomaterials. Metal oxides as well as non-oxides or metallic nanoparticles can be produced by this approach. The process is based on reactions of salts in solvents. A precipitating agent is added to yield the desired particle precipitation, and the precipitate is filtered out and thermally post-treated.

In precipitation processes, particle size and size distribution, crystallinity and morphology (shape) are determined by reaction kinetics (reaction speed). The influencing factors include, beyond the concentration of the source material, the temperature, pH value of the solution, the sequence in which the source materials are added, and mixing processes.

A good size control can be achieved by using self-assembled membranes, which in turn serve as nanoreactors for particle production. Such nanoreactors include micro-emulsions, bubbles, micelles and liposomes. They are composed of a polar group and a non-polar hydrocarbon chain.

Micro-emulsions, for example, consist of two liquids that cannot be mixed with one another in the concentrations used, usually water and oil along with at least one tenside (substance that reduces the surface tension of liquids). In certain solvents this gives rise to small reactors in which nucleation and controlled particle growth take place. Particle size is determined by the size of the nanoreactors and, at the same time, particle agglomeration is prevented.

Micro-emulsion processes are often used to produce nanoparticles for pharmaceutical and cosmetics applications.

An additional process that is based on self-organized growth with templates and coatings is hydrothermal synthesis. Zeolites (microporous aluminum-silicon compounds) are produced from aqueous superheated solutions in autoclaves (airtight pressure chambers).

The partial vaporization of the solvent creates pressure in the autoclaves (several bars), triggering chemical reactions that differ from those under standard conditions, for example by altering the solubility. Nanoparticle formation and cavity shape can be controlled by adding templates. Templates are particles with bonds that enable the formation of certain forms and sizes.

Sol-gel Processes

Sol-gel syntheses (production of a gel from powder-shaped materials) are wet-chemical processes for producing porous nanomaterials, ceramic nanostructured polymers as well as oxide nanoparticles. The synthesis takes place under relatively mild conditions and low temperatures.

The term sol refers to dispersions of solid particles in the 1-100 nm size range, which are finely distributed in water or organic solvents. In sol-gel processes, material production or deposition takes place from a liquid sol state, which is converted into a solid gel state via a sol-gel transformation. The sol-gel transformation involves a three-dimensional cross-linking of the nanoparticles in the solvent, whereby the gel takes on bulk properties. A controlled heat treatment in air can transform gels into a ceramic oxide material.

To start with, adding organic substances in the sol-gel process produces an organometallic compound from a solution containing an alcoxide (metallic compound of an alcohol, for example with silicon, titanium or aluminum). The pH value of the solution is adjusted with an acid or a base which, as a catalyst, also triggers the transformation of the alcoxide.

The subsequent reactions are hydrolysis (splitting of a chemical bond by water) followed by condensation and polymerization (reaction giving rise to many- or long-chained compounds from single-chained ones). The particles or the polymer oxide grow as the reaction continues, until a gel is formed. Due to the high porosity of the network, the particles typically have a large surface area, i.e. several hundred square meters per gram.

The course of hydrolysis and the polycondensation reaction depend on many factors: the composition of the initial solution, the type and amount of catalyst, temperature as well as the reactor- and mixing geometry.

For coatings, the alcoxide initial solution of the sol-gel process can be applied on surfaces of any geometry. After the wetting, the build-up of the porous network takes place through gel formation, yielding thicknesses of 50-500 nm. Thicker layers, suitable as membranes for example, are created by repeated wetting and drying. The sol-gel process can also be used to produce fibers. In all cases, gel formation is followed by a drying step. Figure illustrates the different reaction and processing steps of the sol-gel process.

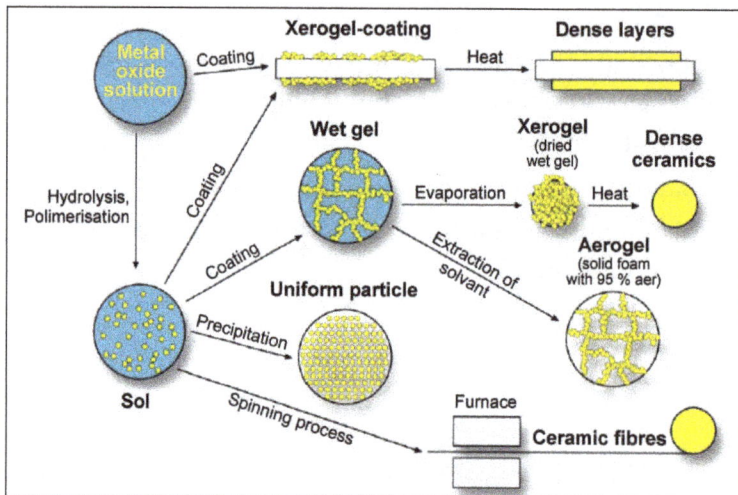

Reaction and processing steps in the sol-gel process.

A distinct advantage of the sol-gel process lies in the processability of the sols and gels, depending on processing step, into powders, fibers, ceramics and coatings. Moreover, highly porous nanomaterials can be produced. Composites can be created by filling in these pores during or after gel production. The low process temperature also enables substances to be embedded into the gel during the synthesis step; these can then be stored or released in a controlled manner.

The disadvantage of the sol-gel process lies in the difficult-to-control synthesis and drying steps, which complicate scaling up the process. Moreover, organic contaminants can remain in the gel. The resulting necessary cleaning steps, drying and thermal post-treatment makes this production process more complex than gas phase synthesis.

The disadvantage of the wet-chemical synthesis of nanomaterials is that the desired crystalline shapes often cannot be configured and that the thermal stability of the product powder is lower. This requires thermal post-treatment with repeated reduction of the particle surface. The advantage is that the liquid phase enables highly porous materials to be produced; this would normally not be possible in gas phase reactors due to the high temperatures.

With a few exceptions, gas phase processes also do not allow the production of organic nanoparticles. Liquid phase processes are particularly suited for the targeted production of monodisperse product powder (with uniform particle size).

Endocytosis and Exocytosis of Nanoparticles

Nano-sized materials have been increasingly used in the medical field to improve the target efficiency of drugs. In order to successfully apply nanoparticles in drug delivery, their physical and chemical properties must first be understood, thereby assisting in controlling the biological responses to their use. Because drug delivery nanosystems transport pharmaceutical compounds in the body, it is important to understand their physiochemical properties to safely achieve a desired therapeutic effect.

However, these drug delivery nanosystems have shown some limitations regarding the toxicity of the nanoscale materials in the body. In order to reduce their toxicity, it is crucial to study endocytosis, exocytosis, and clearance mechanisms for nanoparticles released from the nanoparticle–drug conjugates. Nanoparticles exposed to the bloodstream interact with opsonin proteins. When opsonin proteins attach to the surface of nanoparticles, they allow macrophages of the mononuclear phagocytic system (MPS) to easily recognize the nanoparticles and hence the nanoparticles eventually accumulate in the MPS organs, such as liver and spleen. These phenomena cause low targeting efficiency and severe systemic toxicity of the drug-delivery nanosystems.

Nanoparticle Stability

Nanoparticles have been widely used in the fields of drug delivery and bioimaging because their size, shape, and surface properties can be precisely engineered for specific diseases. The nanoparticle surface can be modified with various targeting molecules (eg, antibody, peptide, aptamer, etc) in order to achieve efficient targeting to disease sites.

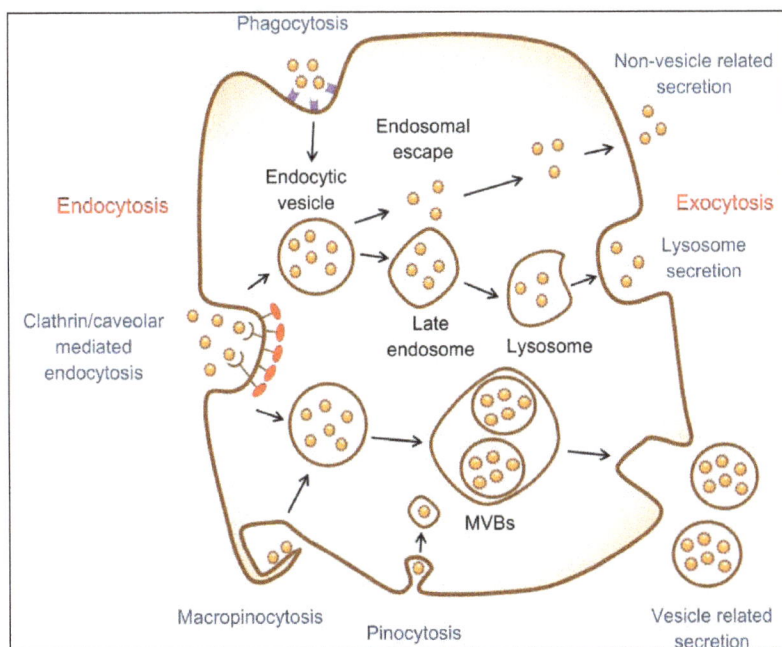

Schematic of endocytosis and exocytosis patterns of nanoparticles. Nanoparticles enter the cell via four types of pathway: clathrin/caveolar-mediated endocytosis, phagocytosis, macropinocytosis, and pinocytosis. Nanoparticles exit the cell via three types of pathway: lysosome secretion, vesicle-related secretion, and non-vesicle-related secretion.

Multivesicular Bodies

Recently, many scientists have begun to investigate the effects of different sizes, shapes, and surface chemistries on endocytosis, toxicity, and gene regulation. However, aggregates (strongly bonded between nanoparticles) and agglomerates (loosely bonded between nanoparticles or aggregates acting under weak forces, eg, van der Waals force) formed by forces between nanoparticles and components in biological media have not been fully considered to optimize their physicochemical properties for biological applications. It was recently suggested that the aggregates or agglomerates occur when the van der Waals attractive forces between nanoparticles are larger than the electrostatic repulsive forces. Aggregated or agglomerated forms of nanoparticles would behave differently within biological systems than would nanoparticles in their single form. Thus, size uniformity of nanoparticles should be considered when the effect of physical and chemical properties of nanoparticles on their interactions with biological systems is examined.

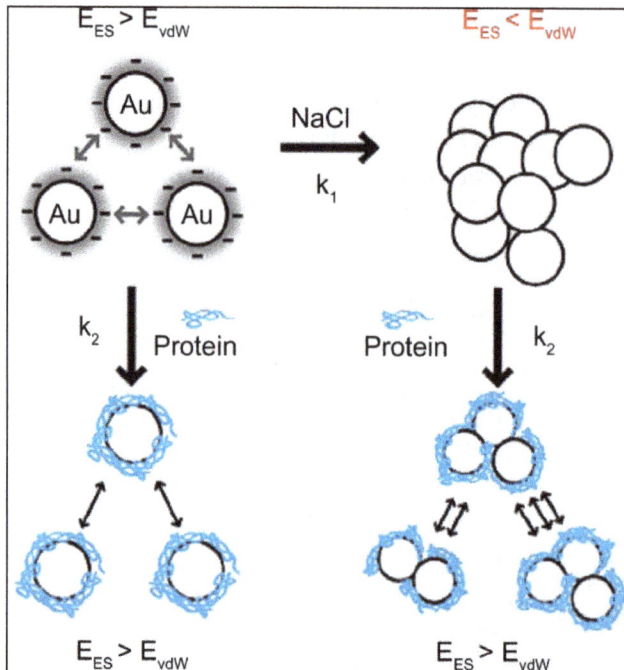

Scheme of aggregation or agglomeration mechanism. The stabilizing electrostatic forces (E_{ES}) on the surface of bare nanoparticles are neutralized by NaCl ions in the biological solution, causing the van der Waals forces (E_{vdW}) to drive formation of aggregation or agglomeration. The protein coating of nanoparticles can reduce the aggregation or agglomeration.

In biological solutions, such as blood, saliva, and cell culture media, the surface chemistry of nanoparticles plays a crucial role in determining their behavior because they are directly related to types and compositions of biomolecules attached to the nanoparticle surface. The surface chemistries on the nanoparticle surface are dynamically changed because various biomolecules are attached and detached based on their binding affinity to the surface. Pre-coating of the nanoparticle surface with stabilizing molecules such as polyethylene glycol (PEG), deoxyribonucleic acid (DNA), and albumin has been utilized to reduce ionic strength and prevent nanoparticles from aggregation or agglomeration in the biological solutions.

Additionally, individual nanoparticles can be naturally coated with various biomolecules, forming the nanoparticle–protein complex, when solubilized in biological solutions. The stability lifetimes of the nanoparticle–protein complexes range from hours to days in the biological solutions. The proteins covering the nanoparticle surface further prevent the individual nanoparticles from aggregation or agglomeration. Because the formation of nanoparticle–protein complexes is mainly determined by surface chemistries of the nanoparticles, it is important to investigate which surface chemistry is the most favorable to form the nanoparticle–protein complex.

Therefore, natural nanoparticle–protein complexes formed in biological environments would allow us to study how individual nanoparticles interact with various types of cells.

When we study the endocytosis and exocytosis of nanoparticles, cells are treated with the nanoparticles in the culture medium containing various serum proteins. Most of the nanoparticles are first coated with the serum proteins and then met with the plasma membrane of cells. If the nanoparticles are already aggregated or agglomerated prior to binding to the membrane, their endocytosis patterns would differ from the endocytosis patterns of individual nanoparticles. The degree of aggregation or agglomeration of the nanoparticles can be determined by measuring time-dependent change of size and surface charge of the nanoparticles in the culture medium. The (hydrodynamic) size has been mainly analyzed using transmission electron microscopy (TEM) and dynamic light scattering (DLS) while the surface charge is determined by zeta potential measurements. In particular, the ultraviolet–visible (UV/Vis) spectrophotometry has also been used to monitor the size of gold nanoparticles because their localized surface plasmon resonance peaks can be shifted to a longer wavelength by increasing their size. The DLS technique has been the most widely used to monitor size change because it directly measures hydrodynamic sizes of protein-coated nanoparticles in the biological solution with nanometer precision. Furthermore, the zeta potentials of protein-coated nanoparticles mostly appeared as a negative surface charge although the nanoparticles had different original surface chemistries. It suggests that most proteins attached on the nanoparticle surface seem to be negatively charged, regardless of their content and composition, although they may alter depending on the surface chemistries of the nanoparticles. Thus, the content and composition of proteins preferentially attached on the nanoparticle surfaces should be studied for the cellular uptake and immune response of nanoparticles.

Table: Classification of endocytosis pathways.

	Pathway	Definition
Specific pathway		
Endocytosis Phagocytosis	Clathrin- and caveolin-mediated Mannose receptor-, complement receptor-, Fcγ receptor-, and scavenger receptor-mediated	Energy-dependent process by which cells internalize biomolecules. Actin-dependent endocytic process by which professional phagocytes (macrophages, dendritic cells and neutrophils) engulf particles with sizes larger than 0.5 µm.
Non-specific pathway		
Macro-pinocytosis	-	Endocytic process by which cells internalize fluids and particles together, and large vesicles (0.2–5 µm) are formed.
Pinocytosis	-	Endocytic process by which cells absorb extracellular fluids, small molecules and small vesicles (~100 nm) are formed.

Nanoparticle Endocytosis

Endocytosis Mechanism

All types of cells in the body use the endocytosis process to communicate with the biological environments. This process is an energy-dependent process through which cells internalize ions and biomolecules. In particular, the cells internalize nutrients and signaling molecules to obtain energy and interact with other cells, respectively. The endocytosis pathways are typically classified into clathrin- and caveolae-mediated endocytosis, phagocytosis, macropinocytosis, and pinocytosis. Clathrin- and caveolae-mediated endocytosis indicates receptor-mediated endocytosis. Many types of cells use the clathrin- and caveolae-mediated endocytosis pathways to internalize nanoscale materials, including viruses and nanoparticles. These endocytosis pathways are the most important pathways for the internalization of nanoparticles into cells because the nanoparticles are directly coated with the plasma proteins when exposed to physiological solutions. The phagocytosis pathway is used when phagocytic cells internalize foreign materials with sizes larger than 0.5 µm 0.16 The phagocytosis pathway is actin-dependent and restricted to professional phagocytes, such as macrophages, dendritic cells, and neutrophils. The macropinocytosis pathway is a non-specific process to internalize fluids and particles together into the cell, whereas the pinocytosis pathway absorbs biological fluids from the external environment of a cell. These pathways are very important to translocate single nanoparticles with sizes below 10 nm into the cell.

When nanoparticles are systemically administered into the body, they are confronted with many types of cells. Since nanoparticles have emerged as effective drug carriers to treat complex diseases, it has become crucial to understand nanoparticle endocytosis mechanisms. It is believed that the endocytosis efficiency of nanoparticles is dependent on the physicochemical properties, such as size, shape, and surface chemistry, as well as cell type.

Factors Affecting Nanoparticle Endocytosis

Nanoparticles circulating in the bloodstream happen to meet and internalize into many types of cells through the plasma membrane. The plasma membrane is a selectively permeable membrane that transfers materials that are essential for sustaining the cell's life. Naturally, materials necessary for the cell's life, such as ions and nano-sized proteins, can pass through the lipid bilayer using specialized membrane-transport protein channels. Thus, the plasma membrane of cells would select the endocytosis pathways of nanoparticles depending on their size, shape, and surface chemistry.

Size

Size-dependent cellular uptake of nanoparticles has been extensively investigated in various cell lines because the nanoparticle size has been known to be a key determinant of the uptake pathways. Many critical in vivo functions of nanoparticles, such as circulation time, targeting, internalization, and clearance, depend on their size.

Much interest has focused on understanding sizedependent internalization of nanoparticles in cancer cells and fibroblasts. The cellular uptake of gold nanoparticles of various sizes was studied in human cervical cancer cells. Researchers demonstrated that uptake mechanism and saturation concentration of nanoparticles was dependent on their size. The 50 nm gold nanoparticles showed the most efficient cellular uptake compared with other sizes. The cellular uptake of polystyrene (PS) nanoparticles of various sizes was also tested on a human colon adenocarcinoma cell line. The PS nanoparticles with a size of 100 nm were taken up into the cells more efficiently than those with sizes of 50, 200, 500, and 1,000 nm. The internalization efficiency of the PS nanoparticles with a size of 50 nm was the lowest of all sizes. These two studies suggest that the patterns of cellular uptake could also vary according to nanoparticle material type. Although different cancer cell lines were used for each experiment, the stiffness of nanoparticles could affect their cellular uptake. Stiffer nanoparticles would interact tightly with the plasma membrane of cell, thereby causing rapid endocytosis.

Single-walled carbon nanotubes (SWNTs) coated with DNA molecules were used to investigate length-dependent cellular uptake. The results demonstrated that long (660 ± 40 nm) and short (130 ± 18 nm) SWNTs have lower uptake efficiencies in the fibroblasts than SWNTs with average lengths of 430 ± 35 nm and 320 ± 30 nm. The SWNTs with an average length of 320 ± 30 nm had the greatest uptake pattern.

In addition to endocytosis in non-phagocytic cells, much attention has recently been paid to understanding interactions between nanoparticles and phagocytic cells such as macrophages, because it could be relevant to the design of nanoparticles to avoid the immune system, thus increasing their target efficiency. Nanoparticles with sizes larger than 0.5 μm have been known to enter phagocytic cells via phagocytosis pathways. Polymeric microspheres with diameters of 2–3 μm exhibited the maximal phagocytosis rate. Interestingly, this size range coincides with the general size of the bacteria that are the most common targets of the MPS.

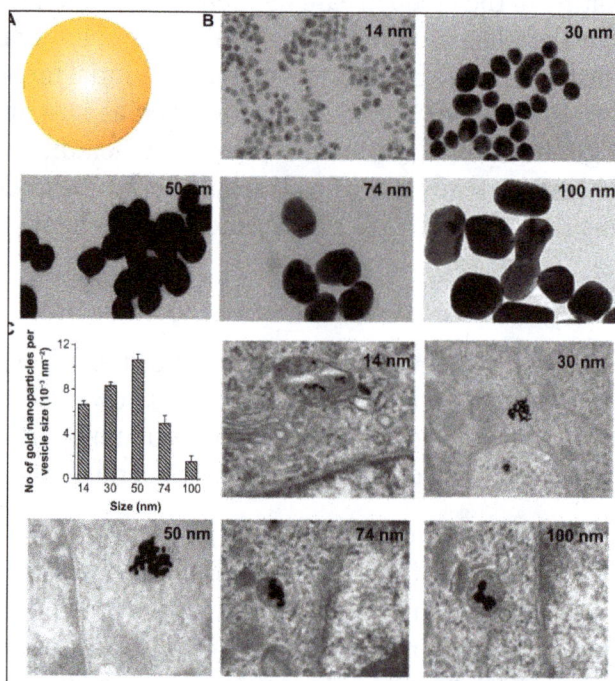

(A) Schematic representation of gold nanoparticles. (B) Transmission electron microscope images of citrate-coated gold nanoparticles with various sizes. (C) Transmission electron microscope images of the gold nanoparticles entrapped in cellular vesicles. Graph showing the number of the gold nanoparticles per vesicle diameter.

On the other hand, there have been some efforts to examine size-dependent phagocytosis of nanoparticles smaller than 0.5 μm. Among many types of nanoparticles, lipid-based nanoparticles, including US Food and Drug Administration (FDA)-approved liposomes have been of particular interest in the drug-delivery field due to their drug-loading capability and biocompatibility. Lipid nanoparticles with sizes of 20, 50, and 100 nm were taken up into the macrophages by complement receptor-mediated phagocytosis. In addition, liposomes with sizes ranging from 100 to 2,000 nm were also tested for their intracellular uptake in the macrophages. The amount of liposomes taken up by the macrophages increased with size over the range 100–1,000 nm, but the uptake rate became constant with sizes over 1,000 nm. Size-dependent phagocytosis of gold nanoparticles was also studied in many research groups. It was reported that gold nanoparticles with sizes below 100 nm were phagocytosed via scavenger receptor-mediated phagocytosis. Tsai et al also demonstrated that gold nanoparticles with a size of 4 nm showed the highest uptake in the macrophages based on the number of nanoparticles taken up per cell, compared with those sized 11, 19, 35, and 45 nm. The 4 nm gold nanoparticles exhibited the highest potency in inhibiting tumor necrosis factor (TNF)-α production related to TNR9-mediated innate immune systems.

Although the experimental results introduced here showed size-dependent cellular uptake of nanoparticles in many types of cells, the physical size would not be fully reflected when they meet the plasma membrane of the cell. Most nanoparticles tend to aggregate in biological solutions, increasing their overall size. The results regarding the size effect would be influenced by nanoparticle aggregation before entering the cell. Thus, it should be tested whether nanoparticles prepared to study their size-dependent endocytosis retain their singularity in the biological media before they enter the cell.

Surface Chemistry

Surface chemistry (or surface charge) of nanoparticles can be determined by the chemical composition on the nanoparticle surface. The surface charge of nanoparticles can affect their efficiency and the pathway of cellular uptake, because biological systems consist of numerous biomolecules with various charges. Therefore, the charge of biomolecules covering the nanoparticle surface can influence the endocytosis patterns of nanoparticles. For polymeric nanoparticles, carboxymethyl chitosan-grafted nanoparticles as negatively charged nanoparticles, and chitosan hydrochloride-grafted nanoparticles as positively charged nanoparticles, were used to test their cellular uptake efficiency. The different surface charges significantly affected their uptake by macrophages. The positively charged nanoparticles exhibited a higher phagocytic uptake than did the negatively or neutrally charged nanoparticles. Moreover, when the uptake efficiency of the positively charged nanoparticles was compared with that of the negatively charged, neutrally charged, and PEGylated nanoparticles, the positively and negatively charged nanoparticles were internalized more rapidly than the neutrally and PEGylated charged nanoparticles. It was also demonstrated that negatively charged polymeric nanoparticles with diameters of around 100 nm were more efficiently phagocytized by macrophages than positively charged nanoparticles. Cellular uptake was also greater in the macrophages than in the monocytes.

Furthermore, the surface functionalization with PEG, poloxamer, and poloxamine polymers prevented phagocytosis because these polymers protect the nanoparticles from ionic strength, promote particle dispersion, and reduce absorption of proteins in blood on their surface. In addition, the uptake efficiency of PEGylated nanoparticles is closely related to PEG grafting density, which can determine the protein absorption. That is, high PEG-grafting density inhibits protein adsorption on the nanoparticle surface in the biological solution. Interaction between nanoparticles and biological media can further lead to surface modification that eventually affects their phagocytosis through the attachment of complementary proteins and immunoglobulins.

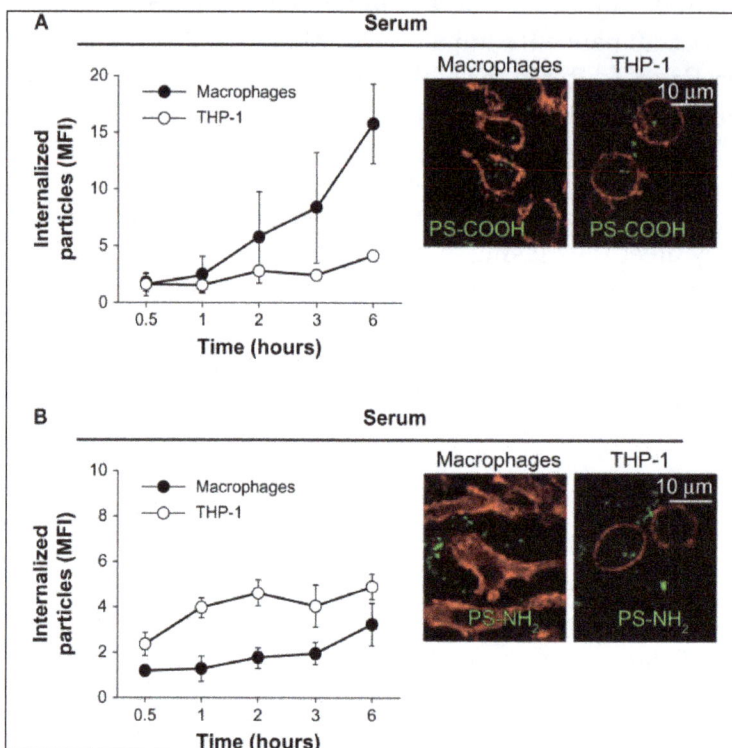

Kinetics of cellular uptake of negatively (–COOH) (A) and positively (–NH2) (B) charged polymeric nanoparticles in macrophages and monocytes (THP-1). Confocal fluorescence images (right) were taken after 2 hours' incubation with the indicated nanoparticles. Cell membrane was stained with red dyes and the nanoparticles were tagged with green dyes.

Recently, there has been much effort to utilize biological nanomaterials in drug delivery applications, due to their biocompatibility and natural cell-binding ability. Apoferritin, a demineralized form of ferritin, was suggested as a drug carrier because it has been known to enter the cell via ferritin receptor-mediated endocytosis. Apoferritin can be further utilized for a switchable delivery system because the endocytosis can be reversibly inhibited in various ways. Additionally, the internal cavity of apoferritin can be used to load therapeutic molecules through channels on the protein shell. Other virus- and protein-based biomaterials also have great potential to serve as biocompatible nano-platforms in the drug-delivery system.

Shape

As mentioned in previous subsections, size and surface properties of nanoparticles play crucial roles in controlling the interaction between nanoparticles and biological systems.

The nanoparticle shape might also be important in determining biological behaviors of nanoparticles. For example, it has been reported that bacteria with various shapes, such as rods, spirals, and ellipsoids, have implications for macrophage recognition. These shapes can directly affect the endocytosis pattern of bacteria. Recent experiments have demonstrated a shape effect of the nanoparticle on cellular uptake. Nanoscale rods exhibit the highest uptake in human cervical cancer cells, followed by spheres, cylinders, and cubes, and the cellular uptake of cylindrical particles depends strongly on their aspect ratio. However, the receptor-mediated endocytosis of gold nanorods was vastly decreased with increases in their aspect ratio. The comparison of intracellular uptake efficiency between rod-shaped and spherical nanoparticles has also been investigated in many types of cells. Interestingly, the results demonstrated that the uptake of rod-shaped nanoparticles by macrophages was more efficient than that of the spherical nanoparticles, while the spherical nanoparticles were taken up by cervical cancer cells and human lung epithelial cells more efficiently than were rod-shaped nanoparticles.

Nanoparticle Exocytosis

An understanding of the cellular uptake and organ distribution of nanoparticles is important when examining their targeting and therapeutic efficiency in drug-delivery applications. Nanoparticles administered into the body are eventually cleared by organs in the MPS, such as the liver and spleen. Nanoparticles remain in these organs for a long time after being taken up by the macrophages, which increases the likelihood of unintended acute or chronic toxicity. Thus, it is also crucial to study exocytosis of the internalized nanoparticles from many types of cells, particularly macrophages, to evaluate their biosafety. However, compared with investigations of nanoparticle endocytosis, relatively little effort has been made to investigate the exocytosis of nanoparticles that may be responsible for their systemic elimination and toxicity.

Many studies have compared the exocytosis of rodshaped nanoparticles with that of spherical nanoparticles. Chithrani and Chan9 examined differences in exocytosis phenomena between spherical and rod-shaped gold nanoparticles using various cell types. The surface of gold nanoparticles was coated with transferrin proteins for their receptor-mediated endocytosis. Importantly, this work described that the cellular uptake could be considered as a result of competition between the thermodynamic driving force for wrapping and the receptor diffusion kinetics. In that respect, the 50 nm gold nanoparticles showed the fastest wrapping time, and therefore the receptor-ligand interaction could produce sufficient free energy to drive the nanoparticles into the cell. On the other hand, smaller nanoparticles with a slower wrapping time exhibited a faster rate of exocytosis. The exocytosis rate of the 14 nm nanoparticles was much faster than that of the 74 nm nanoparticles. In addition, the fraction of the rod-shaped nanoparticles released outside the cells was generally higher than that of the spherical nanoparticles. The exocytosis of peptide-coated gold nanoparticles was also investigated in endothelial cells. Nanoparticles functionalized with KATWLPPR peptides have been known to bind to plasma membrane receptors on endothelial cells for endocytosis, while nanoparticles coated with KPRQPSLP peptides did not interact with the receptors for endocytosis. KATWLPPR peptide-coated nanoparticles taken up by cells were progressively exocytosed up until 6 hours. On the other hand, KPRQPSLP peptide-coated nanoparticles showed a more complex exocytosis profile. Interestingly, it was found that the exocytosed KPRQPSLP peptide-coated nanoparticles were re-taken up by the cells after 4 hours.

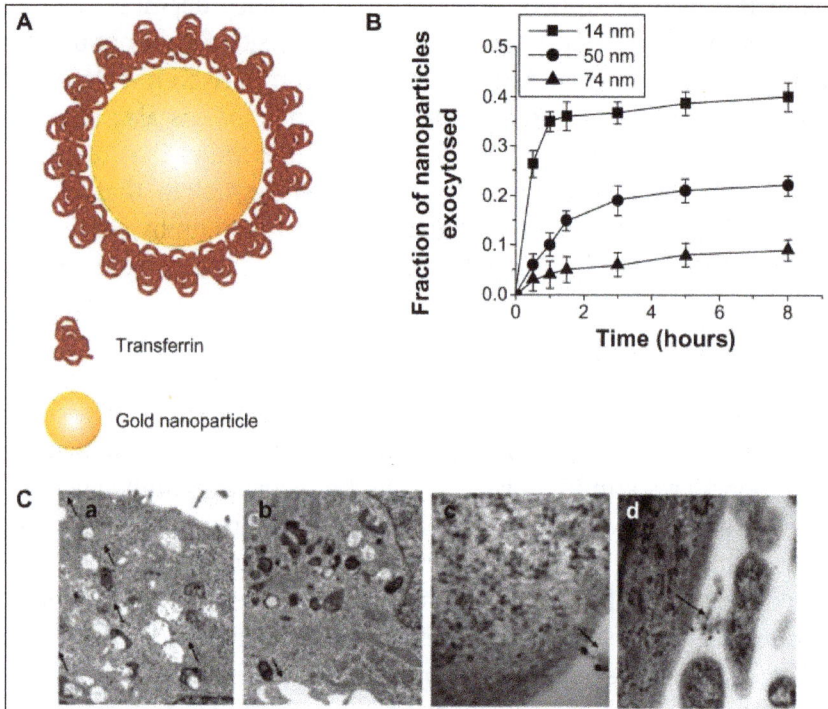

(A) Schematic depicting of transferrin-coated gold nanoparticles. (B) Kinetics of exocytosis patterns of the nanoparticles with different sizes. (C) Different stages of exocytosis patterns: (a) Movement of the vesicles containing nanoparticles toward the cell membrane; (b) Docking of one of the vesicles at the cell membrane; (c) Excretion of nanoparticles; (d) Cluster of nanoparticles after exocytosis.

The exocytosis of poly(D,L-lactide-co-glycolide) (PLGA) nanoparticles was also examined in vascular smooth muscle cells. The size and zeta potential of PLGA nanoparticles coated with bovine serum albumin (BSA) were around 97 nm and -20 mV, respectively. The cellular uptake of nanoparticles increased with incubation time. The exocytosis of nanoparticles increased up to 65% of the internalized fraction within 30 minutes when nanoparticles in the culture media were removed. In addition, the exocytosis of nanoparticles was found to be energy-dependent because it was significantly inhibited with sodium azide and deoxyglucose.

Exocytosis of polysaccharide cationic nanoparticles was also studied in airway epithelium cells. The cationic polymer hydroxycholine was used to coat nanoparticles. After the surface modification, the nanoparticles appeared to be an approximate size of 60 nm. For exocytosis experiments, human bronchial epithelial cells were treated with the nanoparticles for 30 minutes. The amount of exocytosed nanoparticles increased significantly after 1 hour. In addition, cholesterol depletion completely blocked the exocytosis of nanoparticles, indicating that their exocytosis is cholesteroldependent.

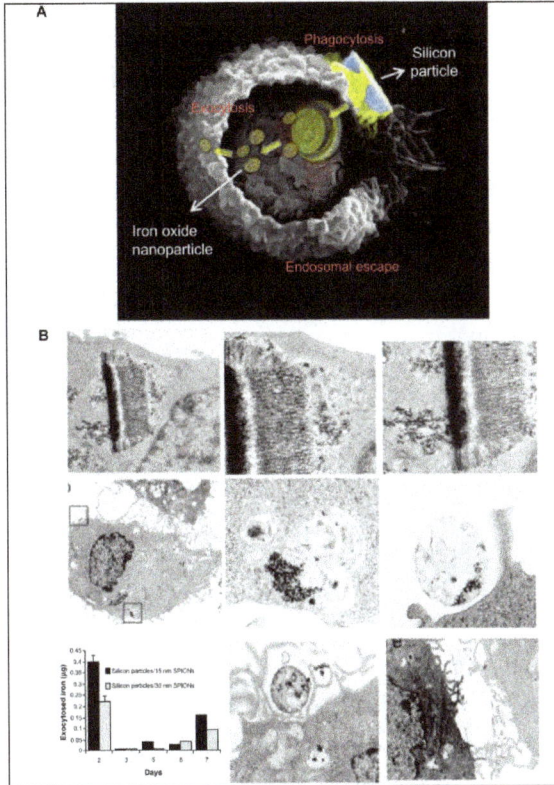

(A) A scanning electron microscopy image of a macrophage showing endocytosis of porous silicon particles incorporated with iron oxide nanoparticles, intracellular partitioning of the particles, endosomal escape of the particles, and exocytosis of the incorporated iron oxide nanoparticles. (B) Exocytosis of iron oxide nanoparticles. Upper column shows transmission electron microscopy images of iron oxide nanoparticles released from the porous silicon carrier in a macrophage. Middle column shows transmission electron microscopy images of the released iron oxide nanoparticles located in the intracellular region of the macrophage 6 days after uptake of the porous silicon particles. Bottom column shows that graph displayed time-dependent iron content in the supernatant and transmission electron microscopy images showed the internalized iron oxide nanoparticles were exocytosed by membrane vesicles.

The exocytosis patterns of superparamagnetic iron oxide nanoparticles and quantum dots were also studied in many research groups. Serda et al investigated the intracellular trafficking of iron oxide nanoparticles delivered into macrophage cells using porous silicon microcarriers. In the exocytosis process, the amine-functionalized nanoparticles enriched in the multivesicular bodies were incorporated into the membrane vesicles in the intracellular region and efficiently secreted from the cells 6 days after intracellular delivery. The exocytosis rate of 15 nm nanoparticles was faster than that of 30 nm nanoparticles, thus indicating that smaller nanoparticles are more favorable for exocytosis. Exocytosis of zwitterionic quantum dots was also examined in human cervical cancer cells. Quantum dots were coated with D-penicillamine to improve colloidal stability in biological solutions. The size of D-penicillamine-coated quantum dots was around 4 nm, which is much smaller than conventional nanoparticles studied for exocytosis. In the endocytosis process, most of these smaller nanoparticles were observed on the plasma membrane prior to internalization. In addition, the quantum dots were found to enter the cells via a clathrin-mediated endocytosis pathway and macropinocytosis. In the exocytosis process, some of the quantum dots trapped in endosomes were actively transported to the cell periphery and exocytosed to the media within 21 minutes after internalization.

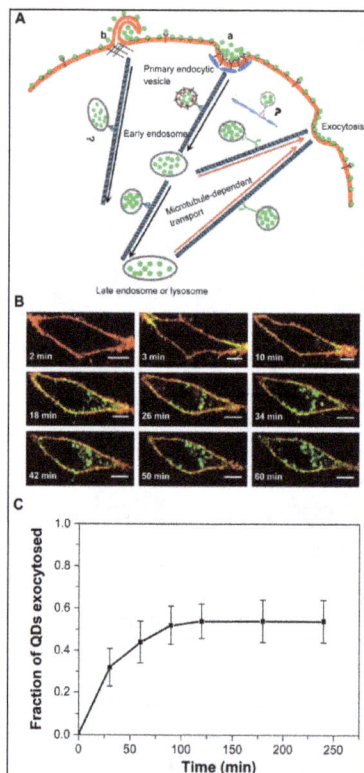

(A) Schematic diagram showing the endocytosis and exocytosis processes of D-penicillamine-coated quantum dots; a: Clathrin-mediated endocytosis; b: Macropinocytosis. (B) Interaction of D-penicillamine-coated quantum dots (green) with plasma membrane of a HeLa cell before internalization. The plasma membrane was stained with the red membrane dye. Scale bar: 10 μm. (C) Kinetics of exocytosis of D-penicillamine-coated quantum dots after removing the nanoparticles in the media.

Magnetic Nanoparticles

Magnetic nanoparticles are a class of nanoparticle that can be manipulated using magnetic fields. Such particles commonly consist of two components, a magnetic material, often iron, nickel and cobalt, and a chemical component that has functionality. While nanoparticles are smaller than 1 micrometer in diameter (typically 1–100 nanometers), the larger microbeads are 0.5–500 micrometer in diameter. Magnetic nanoparticle clusters that are composed of a number of individual magnetic nanoparticles are known as magnetic nanobeads with a diameter of 50–200 nanometers. Magnetic nanoparticle clusters are a basis for their further magnetic assembly into magnetic nanochains. The magnetic nanoparticles have been the focus of much research recently because they possess attractive properties which could see potential use in catalysis including nano-material-based catalysts, biomedicine and tissue specific targeting, magnetically tunable colloidal photonic crystals, microfluidics, magnetic resonance imaging, magnetic particle imaging, data storage, environmental remediation, nanofluids, optical filters, defect sensor, magnetic cooling and cation sensors.

Physical Properties of Magnetic Nanoparticles

Magnetic effects are caused by movements of particles that have both mass and electric charges. These particles are electrons, holes, protons, and positive and negative ions. A spinning electric-charged particle creates a magnetic dipole, so-called magneton. In ferromagnetic materials, magnetons are associated in groups. A magnetic domain (also called a Weiss domain) refers to a volume of ferromagnetic material in which all magnetons are aligned in the same direction by the exchange forces. This concept of domains distinguishes ferromagnetism from paramagnetism. The domain structure of a ferromagnetic material determines the size dependence of its magnetic behavior. When the size of a ferromagnetic material is reduced below a critical value, it becomes a single domain. Fine particle magnetism comes from size effects, which are based on the magnetic domain structure of ferromagnetic materials. It assumes that the state of lowest free energy of ferromagnetic particles has uniform magnetization for particles smaller than a certain critical size and has nonuniform magnetization for larger particles. The former ones are referred to as single domain particles, while the latter are called multidomain particles. According to the magnetic domain theory, the critical size of the single domain is affected by several factors including the value of the magnetic saturation, the strength of the crystal anisotropy and exchange forces, surface or domain-wall energy, and the shape of the particles. The reaction of ferromagnetic materials on an applied field is well described by a hysteresis loop, which is characterized by two main parameters: remanence and coercivity. The latter is related to the 'thickness' of the curve. Dealing with fine particles, the coercivity is the single property of most interest, and it is strongly size-dependent. It has been found that as the particle size is reduced, the coercivity increases to a maximum and then decreases toward zero.

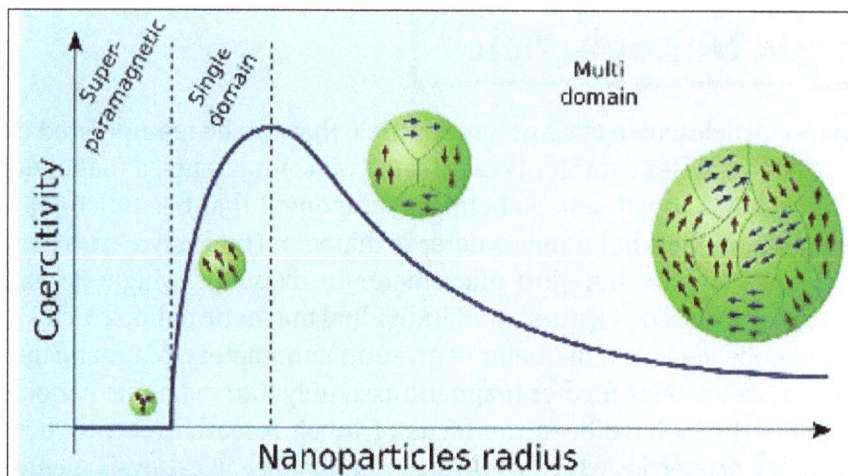

Schematic illustration of the coercivity-size relations of small particles.

When the size of single-domain particles further decreases below a critical diameter, the coercivity becomes zero, and such particles become superparamagnetic. Superparamagnetism is caused by thermal effects. In superparamagnetic particles, thermal fluctuations are strong enough to spontaneously demagnetize a previously saturated assembly; therefore, these particles have zero coercivity and have no hysteresis. Nanoparticles become magnetic in the presence of an external magnet, but revert to a nonmagnetic state when the external magnet is removed. This avoids an 'active' behavior of the particles when there is no applied field. Introduced in the living systems, particles are 'magnetic' only in the presence of an external field, which gives them unique advantage in working in biological environments. There are a number of crystalline materials that exhibit ferromagnetism, among others Fe, Co, or Ni. Since ferrite oxide-magnetite (Fe_3O_4) is the most magnetic of all the naturally occurring minerals on earth, it is widely used in the form of superparamagnetic nanoparticles for all sorts of biological applications.

Magnetic Property (Magnetic Behavior)

Materials are classified by their response to an externally applied magnetic field. Descriptions of orientations of the magnetic moments in a material help identify different forms of magnetism observed in nature. Five basic types of magnetism can be described: diamagnetism, paramagnetism, ferromagnetism, antiferromagnetism, and ferrimagnetisms. In the presence of an externally applied magnetic field, the atomic current loops created by the orbital motion of electrons respond to oppose the applied field. All materials display this type of weak repulsion to a magnetic field known as diamagnetism. However, diamagnetism is very weak, and therefore, any other form of magnetic behavior that a material may possess usually overpowers the effects of the current loops. In terms of the electronic configuration of the materials, diamagnetism is observed in materials with filled electronic subshells where the magnetic moments are paired and overall cancel each other. Diamagnetic materials have a negative susceptibility ($\chi < 0$) and weakly repel an applied magnetic field (e.g., quartz SiO_2). The effects of these atomic current loops are overcome if the material displays a net magnetic moment or has a long-range ordering of its magnetic moments. All other types of magnetic behaviors are observed in materials that are at least partially attributed to unpaired electrons in their atomic shells, often in the 3d or 4f shells of each atom. Materials whose atomic magnetic moments are uncoupled display paramagnetism; thus, paramagnetic materials have moments with no long-range order, and there is a small positive magnetic susceptibility ($\chi \approx 0$), e.g., pyrite. Materials that possess ferromagnetism have aligned atomic magnetic moments of equal magnitude, and their crystalline structures allow for direct coupling interactions between the moments, which may strongly enhance the flux density (e.g., Fe, Ni, and Co). Furthermore, the aligned moments in ferromagnetic materials can confer a spontaneous magnetization in the absence of an applied magnetic field. Materials that retain permanent magnetization in the absence of an applied field are known as hard magnets. Materials having atomic magnetic moments of equal magnitude that are arranged in an antiparallel fashion display antiferromagnetism (e.g., troilite FeS). The exchange interaction couples the moments in such a way that they are antiparallel, therefore, leaving a zero net magnetization. Above the Néel temperature, thermal energy is sufficient to cause the equal and oppositely aligned atomic moments to randomly fluctuate, leading to a disappearance of their long-range order. In this state, the materials exhibit paramagnetic behavior. Ferrimagnetism is a property exhibited by materials whose atoms or ions tend to assume an ordered but nonparallel arrangement in a zero applied field below a certain characteristic temperature known as the Néel temperature (e.g., Fe_3O_4 and Fe_3S_4). In the usual case, within a magnetic domain, a substantial net magnetization results from the antiparallel alignment of neighboring non-equivalent sublattices. The macroscopic behavior is similar to ferromagnetism. Above the Néel temperature, the substance becomes paramagnetic.

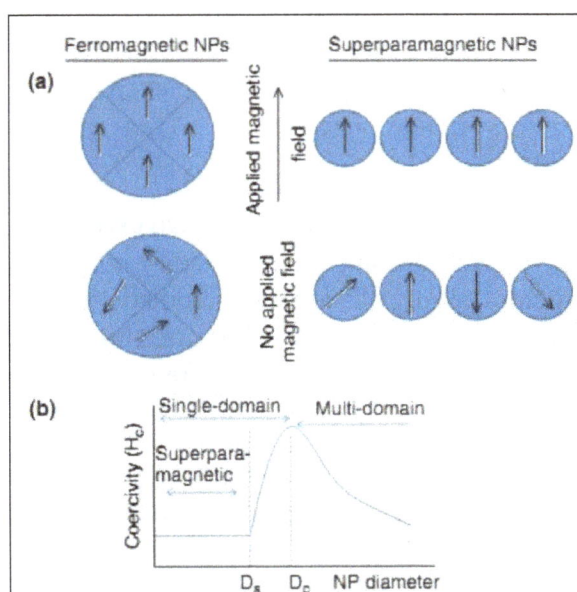

Magnetization behavior of ferromagnetic and superparamagnetic NPs under an external magnetic field. (a) Under an external magnetic field, domains of a ferromagnetic NP align with the applied field. The magnetic moment of single domain superparamagnetic NPs aligns with the applied field. In the absence of an external field, ferromagnetic NPs will maintain a net magnetization, whereas superparamagnetic NPs will exhibit no net magnetization due to rapid reversal of the magnetic moment. (b) Relationship between NP size and the magnetic domain structures. Ds and Dc are the 'superparamagnetism' and 'critical' size thresholds.

Applications

Industrial Applications

Magnetic iron oxides are commonly used as synthetic pigments in ceramics, paints, and porcelain. Magnetic encapsulates may find very important uses in many areas of life and also in various branches of industry. Such materials are interesting from both points of the fundamental study of materials science as well as their applications. Hematite and magnetite have been applied as catalysts for a number of important reactions, such as the preparation of NH_3, the desulfurization of natural gas, and the high-temperature water-gas shift reaction. Other reactions include the Fishere-Tropsch synthesis for hydrocarbons, the dehydrogenation of ethylbenzene to styrene, the oxidation of alcohols, and the large-scale synthesis of butadiene.

Biomedical Applications

Biomedical applications of magnetic nanoparticles can be classified according to their application inside or outside the body (in vivo, in vitro). For in vitro applications, the main use is in diagnostic separation, selection, and magnetorelaxometry, while for in vivo applications, it could be further separated in therapeutic (hyperthermia and drug-targeting) and diagnostic applications (nuclear magnetic resonance (NMR) imaging).

In Vivo Applications

Two major factors play an important role for the in vivo uses of these particles: size and surface functionality. Even without targeting surface ligands, superparamagnetic iron oxide NP (SPIOs) diameters greatly affect in vivo biodistribution. Particles with diameters of 10 to 40 nm including ultra-small SPIOs are important for prolonged blood circulation; they can cross capillary walls and are often phagocytosed by macrophages which traffic to the lymph nodes and bone marrow.

Therapeutic applications. Hyperthermia: Placing superparamagnetic iron oxide in altering current (AC) magnetic fields randomly flips the magnetization direction between the parallel and antiparallel orientations, allowing the transfer of magnetic energy to the particles in the form of heat, a property that can be used in vivo to increase the temperature of tumor tissues to destroy the pathological cells by hyperthermia. Tumor cells are more sensitive to a temperature increase than healthy ones. In past studies, magnetite cationic liposomal nanoparticles and dextran-coated magnetite have been shown to effectively increase the temperature of tumor cells for hyperthermia treatment in cell irradiation. This has been proposed to be one of the key approaches to successful cancer therapy in the future. The advantage of magnetic hyperthermia is that it allows the heating to be restricted to the tumor area. Moreover, the use of subdomain magnetic particles (nanometer-sized) is preferred instead multidomain (micron-sized) particles because nanoparticles absorb much more power at tolerable AC magnetic fields which is strongly dependent on the particle size and shape, and thus, having well-defined synthetic routes able to produce uniform particles is essential for a rigorous control in temperature.

Drug delivery. Drug targeting has emerged as one of the modern technologies for drug delivery. The possibilities for the application of iron oxide magnetic nanoparticles in drug targeting have drastically increased in recent years. MNPs in combination with an external magnetic field and magnetizable implants allow the delivery of particles to the desired target area, fix them at the local site while the medication is released, and act locally (magnetic drug targeting). Transportation of drugs to a specific site can eliminate side effects and also reduce the dosage required. The surfaces of these particles are generally modified with organic polymers and inorganic metals or oxides to make them biocompatible and suitable for further functionalization by the attachment of various bioactive molecules. The process of drug localization using magnetic delivery systems is based on the competition between the forces exerted on the particles by the blood compartment and the magnetic forces generated from the magnet.

Diagnostic Applications

NMR imaging. The development of the NMR imaging technique for clinical diagnosis has prompted the need for a new class of pharmaceuticals, so-called magneto-pharmaceuticals. These drugs must be administered to a patient in order to (1) enhance the image contrast between the normal and diseased tissue and (2) indicate the status of organ functions or blood flow.

In Vitro Applications

Diagnostic Applications

Separation and selection: At present, considerable attention is being paid to solid-phase extraction (SPE) as a way to isolate and preconcentrate desired components from a sample matrix. SPE is a routine extraction method for determining trace-level contaminants in environmental samples. Recently, nanometer-sized particles (nanoparticles, NPs) have gained rapid and substantial progress and have a significant impact on sample extraction. SPE offers an excellent alternative to the conventional sample concentration methods, such as liquid-liquid extraction. The separation and preconcentration of the substance from large volumes of solution can be highly time consuming when using standard column SPE, and it is in this field where the use of magnetic or magnetizable adsorbents called magnetic solid-phase extraction (MSPE) gains importance. In this procedure, the magnetic adsorbent is added to a solution or suspension containing the target. This is adsorbed onto the magnetic adsorbent, and then, the adsorbent with the adsorbed target is recovered from the suspension using an appropriate magnetic separator. For separation and selection, the advantage of using magnetic nanoparticles instead magnetic microparticles is that we can prepare suspensions that are stable against sedimentation in the absence of an applied magnetic field. The applicability of iron oxide magnetic nanoparticles in MSPE is clearly evidenced by the fact that it already exists in the market companies (DYNAL Biotech) that commercialize these products.

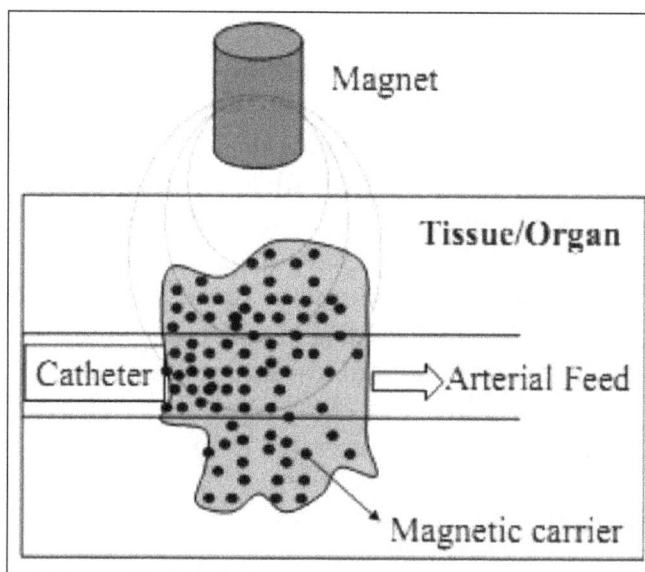

Schematic representation of the magnetically driven transport of drugs to a specific region. A catheter is inserted into an arterial feed to the tumor, and a magnetic stand is positioned over the targeted site.

Magnetorelaxometry: It was introduced as a method for the evaluation of immuno-assays. Magnetorelaxometry measures the magnetic viscosity, i.e., the relaxation of the net magnetic moment of a system of magnetic nanoparticles after removal of a magnetic field. There are two different relaxation mechanisms. First, the internal magnetization vector of a nanoparticle relaxes in the direction of the easy axis inside the core; this is called Néel relaxation. Second, particles accomplish rotational diffusion in a carrier liquid, called Brownian relaxation. Néel and Brownian relaxation can be distinguished by their different relaxation times. Furthermore, Brownian relaxation can take place only in liquids, whereas Néel relaxation does not depend on the dispersion of the nanoparticles. The fact that magnetorelaxometry depends on the core size, the hydrodynamic size, and the anisotropy allows this technique to distinguish between the free and bound conjugates by their different magnetic behavior and therefore can be used as an analytical tool for the evaluation of immunoassays. For this application, the benefits of reducing the particle size to the nanometer size are similar to those described for separation and selection applications.

Magnetic resonance imaging: At the boundary between nanomaterials and medical diagnostics, superparamagnetic iron oxide NPs are proving to be a class of novel probes useful for in vitro and in vivo cellular and molecular imaging. Superparamagnetic contrast agents have an advantage of producing an enhanced proton relaxation in magnetic resonance imaging (MRI) in comparison with paramagnetic ones. Consequently, less amounts of a SPIO agent are needed to dose the human body than a paramagnetic one. To apply the magnetic fluids to a MRI contrast agent, a SPIO should be dispersed into a biocompatible and biodegradable carrier. Recently, Muller et al. comprehensively reviewed the applications of super paramagnetic iron oxide NPs as a contrast agent. However, MRIs are not convenient for in situ monitoring. Thus, a sensitive and simple technique for in situ monitoring of the NPs in living cells is desirable. Compared to conventional organic fluorescence probes, advantages of the nanometer-sized fluorescence probes mainly include their higher photostability and stronger fluorescence. The main problem in cell imaging using the fluorescent nanoprobes is that the fluorescence signal is easily affected by the background noises caused by the cells, matrix, and the nonspecific scattering lights. The high signal to noise (S/N) ratio is not easy to obtain.

Bioseparation: In a biomedical study, separation of specific biological entities (e.g., DNAs, proteins, and cells) from their native environment is often required for analysis. Superparamagnetic colloids are ideal for this application because of their on-off nature of magnetization with and without an external magnetic field, enabling the transportation of biomaterials with a magnetic field. In a typical procedure for separation, the biological entities are labeled by superparamagnetic colloids and then subjected to separation by an external magnetic field. Nanometer-sized magnetic particles, such as super paramagnetic iron oxide particles, have been extensively used for separation and purification of cells and biomolecules in bioprocesses. Due to their small size and high surface area, MNPs have many superior characteristics for these bioseparation applications compared to those of the conventional micrometer-sized resins or beads such as good dispersability, the fast and effective binding of biomolecules, and reversible and controllable flocculation. One of the trends in this subject area is the magnetic separation using antibodies to provide highly accurate antibodies that can specifically bind to their matching antigens on the surface of the targeted species.

Catalysis Applications

In recent years, catalysts supported by MNPs have been extensively used to improve the limitation of heterogeneous catalysis. Magnetically driven, separations make the recovery of catalysts in a liquid-phase reaction much easier than using cross flow filtration and centrifugation, especially when the catalysts are in the submicrometer size range. Such small and magnetically separable catalysts could combine the advantages of high dispersion and reactivity with easy separation. In terms of recycling expensive catalysts or ligands, immobilization of these active species on MNPs leads to the easy separation of catalysts in a quasi-homogeneous system. The various types of transition metal-catalyzed reactions using catalytic sites grafted onto MNPs that have emerged recently include carbon-carbon cross-coupling reactions, hydroformylation, hydrogenation, and polymerization reactions. Other reports on MNP-supported catalysts include enzymes for carboxylate resolution, amino acids for ester hydrolysis, and organic amine catalysts promoting Knoevenagel and related reactions.

Environmental Applications

A similarly important property of nanoscale iron particles is their huge flexibility for in situ applications. Modified iron nanoparticles, such as catalyzed and supported nanoparticles, have been synthesized to further enhance their speed and efficiency of remediation. In spite of some still unresolved uncertainties associated with the application of iron nanoparticles, this material is being accepted as a versatile tool for the remediation of different types of contaminants in groundwater, soil, and air on both the experimental and field scales. In recent years, other MNPs have been investigated for the removal of organic and inorganic pollutants.

Organic Pollutants

There are a few articles about the removal of high concentrations of organic compounds which are mostly related to the removal of dyes. The MNPs have a high capacity in the removal of high concentrations of organic compounds. Dyes are present in the wastewater streams of many industrial sectors such as in dyeing, textile factories, tanneries, and in the paint industry. Therefore, the replacement of MNPs with an expensive or low efficient adsorbent for treatment of textile effluent can be a good platform which needs more detailed investigations.

Inorganic Pollutants

A very important aspect in metal toxin removal is the preparation of functionalized sorbents for affinity or selective removal of hazardous metal ions from complicated matrices. MNPs are used as sorbents for the removal of metal ions. Thus, MNPs show a high capacity and efficiency in the removal of different metal ions due to their high surface area with respect to micron-sized sorbents. These findings can be used to design an appropriate adsorption treatment plan for the removal and recovery of metal ions from wastewaters.

Analytical Applications

Fluorescence techniques. Due to their small size, magnetic luminescent NPs offer a larger surface area-to-volume ratio than currently used microbeads, which result in a good reaction homogeneity and faster reaction kinetics. Thus, the preparation of magnetic fluorescent particles, such as polystyrene magnetic beads with entrapped organic dyes/quantum dots (QDs) or shells of QDs, iron oxide particles coated with dye-doped silica shells, and silica NPs embedded with iron oxide and QDs, is easier. However, their application is limited mostly to biological applications, such as cellular imaging. Only a few papers have reported the use of dual-functional NPs for multiplexed quantitative bioanalysis. The magnetic properties of the MLNPs allowed their manipulation by an external magnetic field without the need of centrifugation or filtration. Their optical characteristics (sharp emission, photostability, long lifetime) facilitated the implementation of an internal calibration in the detection system. This introduced a unique internal quality control and easy quantifications in the multiplexed immunoanalysis. This method developed and enables a direct, simple, and quantitative multiplex protein analysis using conventional organic dyes and can be applied for disease diagnostics and detection of biological threats.

Inorganic and hybrid coatings (or shells) on colloidal templates have been prepared by precipitation and surface reactions. By adequate selection of the experimental conditions, mainly the nature of the precursors, temperature, and pH, this method can give uniform, smooth coatings, and therefore lead to monodispersed spherical composites. Using this technique, submicrometer-sized anionic polystyrene lattices have been coated with uniform layers of iron compounds by aging at an elevated temperature and by dispersions of the polymer colloid in the presence of aqueous solutions of ferric chloride, urea, hydrochloric acid, and polyvinyl pyrrolidone. One of the most promising techniques for the production of superparamagnetic composites is the layer-by-layer self-assembly method. This method was firstly developed for the construction of ultrathin films and was further developed by Caruso et al. for the controlled synthesis of novel nanocomposite core-shell materials and hollow capsules. It consists of the stepwise adsorption of charged polymers or nanocolloids and oppositely charged polyelectrolytes onto flat surfaces or colloidal templates, exploiting primarily electrostatic interactions for layer buildup. Using this strategy, colloidal particles have been coated with alternating layers of polyelectrolytes, nanoparticles, and proteins. Furthermore, Caruso et al. have demonstrated that submicrometer-sized hollow silica spheres or polymer capsules can be obtained after removal of the template from the solid-core multilayered-shell particles either by calcination or by chemical extraction. Their work in the preparation of iron oxide superparamagnetic and monodisperse, dense, and hollow spherical particles that could be used for biomedical applications deserves special mention.

Encapsulation of magnetic nanoparticles in polymeric matrixes. Encapsulation of inorganic particles into organic polymers endows the particles with important properties that bare uncoated particles lack. Polymer coatings on particles enhance compatibility with organic ingredients, reduce susceptibility to leaching, and protect particle surfaces from oxidation. Consequently, encapsulation improves dispersibility, chemical stability, and reduces toxicity. Polymer-coated magnetite nanoparticles have been synthesized by seed precipitation polymerization of methacrylic acid and hydroxyethyl methacrylate in the presence of the magnetite nanoparticles. Cross-linking of polymers has also been reported as an adequate method for the encapsulation of magnetic nanoparticles. To prepare the composites by this method, first, mechanical energy needs to be supplied to create a dispersion of magnetite in the presence of aqueous albumin, chitosan, or PVA polymers. More energy creates an emulsion of the magnetic particle sol in cottonseed, mineral, or vegetable oil. Depending upon composition and reaction conditions, the addition of a cross-linker and heat results in a polydispersed magnetic latex, 0.3 microns in diameter, with up to 24 wt.% in magnetite content. Recently, the preparation of superparamagnetic latex via inverse emulsion polymerization has been reported. A 'double-hydrophilic' diblock copolymer, present during the precipitation of magnetic iron oxide, directs nucleation, controls growth, and sterically stabilizes the resulting 5-nm superparamagnetic iron oxide. After drying, the coated particles repeptize creating a ferrofluid-like dispersion. Inverse emulsification of the ferrofluid into decane, aided by small amounts of diblock copolymer emulsifier along with ultrasonication, creates mini droplets (180 nm) filled with magnetic particles and monomer. Subsequent polymerization generates magnetic latex. A novel approach to prepare superparamagnetic polymeric nanoparticles by synthesis of the magnetite core and polymeric shell in a single inverse microemulsion was reported by Chu et al.. Stable magnetic nanoparticle dispersions with narrow size distribution were thus produced. The microemulsion seed copolymerization of methacrylic acid, hydroxyethyl methacrylate, and cross-linker resulted in a stable hydrophilic polymeric shell around the nanoparticles. Changing the monomer concentration and water/surfactant ratio controls the particle size.

Encapsulation of magnetic nanoparticles in inorganic matrixes. An appropriate tuning of the magnetic properties is essential for the potential use of the superparamagnetic composites. In this way, the use of inorganic matrixes, in particular of silica, as dispersion media of superparamagnetic nanocrystals has been reported to be an effective way to modulate the magnetic properties by a simple heating process. Another advantage of having a surface enriched in silica is the presence of surface silanol groups that can easily react with alcohols and silane coupling agents to produce dispersions that are not only stable in nonaqueous solvents, but also provide the ideal anchorage for covalent bonding of specific ligands. The strong binding makes desorption of these ligands a difficult task. In addition, the silica surface confers a high stability to suspensions of the particles at high volume fractions, changes in pH, or electrolyte concentration. Recently, we have been successful in preparing submicronic silica-coated maghemite hollow and dense spheres with a high loading of magnetic material by aerosol pyrolysis. Silica-coated γ-Fe$_2$O$_3$ hollow spherical particles with an average size of 150 nm were prepared by aerosol pyrolysis of methanol solutions containing iron ammonium citrate and tetraethoxysilane (TEOS) at a total salt concentration of 0.25 M. During the first stage, the rapid evaporation of the methanol solvent favors the surface precipitation (i.e., formation of hollow spheres) of the components. The low solubility of the iron ammonium citrate in methanol when compared with that of TEOS promotes the initial precipitation of the iron salt solid shell. During the second stage, the probable continuous shrinkage of this iron salt solid shell facilitates the enrichment at the surface of the silicon oxide precursor (TEOS). In the third stage, the thermal decomposition of precursors produces the silica-coated γ-Fe_2O_3 hollow spheres. The formation of the γ-Fe_2O_3 is associated with the presence of carbonaceous species coming from the decomposition of the methanol solvent and from the iron ammonium citrate and TEOS. On the other hand, the aerosol pyrolysis of iron nitrate and TEOS at a total salt concentration of 1 M produced silica-coated γ-Fe_2O_3 dense spherical particles with an average size of 250 nm. The increase in salt concentration to a value of 1 M favors the formation of dense spherical particles. Sedimentation studies of these particles have shown that they are particularly useful for separation applications. A W/O microemulsion method has also been used for the preparation of silica-coated iron oxide nanoparticles. Three different non-ionic surfactants have been used for the preparation of microemulsions, and their effects on the particle size, crystallinity, and the magnetic properties have been studied.

The iron oxide nanoparticles are formed by the coprecipitation reaction of ferrous and ferric salts with inorganic bases. A strong base, NaOH, and a comparatively mild base, NH_4OH, have been used with each surfactant to observe whether the basicity influences the crystallization process during particle formation. All these systems show magnetic behavior close to that of superparamagnetic materials. By using this method, magnetic nanoparticles as small as 1 to 2 nm and of very uniform size (standard deviation less than 10%) have been synthesized. A uniform silica coating as thin as 1 nm encapsulating the bare nanoparticles is formed by the base-catalyzed hydrolysis and the polymerization reaction of TEOS in the microemulsion. It is worth mentioning that the small particle size of the composite renders these particles a potential candidate for their use in vivo applications.

Size Selection Methods

Biomedical applications like magnetic resonance imaging, magnetic cell separation, or magnetorelaxometry control the magnetic properties of the nanoparticles in magnetic fluids. Furthermore, these applications also depend on the hydrodynamic size. Therefore, in many cases, only a small portion of particles contributes to the desired effect. The relative amount of the particles with the desired properties can be increased by the fractionation of magnetic fluids. Common methods currently used for the fractionation of magnetic fluids are centrifugation and size-exclusion chromatography. All these methods separate the particles via nonmagnetic properties like density or size. Massart et al have proposed a size sorting procedure based on the thermodynamic properties of aqueous dispersions of nanoparticles. The positive charge of the maghemite surface allows its dispersion in aqueous acidic solutions and the production of dispersions stabilized through electrostatic repulsions. By increasing the acid concentration (in the range 0.1 to 0.5 mol l-1), interparticle repulsions are screened, and phase transitions are induced. Using this principle, these researchers describe a two-step size sorting process in order to obtain significant amounts of nanometric monosized particles with diameters between typically 6 and 13 nm. As the surface of the latter is not modified by the size sorting process, usual procedures are used to disperse them in several aqueous or oil-based media. Preference should be given, however, to partitions based on the properties of interest, in this case, the magnetic properties. So far, magnetic methods have been used only for the separation of magnetic fluids, for example, to remove aggregates by magnetic filtration. Recently, the fractionation of magnetic nanoparticles by flow field-flow fractionation was reported. Field-flow fractionation is a family of analytical separation techniques, in which the separation is carried out in a flow with a parabolic profile running through a thin channel. An external field is applied at a right angle to force the particles toward the so-called accumulation wall.

Superparamagnetic Iron Oxide Nanoparticles

SPIONs are small synthetic γ-Fe_2O_3 (maghemite) or Fe_3O_4 (magnetite) particles with a core ranging between 10 nm and 100 nm in diameter. These magnetic particles are coated with certain biocompatible polymers, such as dextran or polyethylene glycol, which provide chemical handles for the conjugation of therapeutic agents and also improve their blood distribution profile. The current research on SPIONs is opening up wide horizons for their use as diagnostic agents in magnetic resonance imaging as well as for drug delivery vehicles. Delivery of anticancer drugs by coupling with functionalized SPIONs to their targeted site is one of the most pursued areas of research in the development of cancer treatment strategies. SPIONs have also demonstrated their efficiency as nonviral gene vectors that facilitate the introduction of plasmids into the nucleus at rates multifold those of routinely available standard technologies. SPION-induced hyperthermia has also been utilized for localized killing of cancerous cells. Despite their potential biomedical application, alteration in gene expression profiles, disturbance in iron homeostasis, oxidative stress, and altered cellular responses are some SPION-related toxicological aspects which require due consideration.

All great things come in small packages, and products of nanoscience are no exception. Nanoparticles are simply particles in the nanosize range (10^{-9} m), usually, 100 nm in size. Due to their small size and surface area characteristics, they exhibit unique electronic, optical, and magnetic properties that can be exploited for drug delivery. Also known as nanovectors in the field of drug delivery, they are promising new tools for controlled release of drugs because they can satisfy the two most important criteria for successful therapy, ie, spatial placement and temporal delivery.

No drug is free from side effects, and these side effects usually arise from nonspecificity in drug action. For instance, in the case of tumor therapy, it is the side effects of cytotoxic drugs, such as bone marrow depression and reduced immunity, which can be hazardous to the extent that termination of therapy may be required. Modification of the surface characteristics of nanoparticles, such as superparamagnetic iron oxide nanoparticles (SPIONs) with biocompatible polymers, and controlling their size within the desirable range can yield powerful targeted delivery vehicles which can deal with this issue.

Freeman et al were the first to introduce the concept of use of magnetism in medicine in the 1970s. Since then, much research has been done in this area, leading to the design of various magnetic particles and vectors. The main objective is optimization of the properties of these magnetic particles to: provide an increase in magnetic nanoparticle concentration in blood vessels; reduce early clearance from the body; minimize non-specific cell interactions, thus minimizing side effects; and increase their internalization efficiency within target cells, thus reducing the total dose required.

SPIONs are small synthetic γ-Fe_2O_3 (maghemite), Fe_3O_4 (magnetite) or α-Fe_2O_3 (hermatite) particles with a core ranging from 10 nm to 100 nm in diameter. In addition, mixed oxides of iron with transition metal ions such as copper, cobalt, nickel, and manganese, are known to exhibit superparamagnetic properties and also fall into the category of SPIONs. However, magnetite and maghemite nanoparticles are the most widely used SPIONs in various biomedical applications. SPIONs have an organic or inorganic coating, on or within which a drug is loaded, and they are then guided by an external magnet to their target tissue. These particles exhibit the phenomenon of "superparamagnetism", ie, on application of an external magnetic field, they become magnetized up to their saturation magnetization, and on removal of the magnetic field, they no longer exhibit any residual magnetic interaction. This property is size-dependent and generally arises when the size of nanoparticles is as low as 10–20 nm. At such a small size, these nanoparticles do not exhibit multiple domains as found in large magnets; on the other hand, they become a single magnetic domain and act as a "single super spin" that exhibits high magnetic susceptibility. Thus, on application of a magnetic field, these nanoparticles provide a stronger and more rapid magnetic response compared with bulk magnets with negligible remanence (residual magnetization) and coercivity (the field required to bring the magnetism to zero). This superparamagnetism, unique to nanoparticles, is very important for their use as drug delivery vehicles because these nanoparticles can literally drag drug molecules to their target site in the body under the influence of an applied magnet field. Moreover, once the applied magnetic field is removed, the magnetic particles retain no residual magnetism at room temperature and hence are unlikely to agglomerate (i.e., they are easily dispersed), thus evading uptake by phagocytes and increasing their half-life in the circulation. Moreover, due to a negligible tendency to agglomerate, SPIONs pose no danger of thrombosis or blockage of blood capillaries.

The current research on SPIONs is opening up broad horizons for their use in the biomedical sciences. They have been used for both diagnostic as well as therapeutic purposes. In magnetic resonance imaging (MRI), SPIONs have been used as targeted magnetic resonance contrast agents, allowing diagnosis of progressive diseases in their early stages. From a drug delivery point of view, targeting of cancer is the most pursued area, with emphasis on delivery of chemotherapeutics and radiotherapeutics. However, increasing applications of SPIONs have also been found in the areas of gene delivery, cell death with the help of local hyperthermia, and delivery of peptides and antibodies to their site of action.

Toxicity is an important issue which must be dealt with before SPIONs can be considered for widespread use in drug delivery. Much research has been carried out to evaluate the biocompatibility of these magnetic nanoparticles and their possible adverse interactions with cellular and subcellular structures.

Physicochemical Characteristics Essential for Drug Delivery

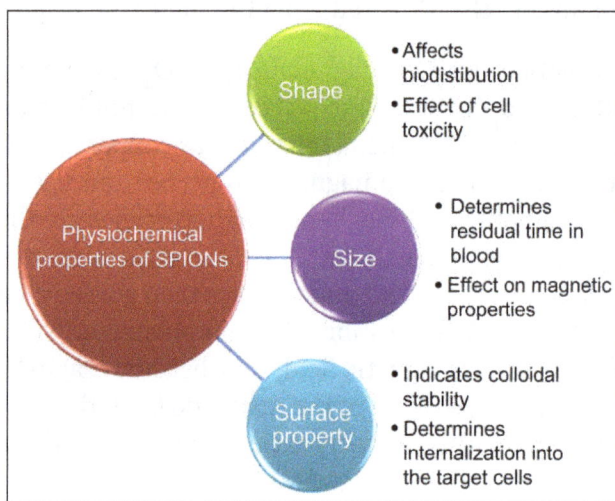

Figure shows the most important physicochemical characteristics of SPIONs, which should be taken into account while designing a successful drug delivery system. Such properties mostly govern the blood distribution profile of these nanoparticles.

Shape

The morphology of Fe_2O_3 nanoparticles has been known to be affected by several factors, including the reaction conditions and chemicals involved. In the presence of surfactants with bulky hydrocarbon chain structures, like oleylamine and adamantane amine, the steric hindrance exerted by surfactants has been shown to affect the shape of growing crystals of iron oxide during synthesis. The shape of magnetic nanoparticles has not been extensively studied as far as its effect on biodistribution of SPIONs is concerned. However, a few researchers studying other nanoparticulate delivery systems have reported that rod-shaped and nonspherical nanoparticles show a longer blood circulation time compared with spherical particles. For instance, Huang et al studied the effect of particle shape on the in vivo behavior of mesoporous silica nanoparticles. They found that the shape of the nanoparticles could affect their biodistribution, clearance, and biocompatibility in vivo. Short-rod mesoporous silica nanoparticles which are more or less spherical were found to accumulate in the liver, whereas long-rod-shaped particles were distributed to the spleen. Moreover, short-rod mesoporous silica nanoparticles showed rapid clearance rates via urine and feces compared with long-rod mesoporous silica nanoparticles. Another reason which may favor a longer blood circulation time for rod-shaped nanoparticles may be the fact that the phagocytic activity of macrophages is stimulated to a lesser extent by rod-shaped particles than by spherical ones. However, spherical magnetite and maghemite particles offer a uniform surface area for coating and conjugation of targeting ligands or therapeutic agents. Mahmoudi et al showed that the shape of SPIONs exerts a direct effect on cell toxicity. Nanobead-shaped, nanoworm-shaped, and nanosphere-shaped SPIONs showed greater cellular toxicity compared with nanorods and colloidal nanocrystal clusters.

Size

The size of nanoparticles largely determines their half-life in the circulation. For instance, particles with sizes smaller than 10 nm are mainly removed by renal clearance, whereas particles larger than 200 nm become concentrated in the spleen or are taken up by phagocytic cells of the body, in both instances leading to decreased plasma concentrations.

However, particles with a size range of 10–100 nm are considered to be optimum, with longer circulation times because they can easily escape the reticuloendothelial system in the body. They are also able to penetrate through very small capillaries. Furthermore, biomedical applications of SPIONs, including MRI, hyperthermia, and magnetic cell separation, depend on the magnetic properties of these particles, which in turn are largely dependent upon size. The small size of SPIONs is also responsible for the enhanced permeability and retention effect, which causes concentration of the particles in target tumor tissue. However, SPIONs with a particle size smaller than 2 nm are not suitable for medical use. This is due to the increased potential of particles in this size range to diffuse through cell membranes, damaging intracellular organelles and thus exhibiting potentially toxic effects. Therefore, control of particle size during preparation of SPIONs is an important concern.

The techniques most commonly employed for measuring the particle size of SPIONs are transmission electron microscopy, dynamic light scattering, and the Scherrer method using x-ray diffractograms. Because only a small number of the particles prepared are of the desired size, it becomes necessary to carry out fractionation of magnetic fluids. Currently applied techniques are centrifugation, size exclusion chromatography, and field flow fractionation. While the first two techniques separate particles on the basis of their density or size, the latter technique utilizes the magnetic properties of SPIONs for their efficient fractionation.

Surface Properties

The surface charge of nanoparticles gives an indication of their colloidal stability.

Nanoparticles having high positive and negative zeta potential show dispersion stability and as a result do not agglomerate on storage. Charge also determines the distribution of these particles in the body and is an important parameter affecting internalization of nanoparticles in their target cells. In one study, it has been reported that uncoated and pullulan-coated SPIONs are internalized into cells by different mechanisms, demonstrating surface-dependent particle endocytosis behavior.

SPIONs having a positive charge are better internalized by human breast cancer cells than are negatively charged particles. However, intake of these nanoparticles also depends upon cell type. Particles with a hydrophobic surface are easily adsorbed at the protein surface (opsonization) and are engulfed by circulating macrophages, resulting in their clearance from plasma. Therefore, they show a low circulation time. However, particles that are surface-engineered with hydrophilic polymers like polyethylene glycol (stealth particles) containing, eg, hydroxyl or amino functional groups, are able to evade engulfment by the reticuloendothelial cells or circulating macrophages, thus having better therapeutic efficacy due to increased residence time in the blood. Surface-engineering of magnetic nanoparticles with different functional groups imparts different surface characteristics, making them suitable for a wide variety of biomedical and other industrial applications.

As the size of iron oxide particles reduces into the nanorange, iron ions on the surface of nanoparticles play an important role in determining the magnetic properties of these particles. The oxidation state of the iron ion on the surface of the nanoparticle is sensitive to its surrounding environment, particularly to surfactant exposure. Analysis of the oxidation state of iron ions on the surface of nanoparticles can help in identifying their structure and chemical environment. It has been reported that the oxidation state of the iron ion can have a potential effect on the morphology of nanoparticles prepared. For instance, iron ions in the trivalent state (+3) favor formation of spherical nanoparticles, whereas metal ions in the divalent state (+2) favor formation of nanorods. The primary objective of research at present is the preparation of uniform and stable SPIONs by controlled synthesis and coating processes. The synthetic route selected not only determines the physical features of SPIONs, but also has a profound effect on their crystallochemical characteristics.

Core Fabrication

Nucleation and crystal growth are the two fundamental steps in preparation of crystals from solution. Taking into consideration the classical model of crystallization proposed by LaMer and Dinegar, monodispersed nanoparticles can be produced by a single short burst of nucleation, which occurs when a solution reaches its critical supersaturation concentration. Nuclei thus obtained then grow as a result of diffusion of solute particles from the solution onto the surface of the nuclei, until a suitable size is reached. For achieving monodispersity, care must be taken to ensure that nucleation does not occur during the crystal growth phase. Multiple nucleations can also result in formation of uniformly dispersed nanoparticles. This occurs as a result of Ostwald ripening, a self-reforming process whereby small nuclei crystals formed get redissolved and deposit onto larger nuclei, forming large uniform crystals. Aggregation of smaller units may also result in uniform-sized nanoparticles.

The most commonly used methods for preparation of uniform iron-based nanoparticles in solution are coprecipitation and microemulsion. Preparation of SPIONs utilizing the coprecipitation method involves two approaches, ie, partial oxidation of ferrous hydroxide suspension by different oxidizing agents such as nitrates, as explored by Sugimoto et al, and addition of base to an aqueous solution containing a mixture of ferrous (Fe^{2+}) and ferric (Fe^{3+}) ions with 1:2 stoichiometry in an oxygen-free environment. Massart et al prepared SPIONs utilizing the second approach and obtained a black precipitate of spherical magnetic nanoparticles in the size range of, 20 nm. On the other hand, particles obtained using the method reported by Sugimoto et al were larger, ranging from 30 nm to 200 nm. The size of magnetic nanoparticles prepared by the co-precipitation method largely depends upon the pH and ionic strength of the precipitating solution. It has been demonstrated that as the pH and ionic strength of the medium increases, the size of the particles decreases. These parameters not only affect the size of the nanoparticles formed, but also determine the electrostatic potential on the surface of these nanoparticles, which is indicative of their dispersion stability.

Smaller and more uniform particles can be synthesized using the microemulsion approach. Water-in-oil (w/o) microemulsions (ie, reverse micelle solutions) are transparent, isotropic, and thermodynamically stable liquids. In these systems, the aqueous phase is dispersed as microdroplets in the continuous oil phase, ie, entrapped within the micellar assembly of stabilizing surfactants. The main advantage of utilizing this approach is that these microdroplets serve as nanoreactors, providing a confined space which limits growth and agglomeration of nanoparticles during their synthesis, thus regulating their size and surface properties. Using this method, iron precursors like ferrous chloride and ferric chloride are precipitated as oxides in the aqueous core. Iron precursors in the organic phase remain unreactive. By controlling the size of the microdroplets, particles in the desired size range can be obtained. A novel nanocomposite consisting of nanometric cores of silver embedded in a matrix of γ-Fe_2O_3 was prepared by sequential reaction of different mixtures of reverse micelles.

An emerging method for preparation of uniform nanoparticles is the polyol technique, whereby fine metallic particles can be made by reduction of their dissolved metallic salts and direct metal precipitation from solution containing a polyol. Iron particles around 100 nm prepared by this process have been reported. High temperature decomposition of iron precursors in the presence of suitable surfactants results in synthesis of uniform magnetic nanoparticles with a desirable size range and surface properties. Sun et al reported synthesis of monodispersed magnetic nanoparticles sized 3 nm to 20 nm by thermal degradation of iron (III) acetylacetonate in phenyl ether in the presence of alcohol, oleic acid, and oleylamine at 265°C. Use of dendrimers as templating hosts for synthesis of magnetic nanoparticles has drawn considerable attention. This is because of the fact that by properly selecting the appropriate dendrimer host, biocompatible SPIONs suitable for in vivo application can be made via a single-step process. High-energy ultrasound waves can also be utilized for the synthesis of magnetite and maghemite nanoparticles. These high-energy sound waves create acoustic cavitations, ie, formation, growth, and implosive collapse of empty cavities, resulting in transient localized hot spots with a temperature of about 5000 K. Formation of these cavities sends out shock waves, leading to particle size reduction with concomitant formation of magnetite nanoparticles. However, large-scale synthesis of magnetic nanoparticles utilizing this approach is not very feasible. Magnetic nanoparticles have also been synthesized by electrochemical deposition of metal on a cathode, produced by reduction of metal ions dissolved from the anode.

Spray and laser pyrolysis are two further emerging approaches for preparation of uniform magnetic nanoparticles, with great commercial scale-up potential. In spray pyrolysis, a solution of Fe^{3+} salt and reducing agents is sprayed through a series of reactors where aerosol droplets undergo evaporation of solvents with solute condensation within the droplets. This is followed by drying and thermolysis of the precipitated product at high temperature, resulting in microporous solids finally sintering to form dense particles. Laser pyrolysis involves heating a flowing mixture of gases with a continuous wave carbon dioxide laser, initiating and sustaining a chemical reaction. Homogenous nucleation of particles results when a critical concentration of nuclei is reached within the reaction zone above a certain pressure and laser power. Usually, iron pentacarbonyl is used as a precursor for synthesis of γ-Fe_2O_3 nanoparticles by the laser pyrolysis method.

Magnetotactic bacteria, a group of Gram-negative prokaryotes, have demonstrated an ability to synthesize fine iron oxide nanoparticles in the size range of 50–100 nm. These bacterial magnetic nanoparticles are covered with phospholipid layers, making them biocompatible and hence useful for a variety of bioapplications. Researchers have also used a variety of natural protein components, such as ferritin, which serve as nanoshells consisting of a central core within which iron oxide nanoparticles 6–8 nm in size can be synthesized.

Coating

The next step after fabrication of SPION cores is their coating. Coating with suitable polymers endows some important characteristics to these nanoparticles that are essential for their use as drug delivery vehicles. Coating of SPIONs is essential because: it reduces the aggregation tendency of the uncoated particles, thus improving their dispersibility and colloidal stability; protects their surface from oxidation; provides a surface for conjugation of drug molecules and targeting ligands; increases the blood circulation time by avoiding clearance by the reticuloendothelial system; makes the particles biocompatible and minimizes nonspecific interactions, thus reducing toxicity; and increases their internalization efficiency by target cells. The presence of amino groups on the coating shell of amino-polyvinyl alcohol-functionalized SPIONs increases their uptake by human melanoma cells.

SPIONs can be coated either during their synthesis or can undergo adsorption after synthesis. Both methods have been reported to produce particles with a uniform coating. However, coating of SPIONs with nonmagnetic polymers like polysaccharides leads to a decrease in saturation magnetization as compared with uncoated SPIONs. Amstad et al have reported a decrease in saturation magnetization of mPEG(550)-gallol-stabilized SPIONs as compared with bare particles, from 58 emu g_{Fe}^{-1} to 50 emu g_{Fe}^{-1}, as measured by a superconducting quantum interference device. A large decrease in saturation magnetization is undesirable because it will result in failure to attain an effective concentration of SPIONs at the target site on application of the external magnetic field. The SPIONs should be coated in such a manner that the coating not only imparts favorable characteristics but also preserves the desirable properties of uncoated SPIONs.

Drug Loading

The primary requirement is that the drug should be loaded in such a manner that its functionality is not compromised. Moreover, these drug-loaded nanoparticles should also release the drug at the appropriate site and at a desired rate. Drug loading can be achieved either by conjugating the therapeutic molecules on the surface of SPIONs or by co-encapsulating drug molecules along with magnetic particles within the coating material envelope.

A number of approaches have been developed for conjugation of therapeutic agents or targeting ligands on the surface of these nanoparticles. They can be grouped under two categories, ie, conjugation by means of cleavable covalent linkages and by means of physical interactions. Covalent linkage strategies involve linkage of the therapeutic agent or targeting molecules directly with, eg, amino or hydroxyl functional groups present on the surface of polymer-coated SPIONs. Alternatively, linker groups such as iodoacetyls, maleimides, and the bifunctional linker, pyridyl disulfide, may be used to attach the drug to the particle. This approach not only leads to enhanced loading capacity, but also results in more specific linkages, protecting the drug's functionality and hence efficacy. Another advantage of using linkers is that this method involves milder reactive conditions for attachment, and hence is suitable for drugs like therapeutic peptides and proteins, which are prone to oxidative degradation.

Physical interactions such as electrostatic interactions, hydrophobic/hydrophilic interactions, and affinity interactions can also lead to coupling of drug molecules on the surfaces of SPIONs. SPIONs coated with polyethylenimine (PEI), a cationic polymer, interact electrostatically with negatively charged DNA, demonstrating their applicability as transfection agents. Similarly, dextran-coated SPIONs functionalized with negatively charged functional groups can couple with peptide oligomers via electrostatic interactions. Because of hydrophobic interactions, lipophilic drugs can easily be attached to SPIONs covered with hydrophobic polymers, from where the drug can be released when the coating degrades. Affinity interactions, such as streptavidinbiotin interactions, can also be utilized for bioconjugation of targeting agents or drugs with SPIONs. Unlike electrostatic and hydrophobic interactions, affinity interactions offer the most stable noncovalent linkages, which are relatively unaffected by environmental conditions, such as changes in pH and ionic strength of the medium.

However, drug delivery by conjugating the drugs onto the surface of SPIONs suffers from some disadvantages. Low entrapment efficiency of drug molecules is one of the major drawbacks of delivering the drug by coating the surface of SPIONs because only a limited amount of drug can be conjugated in this way. Other disadvantages include highly stable linkages as a result of covalent bonding between drug molecules and the surface of SPIONs, leading to failure to release the drug molecule at the target site. Sometimes, a catalyst such as copper used during covalent linking of a drug to the SPION surface may cause in vivo toxicity if it is not purified properly. It has also been shown that it may be difficult to control the orientation of binding ligands when attaching them to the surface of magnetic nanoparticles. This has been observed, particularly in cases where SPIONs decorated with carboxylic acid groups bind with ligands with multiple amine groups, often leading to inactivation of ligands.

Loading of drug molecules along with iron oxide nanoparticles within the coating material represents another approach of delivering a drug to the target site. This approach provides attractive solutions to problems such as low entrapment efficiency and stability. Magnetoliposomes represent a new class of nanocomposites which not only ensures high entrapment efficiency but also better stability and magnetic properties. Magnetoliposomes are nanosized, spherical vesicles consisting of magnetic nanoparticles in a shell composed of a phospholipid bilayer. These magnetoliposomes retain hydrophobic regions that can be used for drug encapsulation. Loading SPIONs with pharmaceuticals within the phospholipid envelope offers several advantages. Firstly, liposomes containing magnetic particles provide simple and easy surface modifications enabling their targeting to a specific tissue. Secondly, magnetoliposomes containing high amounts of SPIONs due to increased entrapment efficiency provide optimum magnetic responsiveness and their nanosize enables exploitation of the enhanced permeability and retention effect for tumor targeting. In addition, a larger amount of both hydrophilic as well as hydrophobic drug can be loaded within the liposomal structure along with SPIONs. Thirdly, encapsulation of SPIONs within liposomes further improves the biocompatibility of SPIONs. Fourthly, the liposomal barrier protects encapsulated pharmaceuticals from the degradative effects of the surrounding environment. SPIONs can be loaded within the liposomes in two ways, ie, by incubating previously prepared liposomes and SPIONs under the influence of an external force resulting in localization of SPIONs within the hydrophilic core of liposomes, or by directly precipitating the SPIONs within the hydrophilic core of liposomes, yielding highly uniform nanoparticles with sizes around 15 nm. However, during coating, agglomerates of SPIONs sometimes get coated instead of discrete SPION cores being coated within a phospholipid shell. This results in poor magnetic and physicochemical properties. Exposing magnetoliposomes to a strong permanent magnetic field over the target tissue leads to alignment of magnetic particles within the bilayer, causing magnetoliposomes to aggregate, fuse, and release the drug only at the target site.

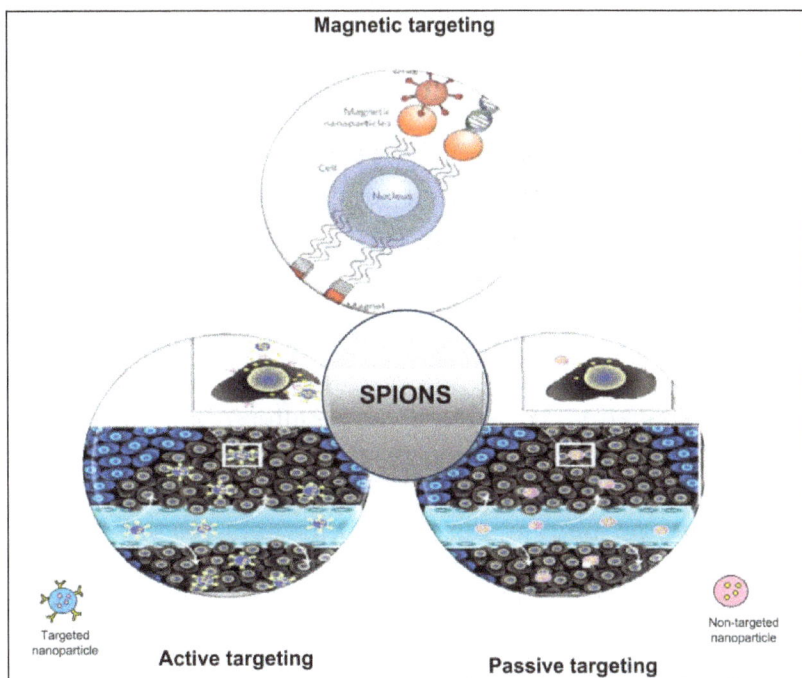

Superparamagnetic iron oxide nanoparticle targeting approaches.

Magnetodendrimers represent another class of nanocomposites which are well suited to imaging of cell trafficking and migration using MRI. Carboxylated polyamidoamine dendrimers are commonly used to coat and stabilize iron oxide nanoparticle suspensions. In general, at an elevated temperature and pH in the presence of dendrimers, oxidation of Fe(II) yields highly stable and soluble SPIONs with dendrimers. Lamanna et al developed dendronized iron oxide nanoparticles for multimodal imaging. SPIONs coated with dendrimers having a hydrodynamic size less than 100 nm and displaying either carboxylate or ammonium groups at the periphery provide a unique opportunity for labeling with a fluorescent dye. Magnetic resonance and fluorescence imaging have been demonstrated to be simultaneously possible.

SPION Targeting Approaches

SPIONs can be properly engineered to reach their target tissue with minimum non-specific cellular interactions. Targeting strategies can be grouped into three classes as shown in figure. Passive targeting takes advantage of the innate size of the nanoparticles as well as the unique characteristics provided by the tumor microenvironment. Tumor cells exhibit leaky vasculature due to incomplete angiogenesis. Nanoparticles enter tumor tissue through these pores, and due to the poor lymphatic drainage system of cancerous tissues are retained for longer periods of time than normal cells. This is also known as the enhanced permeability and retention effect, discovered by Matsumura and Maeda. Thus, particle size in the nanorange itself provides targeted delivery, without any modification. The enhanced permeability and retention effect is considered to be the "gold standard" for developing new anticancer agents.

Active targeting involves targeting ligands which are coupled at the surface of magnetic nanoparticles to interact with receptors that are overexpressed at their target sites. Such particles accumulate in larger quantities in target cells due to "homing" of these ligands onto the receptors and subsequent ease of internalization. Various active targeting agents engineered and attached to the SPION surface not only ensure specific target binding but also minimize the dose required and nonspecific cellular interactions.

Magnetic focusing uses external magnets to create a suitable magnetic field gradient over the targeted area and ensure significant accumulation of drug-loaded SPIONs. The strength of the applied magnetic field can be altered to modulate release of the drug in the desired fashion, resulting in maximum therapeutic benefits. A permanent NdFeB magnet is generally used for magnetic targeting of SPIONs. Mondelak et al increased the permeability of dextran-encapsulated SPIONs over an artificial three-layered membrane on application of an external magnetic field of 0.410 Tesla. Kumar et al synthesized magnetic nanoparticles coupled with plasmid DNA-expressing enhanced green fluorescent protein and coated with chitosan. The particles were successfully directed to the heart and kidney via an external magnetic field when injected into the tail vein of mice. The results demonstrated that application of an external magnetic field is sufficient to target a drug to its specific site of action, excluding the need to develop functionalized nanoparticles. In another study, Lamkowsky et al demonstrated time-dependent, temperaturedependent, and concentration-dependent accumulation of dimercaptosuccinate-coated iron nanoparticles in cultured brain astrocytes. The accumulation of iron oxide nanoparticles was proportionally increased with an increase in the applied magnetic field strength, increasing the cell-specific iron content from an initial 10 nmol/mg of protein within 4 hours of incubation at 37°C to up to 12,000 nmol/mg of protein. The results suggested that application of an external magnetic field enhanced both binding of iron nanoparticles to cell membranes as well as their internalization in cultured astrocytes, demonstrating the utility of SPIONs as potential diagnostic and drug delivery vehicles for imaging and treatment of neurodegenerative diseases. Prijic et al studied the potential increase in cellular internalization of SPIONs by exposure of different cells to an external magnetic field generated by different permanent magnets. They found that exposure to neodymium-iron-boron magnets significantly increased the cellular uptake of SPIONs, predominantly into malignant cells. Accumulation of SPIONs within malignant cells was found to be dependent on duration of exposure to the external magnetic field.

Yolk-shell Nanoparticles

The yolk-shell nanoparticles (YS NPs) are defined as a hybrid structure (mixture of core/shell and hollow) where a core particle is encapsulated inside the hollow shell and may move freely inside the shell, generally represented as a core/void/shell. Among different sub-areas of core/shell nanoparticles, yolk-shell nanoparticles (YS NPs) have drawn significant attention in recent years because of their unique properties such as low density, large surface area, ease of interior core functionalization, a good molecular loading capacity in the void space, tunable interstitial void space, and a hollow outer shell. The YS NPs have better properties over simple core/shell or hollow NPs in various fields including biomedical, catalysis, sensors, lithium batteries, adsorbents, DSSCs, microwave absorbers etc., mainly because of the presence of free void space, porous hollow shell, and free core surface.

Types of YS NPs

The YS NPs can be broadly classified into 'spherical' and 'nonspherical' structures depending on their core and shell morphologies, without considering their material properties. In the first case, both the core and the shell are spherical in shape, and in the latter case at least one structure should be non-spherical. Further, the YS structures under the head of spherical shape can be classified into five distinct sub-categories such as: (i) single core/shell, (ii) multi-cores/single shell, (iii) single core/ multi-shell, (iv) multi-cores/shells, and (v) multi-shells or shell in shell. Different types of reported spherical YS NPs are shown in figure. Non-spherical YS NPs are also divided into two categories based on their geometry: (i) complete non-spherical YS NPs, where both the core and the shell have a non-spherical shape, (ii) partially non-spherical YS NPs, where either the core or the shell of the YS has a non-spherical shape.

Approaches for YS NPs Synthesis

Synthesis approaches play a crucial role in the core size and shape, void space, outer shell thickness, shell porosity and shape to fabricate the YS NPs. Generally, all types of YS NPs are synthesized by the bottom-up approach because of the complexity of the structure. However, based on the synthesis sequences, the approaches are of two types, such as (i) core-toshell, where core/shell/shell type NPs are synthesized first and then a sacrificial shell layer is removed by a suitable method to obtain the final structure. In this approach, selective etching or calcination of the sacrificial template, galvanic replacement, Kirkendall effect, and Ostwald ripening are used. (ii) Shell-tocore, such as the ship-in-bottle method. In the second approach, hollow shells are synthesized first, and then core NPs are encapsulated inside the hollow shell. The first approach is widely used by several researchers while the second approach is fewer in number. As per the present status of this area, significant research efforts have already been made to develop maximum possible structures of YS NPs. However, still there are huge possibilities ahead to control the size and shape of movable cores and shells with appropriate synthesis methods, suitable functionalization of core and shell surfaces for desired applications, and selection of appropriate materials to obtain improved desired properties.

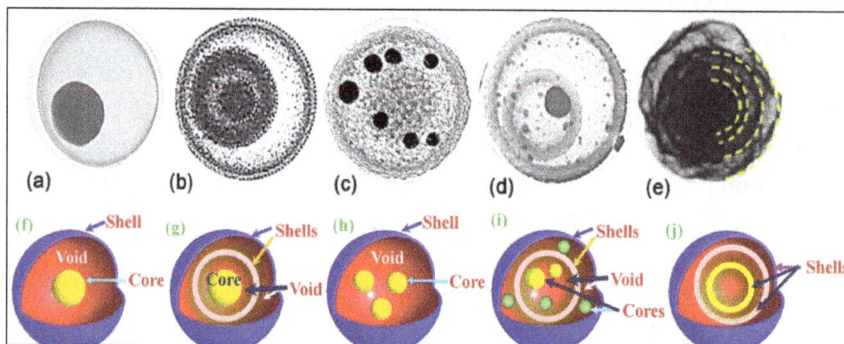

Different classes of spherical YS NPs (a) Single core/shell, (b) Single core multi shells.
(c) Multiple cores single shell. (d) Multi-cores/shells. (e) Multi shells.

Importance of YS Structure

The YS NPs are under a specific class of CS NPs, where the presence of void space makes them so special in several nanotechnology applications. The unique properties arise mainly from the movable core, hollow shell, and void space of the YS NPs as mentioned before. In general, the final YS structure shows individual or synergistic properties of core and shell materials. Additionally, the properties of YS NPs can also be modified via selection of desired core and shell materials, and tuning of other parameters such as core size, void space, shell thickness and porosity. The importance of YS NPs for selective applications is also summarized in some recent reviews. The YS NPs are important in various fields including biomedical, catalysis, sensors, lithium batteries, adsorbents, dye-sensitized solar cells (DSSCs), microwave absorber etc. In catalytic applications, the presence of a mesoporous shell serves as a protective shield to protect the core catalyst NPs from the harsh environmental conditions (chemical and thermal) and also prevents the core from agglomeration. Similarly, as an electrocatalyst, the metallic shell on an electrocatalyst core (bimetallic YS NPs) exhibits high activity, high CO tolerance, and reduced loading of Pt catalysts as compared to conventional Pt catalysts with improved stability, electrical conductivity, reactivity, and optical and electronic properties. In biomedical applications, the large void space between the core and hollow shell is particularly suitable for encapsulation of guest species such as fluorescent and drug molecules with a specific target functionality. The void space also provides a buffering space for electroactive core material during the charging of lithium batteries.

Synthesis Methods for YS NPs

A majority of YS nanostructures were synthesized via different core/shell structures as templates. However, the extra and important feature to design the YS nanostructure is the creation of void space between the core and shell, which is an essential step to fabricate the YS nanostructures. In this topic, we present a comprehensive analysis of synthesis methods of the YS nanostructures to create void space between the core and shell. Generally, synthesis approaches of the YS nanostructures are classified into two types: (i) core-to-shell and (ii) shell-to-core. In the former approach, first the core and then successive shell layers are synthesized; among the shell layers one or more layers may present as a sacrificial shell. In the latter approach, the hollow shells are synthesized first and then core nanoparticles are encapsulated inside the hollow shell. For a better understanding, we can divide the synthesis approach into 4 parts on the basis of the method.

Sacrificial Template-assisted Synthesis

From the introduction of the YS nanostructures, the sacrificial template based approaches have been receiving a lot of attention because of the simplicity of the process, ease to retain the core shape, control over the void space, and so on. The generation of void space by the removal of sacrificial layers can be classified into three types: (i) complete removal of the sacrificial layer, in which the sacrificial layer is made of different materials than that of the core and the shell, (ii) partial dissolution of the core layer, where the core surface is partially etched after the formation of the core/shell structure, and (iii) partial dissolution of the inner shell layer, where the inner shell layer is partially etched after the formation of the core/shell structure. Thus, one can easily understand that the template removal or dissolution step is very important for the formation of YS nanostructures. Now based on the physical states of the materials, the sacrificial layers can be classified into (a) hard, where solid and rigid particles are used as a template, (b) soft types, the self-assembly of surfactant and organic molecules provides the template, (c) galvanic replacement, the redox reaction between two metals helps to create void space because of the difference in electrode potential, (d) the Kirkendall reaction, the void space is created because of a nonreciprocal mutual diffusion process through the interface of two materials.

Hard Templates

Generally, the hard templates refer to solid rigid material layers, which can be easily removed through calcination, dissolution by solvents, and etching by acid or alkali solutions depending on the material properties. When removing the hard template by dissolution or calcination methods, special care is required to retain the external shell layer without collapsing. In some cases such as partial removal of core or shell layers and for the synthesis of multiple hollow shells, special care in the removal technique is very important. So far, different materials as hard templates have been reported, which include silica, carbon, polymers, Fe_3O_4, gold, Ni, V_2O_5, MnO, Cu_2O, sulphur, Ag, AgBr, β-FeOOH etc., to synthesise YS nanostructures.

Silica template. Silica is the most widely used hard template to design YS nanostructures because of its easy synthesis and removal technique. The SiO_2 layer can be selectively or partially etched in the presence of hydrofluoric acid or alkaline solutions.

The dissolution of SiO2 in HF is a well-known process in the glass and integrated circuit industries. The SiO2 hydrolyze in the presence of HF and form silanol on the surface and then form a water soluble H2SiF6 compound because of the high electronegativity of fluorine. The dissolution of silica by aqueous HF can be represented by the chemical reaction described in equation $SiO_{2(s)} + 6HF_{(1)} \rightarrow H_2SiF_{6(aq.)} + 2H_2O$. On the other hand, silica also dissolves in the presence of alkaline solutions such as soluble hydroxides and carbonates, and forms water soluble alkali silicates as shown in the chemical reaction in equation $SiO_{2(s)} + 2NaOH \rightarrow Na_2SiO_{3(aq.)} + H_2O$. Comparing two dissolution methods, the HF mediated process is faster than that of alkali; however, handling of HF required several precautions and safety norms.

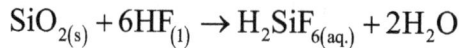

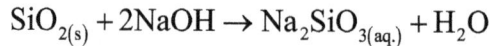

$$SiO_{2(s)} + 6HF_{(1)} \rightarrow H_2SiF_{6(aq.)} + 2H_2O$$

$$SiO_{2(s)} + 2NaOH \rightarrow Na_2SiO_{3(aq.)} + H_2O$$

The most commonly used sol–gel and Stöber methods are generally used for the synthesis of SiO_2 layers. Using the Stöber method, two types of silica layers are synthesized: (i) pure inorganic silica and (ii) organic–inorganic hybrid silica. Generally, a pure silica layer is synthesized using a silica precursor (TEOS) by the Stöber method. On the other hand, organic–inorganic hybrid silica (organosilica) can be synthesized from the combined use of a silica precursor (TEOS) and organosilanes (C_{18}TMS, APTES, APTMS, 3-MPTS, HDTES, EPTMS etc.) by the Stöber method. Thus, both types of silica have specific importance as a sacrificial template for different purposes. The organosilica layer is used when either the core or shell is formed with pure silica. Thus, during selective etching, organisilica will react faster with HF or alkaline solution than pure silica due to less dense and cross-linked structures and gives void space as shown in figure. In addition, the organosilica can also be calcined at high temperature to obtain the YS structure by the removal of the organic group. When the pure form of silica is used as a sacrificial template, it can be partially or completely removed by HF and alkaline solution.

Similarly, the surface protecting etching of the silica layer is also used for the synthesis of porous YS NPs, where a SiO_2 layer is pre-adsorbed with a protecting layer of PVP or other ligands as shown in figure. Here, PVP is used as a surface protecting agent of silica because its abundant carbonyl groups can interact with the hydroxyl groups on the SiO_2 surface via strong hydrogen bonding. Thus, the interior part of the SiO_2 layer only etches and the surface of shell is protected during the dissolution, which creates porosity on the silica shell.

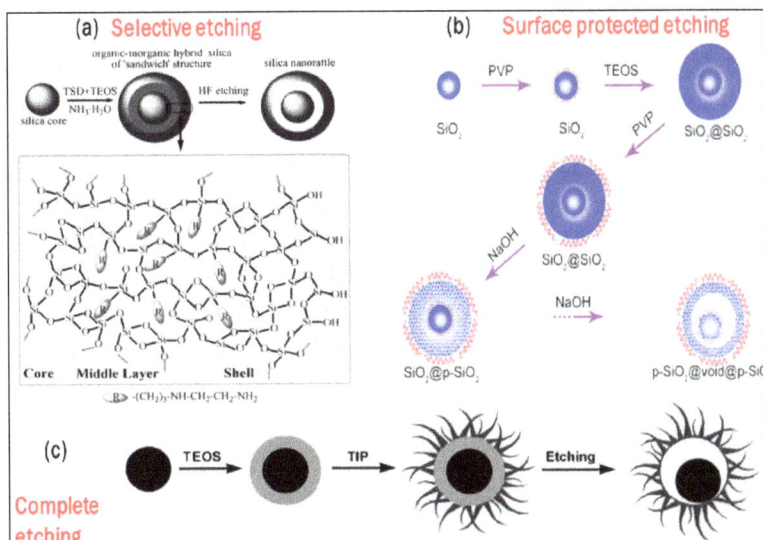

(a) Selective etching of an organosilicate layer. (b) Surface protected etching of silica using PVP. (c) complete etching of a silica layer.

A partial or complete removal technique of the silica layer depends on the nature of core and shell materials as shown in figure. The silica layer is removed completely to obtain the hollow space for the synthesis of nonsilica-based YS NPs such as Au/polymer, γ-Fe_2O_3/Y_2O_3, Au/ZrO_2, Fe_3O_4/TiO_2, Fe_3O_4/YPO_4, Fe_3O_4/SnO_2, Au/carbon, Si/carbon, C/C, Pd/C, Au/TiO_2, Pd/CeO_2 etc. On the other hand, for the synthesis of silica-based YS NPs, partial etching of silica is a useful method. Many researchers used partial etching of silica for the synthesis of Au/SiO_2, Ni/SiO_2, Pd/SiO_2, SiO_2/SiO_2, Fe_3O_4/SiO_2, SiO_2/TiO_2 YS NPs. The silica template is also useful to make silicate shells in YS nanostructures. During the silica etching, the noble metal and metal-oxide partially react with the silica to form products of silicates.

Carbon template: The sacrificial hard shell layer of carbon is used as another important material for synthesizing the YS structure. Generally, the carbon template layer can be completely removed by calcination at high temperature in the presence of air. Several YS nanostructures such as Fe_3O_4/SiO_2, Fe_2O_3/SiO_2, Au/SiO_2, SnO_2/C, Fe_3O_4/C, TiO_2/TiO_2 and SnO_2/TiO_2 have been synthesized by using the carbon layer as a sacrificial template and removed completely by calcination. In some approaches, presynthesized hollow carbon spheres are used as a sacrificial template, where core precursors diffuse inside to form the core particles and then the shell material is coated on the carbon shell. Finally, the carbon layer is removed by heat treatment to obtain the YS nanostructure. Utilization of pre-synthesized carbon spheres has the advantage of a tuneable size which can be controlled by tuning the reaction time, temperature, and the concentration of the carbon precursor. On the other hand, the carbon layer can also be formed by carbonization of organic compounds under hydrothermal conditions. In general, the carbon layer is synthesized easily by the hydrothermal treatment of the carbonaceous water soluble precursors such as glucose and urea, where the transformation of the carbon source to elemental carbon involves dehydration, polymerization, condensation, and carbonization steps. Initially intermolecular dehydration occurs under the hydrothermal conditions and forms organic compounds and organic acids. Thus, the hydronium ions are formed in acidic pH, which acts as a catalyst for degradation of the organic compound. Then, subsequent polymerization and condensation reactions form the final carbon nanosphere.

Polymer template: Different polymers are also used as sacrificial templates for the design of YS nanostructures. The polymeric sacrificial layers are generally removed by calcination or by dissolution in the presence of an appropriate solvent. Here, the choice of the removal technique totally depends on the shell material as shown in figure. If the shell material is a polymer or organic, then the removal by calcina-tion can be avoided; only a dissolution method with organic solvents is used. On the other hand, if the shell is made up of an inorganic layer, then the polymer template can be easily removed by the calcination process. Thus, many different YS materials such as SiO_2/TiO_2 (polystyrene), Au/SiO_2 (polystyrene-co-poly(4-vinylpyridine), magnetic silica/silica (polystyrene), SiO_2/polypyrrole (polystyrene), Fe_3O_4/SiO_2/poly- (MBAAm-co-MAA) (PMMA), poly(DVB-co-AA)/poly(DVB-co- AA) (poly(acrylic acid)), PS-co-PAEMAco-PVTES-co-PMAA (polystyrene), Fe_3O_4/silica (polystyrene), TiO_2/TiO_2 (polystyrene), SiO_2/TiO_2 (PMMA, polystyrene), Au/SiO_2 (PS-co- P4VP), Fe_3O_4/PANi (polystyrene), TiO_2/SiO_2 (PMMA) and silica/PDVB (PMAA) were also reported using the polymer as a sacrificial template (mentioned in parentheses). Mostly, the polymer layer was synthesized by the emulsion polymerization method.

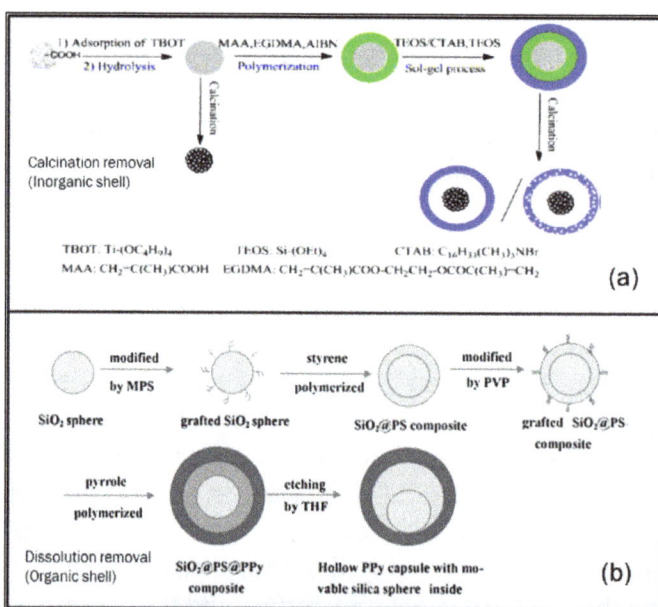

(a) Formation of TiO_2/SiO_2 YS NPs through removal of a sacrificial layer (PMMA) by calcination. (b) synthesis of SiO2/PPy YS NPs through removal of a sacrificial layer (PS) by the dissolution method.

Other materials: While silica, polymer, and carbon materials are mostly used as a sacrificial template for YS synthesis, many other inorganic materials such as Fe_3O_4, gold,

Ni, V_2O_5, MnO, Cu_2O, sulphur, Ag, AgBr, and β-FeOOH have also been reported. When metal oxides and metals are used as a sacrificial template, mostly acid (HNO_3, H_2SO_4, HCl) and alkali leaching (NH_4OH, NaOH) techniques are used for complete removal of the sacrificial layer. The metal and metal oxides react with acid to form a soluble metal salt and hydrogen as shown in equation Metal + acid → metal salt + hydrogen and Metal oxides + acid → metal salt + water. Apart from acidic dissolution, reductive dissolution can also be used to remove a metal oxide layer with a reducing agent as presented in equation $Fe_3O_4 + 2e^- + 8H^+ \rightarrow 3Fe^{2+} + 4H_2O$ and figure. For example, the Au/SiO2 YS NPs were prepared by reductive dissolution of Fe3O4 in the presence of HCl and NaBH4 as shown in equation Metal oxides + acid → metal salt + water. Similarly, MnO was also selectively etched by NH2OH-based solutions (hydroxylamine hydrochloride), where dissolution of MnO occurs by reduction of Mn(IV) or Mn(III) to soluble Mn(II) as presented in equation $MnO_2 + (Mn(iv)) + 2NH_3OH^+ \rightarrow Mn^{2+} + N^{2+} + H_2O$ and $Mn_2O_3(Mn(iii)) + 2NH_3OH^+ + 2H^+ \rightarrow Mn^{2+} + N_2(g) + 5H_2O$.

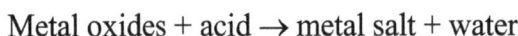

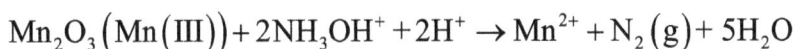

Metal + acid → metal salt + hydrogen

Metal oxides + acid → metal salt + water

$$Fe_3O_4 + 2e^- + 8H^+ \rightarrow 3Fe^{2+} + 4H_2O$$

$$MnO_2 + (Mn(IV)) + 2NH_3OH^+ \rightarrow Mn^{2+} + N^{2+} + H_2O$$

$$Mn_2O_3(Mn(III)) + 2NH_3OH^+ + 2H^+ \rightarrow Mn^{2+} + N_2(g) + 5H_2O$$

Some specific reagents are also used, such as leaching of gold by cyanide (cyanidation), which is a common process to dissolve Au as shown in figure. In this process, cyanide ions oxidize metallic gold and form a water soluble Au–CN coordination complex as shown in equation $Mn_2O_3(Mn(iii)) + 2NH_3OH^+ + 2H^+ \rightarrow Mn^{2+} + N_2(g) + 5H_2O$. Using this reaction, the thickness of the dissolution layer can be controlled by controlling the concentration of added KCN as per the stoichiometry mentioned in eqution.

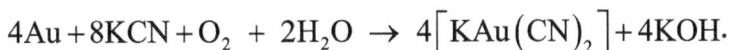

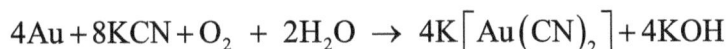

$$4Au + 8KCN + O_2 + 2H_2O \rightarrow 4[KAu(CN)_2] + 4KOH.$$

$$4Au + 8KCN + O_2 + 2H_2O \rightarrow 4K[Au(CN)_2] + 4KOH$$

Among non-metallic sacrificial layers, sulfur is also an important material, as the sulfur layer can be removed by both dissolution (CS2 and toluene) and calcination at low temperature. Recently some studies also reported the synthesis of the YS nanostructure by using the sulfur core as a sacrificial template, where the sulfur core dissolves partially in the presence toluene solution as shown in figure. The sulfur reacts with toluene and forms H2S gas as shown in equation $2C_6H_5CH_3 + 2S \rightarrow C_6H_5CH : CHC_6H_6 + 2H_2S$. Besides toluene, sulphur can also be removed using carbon disulfide, carbon tetrachloride, xylene, and benzene solution. We can also eliminate sulphur via the calcination at elevated temperature. The Ag template was also reported for the synthesis of Au/PPy YS NPs, where the Ag template was removed using NH3 solution by forming a soluble metal–amine complex compound as shown in equation $Ag + 2NH_3 \; (aq.) \rightarrow \left[Ag(NH_3)_2\right]$. Silver halides such as AgBr were also reported as a sacrificial template, where AgBr/PPy core/shell transforms into Ag/PPy YS through photoreduction of AgBr in the presence of UV light. AgBr is a photoactive material, which absorbs photons in the UV-Vis wavelengths to generate electrons and hole pairs. The generated electron associated with interstitial Ag ions to produce Ag atoms as shown in equation $AgBr + hv \rightarrow AgBr \left(ht^+ + e^-\right) - AgBr + hv \rightarrow Ag + \frac{1}{2}Br_2$. The AgBr can also be dissolved in the presence of NH3 solution by forming silver– amine complexes as shown in equation $AgBr(s) + 2NH_3(aq.) \rightarrow Ag(NH_3)_2^+(aq.) + Br^-(aq.)$. The β-FeOOH was also used as a template, where it was converted into Fe_3O_4 by reduction.

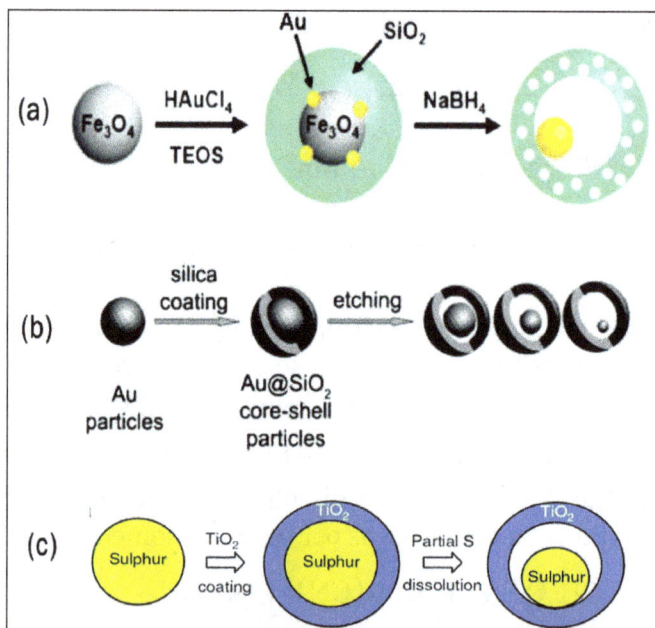

(a) The synthesis of Au/h-SiO$_2$ YS NPs by using Fe$_3$O$_4$ as a sacrificial layer. (b) synthetic procedure of Au/SiO$_2$ nanoreactor framework by using Au as a sacrificial template. (c) schematic of the synthetic process that involves coating of sulphur nanoparticles with TiO$_2$ to form sulphur/TiO$_2$ core/shell nanostructures, followed by partial dissolution of sulfur in toluene to achieve the YS morphology.

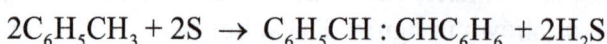

$$2C_6H_5CH_3 + 2S \rightarrow C_6H_5CH : CHC_6H_6 + 2H_2S$$

$$Ag + 2NH_3 \ (aq.) \rightarrow \left[Ag(NH_3)_2 \right]$$

$$AgBr + h\nu \rightarrow AgBr \left(h^+ + e^- \right)$$

$$e^- + Ag_i^+ \rightarrow Ag_i^0$$

$$AgBr + h\nu \rightarrow Ag + \frac{1}{2} Br_2$$

$$AgBr(s) + 2NH_3(aq.) \rightarrow Ag(NH_3)_2^+ (aq.) + Br^- (aq.)$$

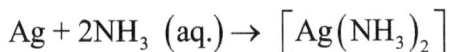

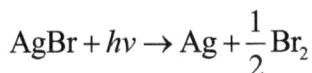

In general, conventional hard template methods are usually time consuming because of the requirements of pre-designing or multiple step control, which may finally lead to difficulty to obtain more complex structures. Besides, these templates are also advantageous for many reasons such as easy synthesis, monodispersity, size variety, tuneability of components, bulk synthesis, and being easy to scale up for large-scale production.

Soft template based method. The hard template based processes are probably the most effective and are widely used over the soft templates to design YS nanostructures. However, these methods have several issues such as stability of core and shell structures during removal of the sacrificial layer (calcinations or dissolution) and a multistep process. Many applications including drug delivery, therapeutic delivery, bioimaging, and catalysis require a facile way to access void space to encapsulate the guest molecules. In the hard templated method, encapsulation of the guest molecules in the void space during shell formation is very difficult. These problems can be solved by the use of a soft template based strategy. The soft template based approach allows easy encapsulation of guest molecules inside the hollow shell. The soft templates are considered adsorbed surfactant layers and microemulsions for the synthesis of YS nanostructures. Usually, a soft template can be completely removed by various routes such as the acid extraction method, water or alcohol washing, and thermal annealing or calcination.

The surfactants are organic amphiphilic molecules used in the NP synthesis process as template or capping agents to control the size and shape of the NPs. In general, adsorbed or self-assembled layers of surfactant molecules on the core surface, surfactant micelle, reverse micelle, vesicle, and microemulsions which contain a huge surfactant are generally used as a template. In the surfactant based template method, surfactant molecules are adsorbed on the core surface by electrostatic or van der Waals interactions and then the outer shell layer is deposited on that surface because of similar interactions. For example, a detailed process of the formation of YS nanostructures is shown in figure. First, the core NPs were dispersed in the aqueous solution of a suitable surfactant to induce the formation of vesicles with core NPs with the assistance of a vesicle-inducing agent such as a short alkyl amine or other small organic molecules.

Then, the co-structure-directing agents (CSDAs), typically aminosilane such as APTES or TMAPS, were attached to the surface of the vesicles through electrostatic attraction. Finally, the silica shells were formed through the sol–gel process by aminosilane and TEOS. In the whole process, the aminosilane acts as both a vesicle inducing agent and a co-structure directing agent simultaneously. Thus, the core NPs are accommodated into the vesicles and the YS structure formation occurs through the replication of vesicles by silica shells. Finally, the surfactant layer was removed by the ethanol washing and acid extraction (acetonitrile solution containing ~35% HCl) method.

Similarly in another approach, FC4 (fluorocarbon surfactant) surfactant was used as the vesicle–core complex template for the synthesis SiO_2/SiO_2, Au/SiO_2, Fe_3O_4/SiO_2 YS nanostructures with a tuneable shell thickness as shown in figure, where the organic template was removed via calcination to obtain the hollow structure. In the case of surfactant-based soft templated methods, researchers are continuously exploring various surfactants to obtain their desire advantages. As another example, an aqueous mixture of lauryl sulfonate betaine (LSB, a zwitterionic surfactant) and sodium dodecyl benzenesulfonate (SDBS, an anionic surfactant) was also used as a soft template to synthesize YS nanostructures. For designing this process, the molecular structure of the surfactant, the concentration, and the addition of a co-surfactant or other molecules are extremely important to control the morphological parameters of YS structures.

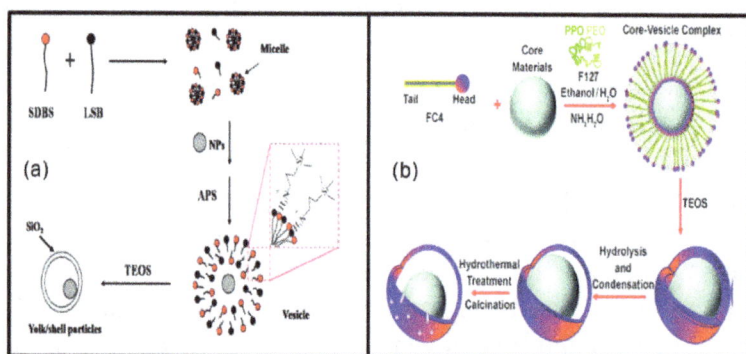

(a) Schematic procedure used to generate YS nanostructures through a soft template (b) procedure for the preparation of YS nanostructures with a mesoporous shell. FC4 = fluorocarbon surfacetant, PPO = poly (propylene oxide), PEO (poly (ethylene oxide), TEOS (tetraethoxysilane).

Emulsions can also be used as a soft template to synthesize the YS structure. Generally, microemulsions are used as a template, since nanometer size droplets are formed in the microemulsion. A microemulsion is a thermodynamically stable phase which contains a mixture of oil, water, surfactant, and co-surfactants (if required). Depending on the process and selection of materials, both oil in water (O/W) and water in oil (W/O) emulsions are used. The synthesis of YS can be achieved in two steps by the emulsion process. Firstly, the core NPs were dispersed in the emulsion and then the shell layers were deposited around the interface between the emulsion droplets and the continuous phase of the microemulsion to generate YS structures. Many researchers have also demonstrated water in oil microemulsion as a soft template to prepare YS nanostructures. For example, the Au/SiO$_2$ YS nanostructure was prepared using W/O microemulsion by encapsulating a gold precursor in a SiO$_2$ template and further encapsulating the gold nanoparticles inside the SiO$_2$ shell via reduction of the gold precursor by NaBH$_4$. In another approach, the Fe$_3$O$_4$/SiO$_2$ YS nano-structure was obtained by using Triton X-100 surfactant based W/O microemulsion shown in figure. However, this process is limited to the selective surfactant for the preparation of YS nanostructures.

Galvanic replacement. Galvanic replacement reaction offers an efficient route for the synthesis of noble metallic or alloy type YS nanostructures with controllable hollow interiors, size, shape, composition, and morphologies; this method is also called transmetalation. This process is a sacrificial template type technique based on the electrochemical potential of the used metals, where the template also acts as a reducing agent. The key point of galvanic replacement is the electrical potential difference between two metals, where one metal acts as the cathode and other metal as the anode. Galvanic replacement reaction occurs spontaneously when atoms of a metal (zero valance state) react with ions of another metal having a higher electrochemical potential in a solution phase. Generally, in the galvanic replacement process the lower standard electrode potential metal (A) is used as a sacrificial layer and the higher standard electrode potential metal (B) as a shell. In this process metal A is oxidized and B is reduced. The reaction can be expressed as follows:

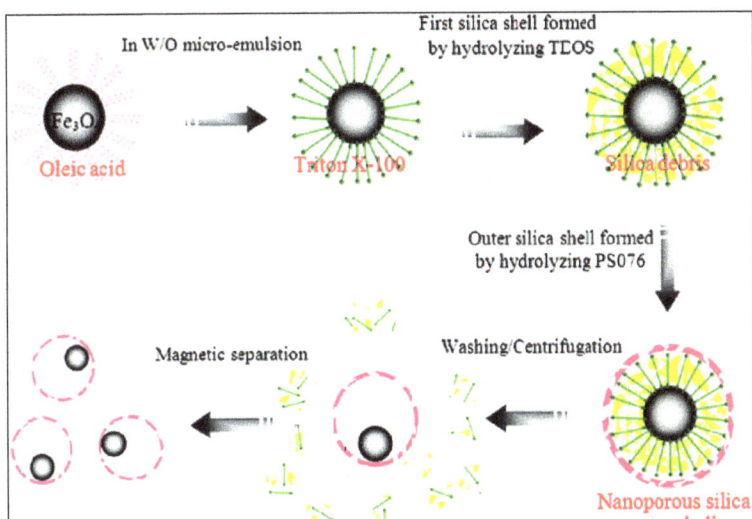

Schematic diagrams of the formation mechanism of Fe$_3$O$_4$/Silica YS NPs via W/O microemulsion.

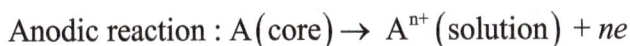

Anodic reaction : $A(\text{core}) \rightarrow A^{n+}(\text{solution}) + ne$

Cathodic reaction : $B^{m+} + me \rightarrow B^0(\text{shell})$

Overall reaction : $mA^0(\text{core}) + nB^{m+} \rightarrow nB^0(\text{shell}) + mA^{n+}(\text{solution})$

The core material is synthesized using a chemical reduction method. In this electrochemical process, partial or complete removal of the sacrificial template is dependent on the final YS structure. When the removal is partial the core material is synthesized first, and the finally A/B type YS structure is formed. In the case of complete removal, A/C type core/shell material is synthesized first and then after removal of the C layer it gives the A/B type YS structure. However, in this case, if there is an incomplete removal of the C layer, then the A/C type core/shell will be encapsulated inside the hollow shell layer of B. Similar to other methods, here also shell thickness, void space etc., can be controlled by changing the process parameters (reactant concentration, standard electrode potential, etc.). In this process, the properties of the shell depends on various parameters such as: (i) when more than one metal precursor is present in the presynthesized core media for galvanic replacement, the distribution of metals in the shell can be controlled by altering the order of addition of precursors. (ii) The core dissolution rate from different positions depends on the chemical reactivity of the respective sites and the shielding effect of the deposited shell layer, and the finally formed shell structures are in general of porous type. During the dissolution process there is a chance alloying too in the galvanic replacement. When metal B is deposited on A, in the initial period of the process alloying started within the thin shell layer. It has been reported that the kinetics of alloying obeys Fick's second law of diffusion.

The standard reduction potentials of some metals are shown in table.

Reduction reaction	E° (V vs. SHE)
$Co^{2+} + 2e^- \rightarrow Co$	−0.28
$Cu^{2+} + 2e^- \rightarrow Cu$	0.34
$Ag^+ + e^- \rightarrow Ag$	0.80
$Pd^{2+} + 2e^- \rightarrow Pd$	0.95
$Pt^{2+} + 2e^- \rightarrow Pt$	1.18
$Au^{3+} + 3e^- \rightarrow Au$	1.50

$$C_{x,t} = C_i - C_i \mathrm{erf}\left(\frac{x}{2DT}\right)^{1/2}$$

where $_{x,t}$ is the atomic concentration of metal B as a function of time (t) and distance (x), C_i is the initial concentration, and D is the diffusion coefficient. Finally, the degree of alloying increases with increasing reaction temperature, thickening of the shell layer, and depression of the energy barrier to diffusion.

Additionally, dealloying is also an important fact in the later stage of this process, to control porosity of the shell layer by removing one metal. In general, dealloying occurs in the presence of more metal ions involved in the galvanic replacement reaction or by etching (acid or alkali). The alloying and dealloying process depends on the stoichiometry of the oxidation number or the valence difference between A and B materials. As an example of the Au–Ag system, when $HAuCl_4$ is used as a Au-precursor, one Au atom formed (reduced) by oxidation of three Ag atoms according to equation

$$3Ag_{(s)} + AuCl_4^-{}_{(aq.)} \rightarrow Au_{(s)} + 3Ag^+{}_{(aq.)} + 4Cl^-{}_{(aq.)}.$$ In this process, alloying continues

till the lower electrochemical potential metal or sacrificial layer molecules (Ag) are present in excess or equivalent moles compared to that of the higher electrochemical potential (Au^{3+}) metal ions. If a larger amount of $HAuCl_4$ is added, in the presence of more Au^{3+}, dealloying of the Au/Ag shell occurs by selective removal of Ag atoms from the alloyed shell, and finally pure Au shell forms. During the dealloying process, many small lattice vacancies are also generated, which form small holes on the shell as shown in figure. As an example of the valance effect, if Au+ ions are used in place of Au^{3+} ions with an Ag core, a thicker Au–Ag shell will form, as one Ag atom will dissolve per Au^+ ion. Thus, the composition or removal of the sacrificial layer can be controlled by tuning the stoichiometric ratio or valence ratio of the molecules present in the sacrificial template and the outer layer precursor material.

The first example of the synthesis of the YS nanostructure via a galvanic replacement reaction was reported for the synthesis of Au/Ag alloy YS nanostructures via a reaction between Ag and $HAuCl_4$. In this study, Au was formed after reduction on the surface of the Ag template, as the reduction potential of $AuCl_4^-/Au$ (0.99 V SHE) is higher than that of Ag^+/Ag (0.8 V SHE). The shell layer was formed through the processes of nucleation and growth; the reaction is shown in equation

$$3Ag_{(s)} + AuCl_4^-{}_{(aq.)} \rightarrow Au_{(s)} + 3Ag^+{}_{(aq.)} + 4Cl^-{}_{(aq.)}.$$ In the Ag–Au system, the thickness

of the void space was also controlled by varying the thickness of sacrificial Ag coating on the Au/Ag alloy core by changing the concentration of $AgNO_3$ in the electrolysis plating bath. In this case, the morphology of the Au NPs is similar to the Ag template as shown in figure. Similarly, many research groups have also used a galvanic replacement reaction to design Au/Au, Au/Pt–Ag, Au nanorod/Au,288 Au–Ag/Au, Au/Pt, Au/Pd, and Au/SiO_2 YS NPs. Other than these, $Pd / M_x Cu_{1-x}$ alloy (M = Au, Pd, and Pt) YS nanostructure was also synthesized using Pd/Cu core/shell nanocubes as a template. While most of the studies used the Ag sacrificial layer, in contrast, Cu also offers a much lower reduction potential (0.34 versus 0.80 V for Ag) to facilitate the galvanic reaction more easily, even at room temperature.

$$3Ag_{(s)} + AuCl_4^-{}_{(aq.)} \rightarrow Au_{(s)} + 3Ag^+{}_{(aq.)} + 4Cl^-{}_{(aq.)}$$

The Kirkendall reaction: The Kirkendall effect is a classical phenomenon in metallurgy, where vacancy (absence of atoms in a crystal structure) diffusion occurs at the interface between two metals because of the difference in diffusion rate between them. According to this concept, atomic diffusion in a metal does not occur directly, but through the vacancy exchange. The material with the higher diffusion coefficient will have a larger associated vacancy flux in it, so the net movement of vacancies will be from the material with the lower diffusion coefficient to the material with the higher diffusion coefficient. Finally, the condensation of excess vacancies can give rise to void formation near the original interface and within the faster diffusion side called 'Kirkendall voids'.

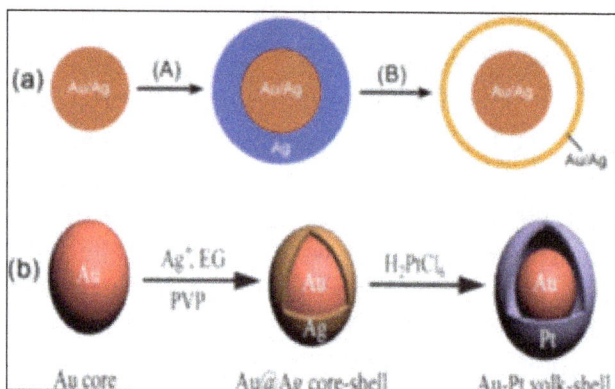

(a) Schematic illustration of the procedure for preparing YS NPs consisting of a Au/Ag alloy core and a Au/Ag alloy shell, where two major steps are involved: (A) electroless deposition of pure Ag on the surface of an Au/Ag alloy NP; and (B) reaction of the resultant particle with an aqueous $HAuCl_4$ solution to transform the coating of pure Ag into an Au/Ag alloy shell larger in size.

The Kirkendall effect is based on two principles: (i) atomic diffusion occurs via vacancies and (ii) each metal diffuses at a different mobility. This effect was first observed by Smigelskas and Kirkendall on the diffusion between zinc and brass. During the last decade, the Kirkendall effect has been recognized as one of the most useful methods to prepare hollow and YS NPs of metal oxides, sulfides and phosphides after it was first reported by Yin's group in 2004, where Co NPs were converted into hollow cobalt oxide and chalcogenide (S, Se) NPs through a reaction with sulfur, oxygen, and selenium. This study demonstrated that at the metal/oxide or metal/sulphide interface, the outward diffusion of the cobalt atoms is much faster than the inward diffusion of oxygen or sulphur at high temperature, which gives rise to voids or vacancies during oxidation and sulfidation. In this case, there is formation of material bridges between the core and the shell, and this persists until the core is completely consumed. These bridges provide a fast transport path for outward diffusion of cobalt atoms that can then spread to the inner shell surface.

This approach inspired several researchers to design oxide, phosphide, and chalcogenide YS nano-structures, where the void can develop through the Kirkendall effect. In this approach, when a metal is exposed to oxygen, phosphorus, sulfur, or selenium precursors at elevated temperature, it gives a diffusion couple. Thus, outward diffusion of metal NP cation species is quicker than inward diffusion of the anions species; then an inward flux of vacancies balances the diffusivity difference of the outward metal cation flux. Finally, condensation of excess vacancies can give rise to void formation at the interface of the faster diffusion side. The YS synthesis through the Kirkendall effect involves three major steps: (i) synthesis of core NPs, (ii) deposition of shell material on the core NPs, (iii) oxidation, phosphidation, or chalcogination of the shell to form a hollow shell by the Kirkendall reaction for diffusion of the outer layer to create void space at the interface. For example, the Kirkendall effect can be explained by the oxidation (a chemical conversion reaction) of the metal (Ni) NP process as shown in figure. During the oxidation process, a thin layer of oxide forms on the surface of metal NPs, which allows metal ions to diffuse more quickly towards the outside than oxygen atoms because of the larger ionic radius of anions than cations and finally leaves a void space inside. Using the same approach, a $FePt/CoS_2$ YS nanostructure was synthesized, where the YS structure was formed after wet sulfidation of FePt/Co nanoparticles. Mostly, researchers have used the thermal oxidation process to synthesize YS nanostructures, which is a type of non-equilibrium inter-diffusion process. Similarly, multifunctional Au/Fe_2O_3, Ag/Fe_2O_3, and $FePt/Fe_2O_3$ YS nanostructures were also prepared from the oxidation of the iron shell on the respective core materials by means of the Kirkendall effect.

Template Free Approach (Ostwald Ripening)

The template-mediated synthesis methods of NPs have been used for several years to synthesize different size and shape nanoparticles to obtain proper control on different parameters such as size, shape, shell layer thickness etc. However, in many cases, the synthesis process required several steps based on the complexity of the desired structure. More specifically, while coming to the YS structure, removal of the sacrificial layer and its purification required a few more steps. Needless to say, the sacrificial layer material is completely undesirable and finally becomes a waste at the end of the process. In many cases, this sacrificial layer is either made of some precious materials (Au, Ag), or the removal process is environmentally harsh. A sacrificial layer removal process finally complicates the overall process and increases the chance of impurity, agglomeration, damage of core or shell nanoparticles, and chances of shell collapsing. Finally, because of these reasons the overall costs of template-mediated methods are higher, while commercialized for the practical applications. So many researchers would prefer a one-step template-free process to synthesize YS nanostructures.

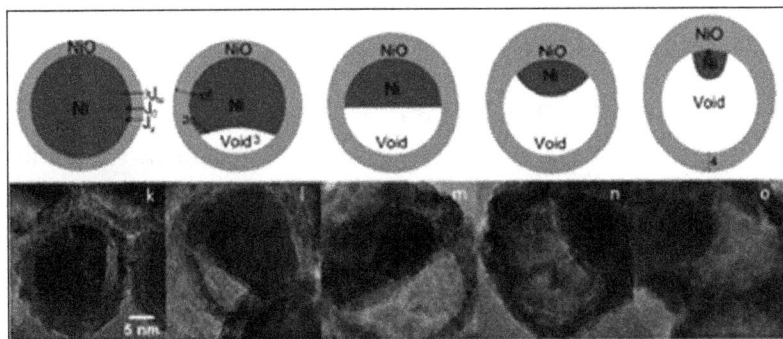

Formation of a single void at the core/shell interface causes asymmetrical outward diffusion of Ni, resulting in a non-uniform NiO shell thickness.

As a result, many research groups have reported one step template-free methods for the synthesis of YS nano-structures based on the Ostwald ripening process. Ostwald ripening is a physical phenomenon that refers to "the growth of larger crystals from those of smaller size which have a higher solubility than the larger ones". In this process, small nanoparticles dissolve in solution and large nanoparticles because of minimization of surface energy. In this process, a void space will be generated when the core is made of aggregates of small NPs and the shell is made of larger NPs. When the small NPs are located at the central part of the spherical aggregates, it is called symmetric Ostwald ripening. On the other hand, if small nanoparticles are located asymmetrically on the spherical aggregates, then it is termed as an asymmetric Ostwald ripening process. To date, a wide variety of YS nanostructures have been synthesized using this template-free method. For instance, semiconductor ZnS and Co_3O_4 YS nanostructures are synthesized based on the symmetric and asymmetric Ostwald ripening processes as shown in figure. In this process, firstly, small nanoparticles were aggregated into a solid sphere. Then, during the recrystallization process, small crystallites of the solid sphere were dissolved and move towards the outer shell to make a void space. With the progress of time, the void space becomes larger and divides the solid sphere into two parts: one is movable core, and other is a hollow outer shell. Similarly, a simple one-pot template-free synthesis of Cu_2O YS NPs was also reported by symmetric and asymmetric Ostwald ripening. The synthesis was performed in aqueous $Cu(CH_3COO)_2$ solution, where the precursor was reduced by ascorbic acid at room temperature to form the Cu_2O nuclei clusters. Firstly, these clusters are self-assembled into a solid sphere and then through the recrystallization process they form void space inside the solid sphere. The rare earth fluoride (SmF_3) YS NPs were also synthesized by symmetric and asymmetric Ostwald ripening processes. The formation of α-Fe_2O_3/SnO_2 YS nano-structures was also reported based on the inside-out Ostwald ripening mechanism, where the SnO_2 nanoparticles were synthesized by hydrolysis of $K_2SnO_3 \cdot 3H_2O$ around the α-Fe_2O_3 NPs to form a solid core/shell structure. Further, during the hydrothermal Ostwald ripening process, the small SnO_2 nanoparticles dissolve and relocate to the outer shell because of a recrystallization process, and form the void space. Similarly, other researchers have also reported Fe_3O_4/Co_3O_4, Fe_3O_4/SnO_2, Pt/CeO_2, and Au/Cu_2O YS nanostructures.

Ultrasonic Spray Pyrolysis

Ultrasonic spray pyrolysis (USP) is a one-step, scalable and continuous approach for the fabrication of a diverse range of nanostructured materials including hollow and YS nano-structures by the evaporation and decomposition of precursor droplets in a reactor. It is a solution based process, where high-frequency ultrasound is passed through the liquid precursor solution to generate a liquid–gas interface (aerosol), which nebulized into micro-droplets. Finally, the generated droplets are carried by a gas flow into a furnace, where solvent evaporation and precursor decomposition take place to form the YS nanostructures. The volatile organic compounds form a layer on the NP surface and leave a void space to generate a YS structure during the heating process. The role of ultrasound in USP is to provide the phase isolation of one microdroplet reactor from another. Generally, this process involves the following steps: (i) micro-droplet generation using USP, (ii) evaporation of solvents in the heated zone, (iii) diffusion of reactants, (iv) reaction or precipitation, and (v) escape of any volatile components. This method has several advantages over conventional methods for the synthesis of YS nanostructures such as a one step process, no acid or alkali treatments for sacrificial layer dissolution, an inexpensive continuous process, environmentally friendly, low-cost initial reactants, control on the composition, and scope for scalability of production of YS/hollow NPs. In addition, the facile control over chemical and physical compositions in the USP method makes it useful in the preparation of multicomponent or composite materials.

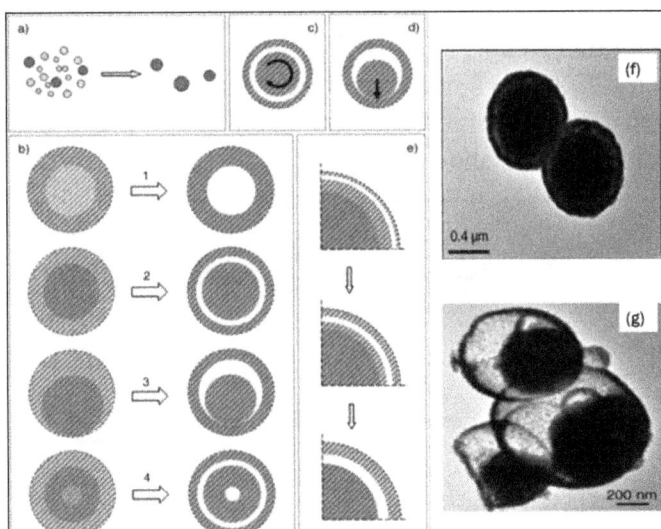

(a) General process of Ostwald ripening. (b) Various schemes of Ostwald ripening for spherical colloidal aggregates: (1) core hollowing process; (2) symmetric Ostwald ripening for formation of a homogeneous core–shell structure; (3) asymmetric Ostwald ripening in formation of a semi-hollow core–shell structure; (4) a combination of 1 and 2. (c) A rotating core. (d) Mass center of a semi-hollow core–shell structure. (e) Proposed model for the formation of the void space between a core and a shell in scheme 2 of (b). Hashed lines: the cross-sectional plane of a sphere. Darker areas: larger and closely packed crystallites. Lighter areas: smaller and loosely packed crystallites. White areas: void space. Note that the above area illustrations are simplified, as actual transitions between two different areas should be much more gradual. (f) TEM images of symmetric Ostwald ripening in Zns. (g) TEM images of symmetric Ostwald ripening in Co_3O_4.

For example, M/C (where M = multiple Sn, Pt, Ag, or Fe–FeO NPs) YS nanostructures have been prepared using ultrasonic spray pyrolysis of aqueous solutions containing sodium citrate and corresponding inorganic metal salts. Initially, the process involves the preparation of aqueous solutions containing sodium citrate and inorganic metal salts used as a precursor. Then during the USP process, first inner nanoparticles are generated via reduction of metal salts with an organic reducing agent in the hot liquid droplets. The sodium citrate outer shell is formed due to the tendency of free sodium citrate molecules to move towards the periphery of the hot liquid droplets; then an outer shell (carbon) is formed via carbonization of the sodium citrate shell. Finally, the resultant M/C YS NPs are formed via removal of water-soluble by-products. In this approach, a carbon precursor/inorganic template composite is first formed, followed by carbonization, and then chemical leaching of the template material. Similarly, double shell SnO_2 YS NPs were also synthesized by the decomposition of tin salt, and carbon was formed by the polymerization and carbonization of sucrose. Finally, after combustion of carbon layers at high temperature, a multi-shell YS nanostructure was produced as shown in figure. More complex structures such as binary (TiO_2, TiO_2/Al_2O_3), ternary ($TiO_2-Al_2O_3-ZrO_2$), quaternary ($TiO_2-Al_2O_3-ZrO_2-CeO_2$), and quinary ($TiO_2-Al_2O_3-ZrO_2-CeO_2-Y_2O_3$) multi-shell YS nanostructures were also synthesized by using different metal salt precursors with the ultra- sonic spray pyrolysis process, where carbon was removed by calcination as shown in figure.

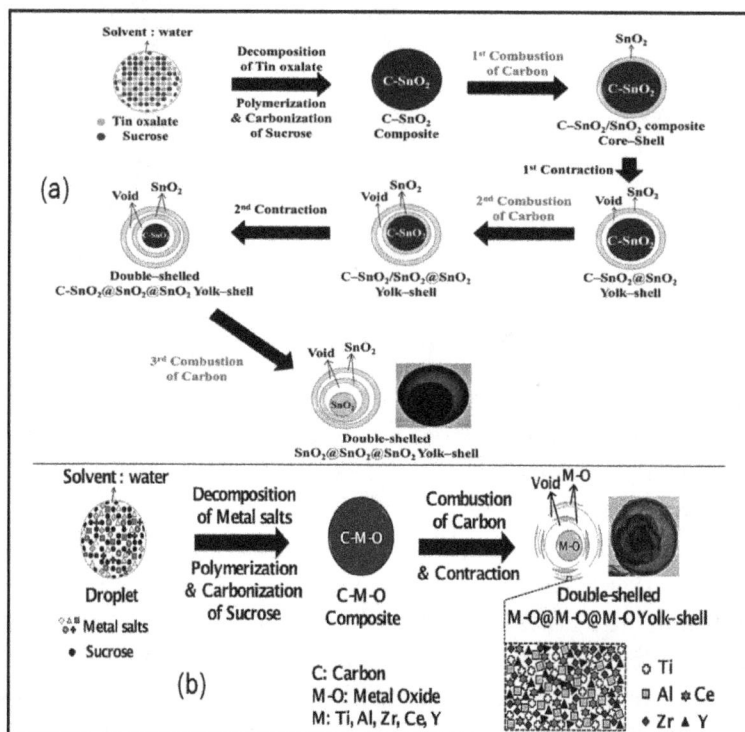

(a) Mechanism of the formation of the double-shelled SnO_2 YS nanostructured powders. (b) schematic diagrams of the formation of YS composite powders with quinary composition. Firstly, drying and decomposition of the droplets generated by the ultrasonic nebulizer in air produced the carbon–metal oxide particles. Further, sequentially combustion of the carbon–metal oxide powder produced the metal oxide YS structure.

Ship in Bottle Method

Generally, in the sacrificial template-mediated methods, a core particle is sequentially coated with a template and then the shell material. Further, the sacrificial template is removed by dissolution or calcination to obtain YS nanostructures.

However, core materials may be affected by the dissolution or calcination depending on the sensitivity of the material, and the number of steps also increases depending on the complexity of the structure as mentioned before. In contrast, many researchers have used the ship in bottle approach for the synthesis of YS nanostructures in recent years. This is a simple approach for the synthesis of YS nano-structures. In this method, the core material precursor is encapsulated inside the pre-synthesized hollow shells first and then core particles are generated inside the hollow shells through a chemical reaction or by the self-assembly process to form a final YS structure as shown in figure. This approach can be used for many materials such as metals, metal oxides, polymers, drugs molecules, and proteins. For instance, Au nanorod/SiO_2 YS nanostructures were synthesized through the ship-in-bottle method as shown in figure. The synthesis method was based on encapsulation of gold seed inside the silica hollow shell, where the hollow silica shell was dispersed in an aqueous solution of gold precursor ($HAuCl_4$), and further reduced to the Au nanoparticle in the presence of an ice-cold aqueous $NaBH_4$ reducing agent. The Au nanoparticles formed outside the hollow shells were removed by centrifugation with water and the remaining Au nanoparticles inside the hollow shells were used as seed to induce the growth of Au nanorods. Similarly, Fe_2O_3/C, Fe_2O_3/silica, Au/silica, Pt/silica, and SnO_2/SiO_2 YS nanostructures were also synthesized by the ship-in-bottle process by encapsulating a metal precursor inside the pre-existing hollow shell and a subsequent reduction of the metal precursor.

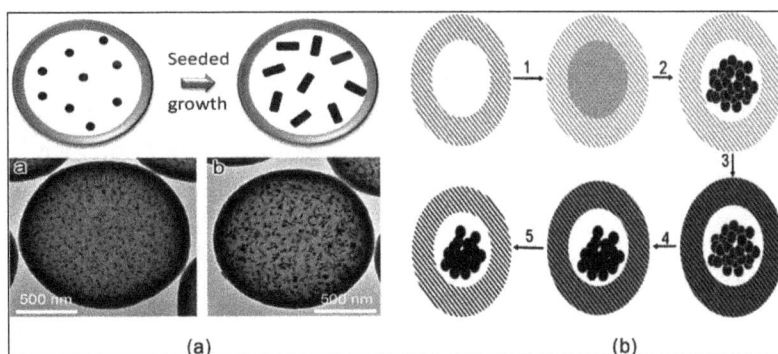

(a) An illustration of the in situ confined growth of Au nanorods from Au nanosphere seeds inside hollow mesoporous silica shells with corresponding TEM images (b) illustration of the ship-in-bottle synthesis procedure of YS NPs: (1) introduction of the iron nitrate solution into the hollow core by vacuum nanocasting, (2) calcination into hematite particles of the core, (3) the introduction of the furfuryl alcohol into the mesoporous channels, and polymerization, (4) carbonization of the polyfur-furyl alcohol, and the reduction of hematite particles into magnetic particles, (5) removal of the silica template.

Emerging Properties with Applications

Biomedical Applications

In recent years, applications of nanoparticles are extended to almost all fields of science and engineering. Among several fields, applications of nanoparticles in different areas of bio-medical fields are rapidly growing and show significant future challenges. In fact, applications of YS NPs are not an exception in this area. Unlike other nanoparticles, multifunctionality can be developed in one particle by using the YS morphology, which is eventually very important for biomedical applications. In the case of YS NPs, their structure, tuneable interior void space, appropriate combination of the core and the shell, and multifunctionality make them superior for bio-medical applications including targeted drug/gene delivery, controlled release, bio-imaging, diagnostic, therapeutic agent, biosensor, antimicrobial activity, and so on.

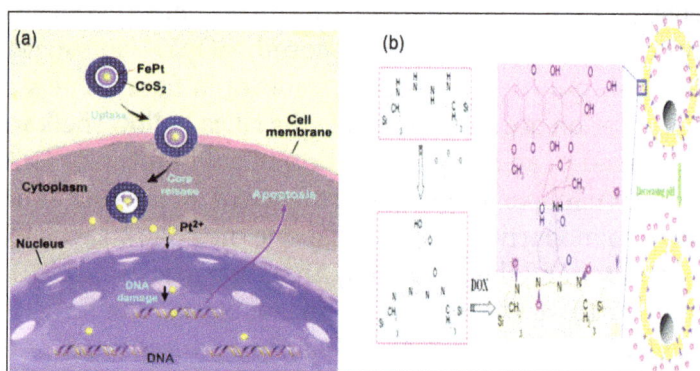

(a) The cytotoxicity mechanism of FePt/CoS$_2$ YS NPs for HeLa cells. (b) conjugation and release scheme of DOX molecules from pH stimuli-responsive YS NPs.

The NP-based drug delivery technique is a promising treatment for many diseases and therapy in bio-medical science. In this regard, silica is the most popular material for drug delivery applications because of its good bio-compatibility and easy bio-functionalization ability. Because of these reasons, mostly silica-based YS NPs are used for drug delivery and drug carrier applications with an enhanced drug loading capacity. Additionally, the magnetic and optical imaging properties are also achieved in, in vivo and in vitro biological specimens by encapsulating magnetic materials, fluorescent molecules, and quantum dots within the hollow silica shells, where magnetic NPs are used as MRI agents, and fluorescent dye or quantum dot NPs are used as optical agents. In some cases, dual imaging is also used for simultaneous magnetic and optical imaging. In the field of drug delivery, magnetic/silica YS NPs have tremendous advantages for magnetic field induced drug targeting or triggering and for use as an MRI agent because of additional magnetic properties and functionalities. Under this class, Fe$_3$O$_4$/SiO$_2$ YS NPs have been reported for the drug delivery applications, where both drug loading capacity and a significant magnetization strength make it suitable as a multifunctional drug carrier. The FePt/CoS$_2$ and FePt/Fe$_2$O$_3$ YS NPs were also reported as potential nanostructured materials for anticancer drug delivery with MRI functionality; in this case, the presence of Pt acts as a potential anti-cancer drug like the well-known cancer drug cisplatin (platinum-containing anti-cancer drug). In the case of the acidic environment of the secondary lysosomes, FePt cores are oxidized to metal ions in the presence of O_2 inside the cells, and the unprotected iron promotes the disintegration of FePt to release platinum ions (Pt^{2+}). The permeability of shells allows Pt ions to diffuse to the nucleus and mitochondria of the cell to damage the DNA helix chains and lead to the apoptosis of the cell as shown in figure.

Recently, smart stimuli-responsive polymer based YS NPs have been reported as a specific multifunctional drug delivery system, where the release of drug molecules can be triggered in the targeted organs and tissues with the desired parameters such as pH, temperature etc. For example, the Fe_3O_4/SiO_2 YS NPs are particularly suitable for drug carriers. When the silica shell is functionalized with the carboxylic group, the NPs can easily react with the amine groups of DOX drug molecules and form amide linkers, which act as a pH triggered switch as shown in figure. At low pH, they released drug molecules into the solution. On the other hand, the superparamagnetic core responds in the presence of an external magnetic field, which provides an additional advantage for biomedical applications. Similarly, polymeric (polymer/polymer) YS NPs were also reported for both pH and temperature stimuli-responsive drug delivery applications. On the other hand, the Co/Au and Fe_3O_4/SiO_2 YS NPs were also reported for non-viral gene delivery to target the HeLa cells with optical imaging and magnetic tracking.

Antimicrobial Applications

Bacterial infections are a major problem in the textile industry, hospitals, water disinfection, medicine, and food packaging, which in turn led to many serious human diseases and effects on human health. The YS NPs are also providing an effective solution for antibacterial activity with multi-functionality. As an example, multifunctional Ag/Fe_2O_3 YS NPs conjugated with glucose are effective for bacterial capturing, killing, and elimination applications, where the magnetic shell allows magnetic removal of attached bacteria from the media and the silver nano core helps to kill the pathogen. Similarly, the combination of Ag and TiO_2 as the YS structure exhibits a strong bactericide property in the presence of both light and dark conditions.

Catalysis

Catalysis is an evergreen field for many chemical industries for the synthesis of different chemicals. So far, many strategies such as metal support on metal oxide, noble metal alloy, and core/shell structures have been developed for high performance chemical, photocatalysis, and electrocatalysis applications. Transition metal NPs such as Au, Ag, Pd, Pt, and Ni are well known because of their catalytic properties. Pure metal NPs may improve the catalytic performance to some level, but cannot avoid the problems of aggregation, chemical/thermal stability, and separation after the completion of reactions for reuse. To solve these problems, researchers have encapsulated many different materials such as metal NPs (Au, Ag, Pt, Pd, and Fe), metal oxides (SiO_2, TiO_2, ZrO_2, CeO_2), and polymers in the form of YS structures to combine the individual properties of core and shell materials, which exhibit collective and synergistic effects to improve the catalytic efficiency. The YS structures show improved catalytic performance because of the following advantages: (i) the movable cores exhibit a high surface area and more active sites because of the presence of an unblocked surface, (ii) the porous shell prevents core NP agglomeration, provides thermal/chemical stability to the core, and acts as a selective membrane for the diffusion of reactants inside the YS, (iii) the void space accommodates more reactants inside the YS structure, (iv) encapsulation of noble metals in the mesoporous metal oxide shell acts as a support catalyst, and (v) replacement of single noble metals with multi or alloy noble metals. The unique properties of YS have great importance in catalytic applications, which will be discussed below for different catalytic systems.

Chemical catalysis. In the case of nanoparticle based heterogeneous catalysis, different oxidation and reduction reactions such as the catalytic reduction of nitro compounds, oxidation of alcohols and CO have been employed as model reactions for the determination of catalytic activity. Generally, Au based YS NPs are used for the reduction of p-nitrophenol, 2-nitroaniline, nitrobenzene; and catalytic oxidation of o-phenylenediamine, CO, and alcohols. Similarly, Ag based YS NPs are also used in the reduction of the nitro compounds. The Pd base YS NPs also exhibit high activity in Suzuki-coupling reaction and oxidation of alcohols. The Ni metal based YS structures are also used for methane reforming and hydrogen transfer for ketone reactions, which is advantageous over the expensive metal NPs. Similarly, Au NPs encapsulated inside metal oxides such as ZrO_2, SiO_2, CeO_2, TiO_2 etc. have been used as high temperature or sinter stable catalytic reactions such as hydrogenation and oxidation–reduction, where a metal core catalyst exhibits high catalytic activity. The incorporation of metal NPs inside the stimuli-responsive polymers has also been reported. Such YS structure's catalytic properties could be tuned by the stimuli responsive polymer functionality.

Photocatalysis. Semiconductor metal oxides such as TiO_2, ZnO, CeO_2 etc. have been widely used for photocatalytic applications to degrade organic pollutants and for H_2 production. The metal oxide/metal oxide combination of YS structures exhibits an enhancement in photocatalytic efficiency because of multiple light scattering in void space and a high available surface area.

Interestingly, plasmonic metal encapsulation in semi-conductor metal oxides called plasmonic catalysts has also been explored for enhanced photocatalytic applications, and show better absorption of visible light and the charge separation property. Similarly, some elements (Bi, Eu, N, Ag) encapsulated in semiconductor metal oxides also help in the enhancement of visible light absorbance and in the increase of charge carrier lifetime. Plasmonic metal/semiconductor metal oxide YS structures have been extensively employed for photocatalytic degradation of organic pollutants and H_2 production. The magnetic material encapsulation inside the semiconductor metal oxide provides additional magnetic separation properties in the presence of an external magnetic field, which provides recyclability of the catalysts.

Electro catalysis. So far, platinum has been mostly used as an electrocatalyst because of its superior catalytic activity and long-term operation stability. However, its expensiveness and poor CO-tolerance are the major concern in practical applications. Recently, intensive efforts have been devoted to the synthesis of bi-metallic or metal oxide YS elec-trocatalysts for fuel cells and ORR reactions. In comparison with the conventional electrode, the YS structure can provide high electrocatalytic activity, enhance the CO-tolerance, and also reduce the loading of a Pt catalyst. There are many reports on bimetallic YS NPs with remarkable activity as an electrocatalyst for the oxidation of methanol to CO_2 at low temperature. For example, Au/Pt YS NPs exhibit superiority in catalytic activity compared to Au/Pt core/shell, hollow Pt, solid Pt nanoparticles, commercial Pt/C, and Au nanoparticles because of bimetallic synergetic effects and the presence of void space.

SERS

The SERS has been one of the most powerful analytical techniques for sensitive and selective identification of molecular species since its discovery in the late 1970s. In recent years, many advanced noble plasmonic metals have been reported for biological and chemical SERS sensing, where the binding agent can be sensed by the change in the local refractive index around the NPs and the shift in the LSPR wavelength. Among all noble metals, Au, Ag, and Cu have attracted more attention in optical sensors because of their biocompatible nature and strong excitation of light in the visible to near infrared region because of the excitation of their plasmon oscillation.

Generally, plasmonic properties of noble metal NPs are affected by the change in size or shape of the metal nano-particles. So, when the nanoparticles are agglomerated in the suspension, it can affect the signal and intensity of SERS. In that case, to maintain the electromagnetic properties of metals and prevent aggregation of nanoparticles, the YS nano-structure with a noble metal core and a porous protective shell is a useful approach. Here, noble metal/silica YS NPs provide many advantages such as optical transparency, reduced electromagnetic coupling between the metallic cores, decreased oxidation rate of noble metals, porous permeable nature of the shell to allow reactant molecules inside, and protection of the core from degradation under harsh environmental conditions. The outer protective layer also prevents the adsorption of the unwanted molecules on their surface or extra signals in the spectrum. More interestingly, metal/SiO_2 YS structures can produce active and stable SERS signals. For example, Au/SiO_2 YS NPs demonstrate good optical stability by the addition of small amounts of quinolone for SERS. Besides silica, many polymers and metal oxides are also used as shell materials, but silica provides more advantages because of its easy tuneability of thickness compared to other oxides or polymers.

Lithium Batteries

Lithium batteries are presently the most dominant power source used to run small electronic devices in our daily life because of their high energy density. Initially, the Li–S battery was considered a promising candidate as a power source since 1960 after its discovery. However, since 1990, Li-ion batteries have dominated the battery market because of their improved stable electrochemistry and longer life span over Li–S batteries. However, the limited electrochemical stability of the electrolyte makes it difficult to increase the cathode operating voltage and capacity. Thus, research interest in Li–S batteries is rising again because sulphur as a cathode has a 5 times higher theoretical capacity than existing commercial graphite materials and has the capability of accommodating more ions.

In recent years, extensive efforts have been devoted to the design of YS NP-based electrode materials in lithium batteries. With the unique structure and composition, YS structures have also shown dominant importance in the field of Li batteries because of their low pulverization and good conductivity compared to pure NPs, core/shell, and hollow NPs. The YS structures show an overall improved performance of lithium batteries because of the following advantages: (i) the core of such structures improves the rate capability as well as the energy density of the powders by increasing the weight fraction of the electrochemically active component, (ii) the void space between the core and shell can also serve as a buffering space for the electroactive core material during lithium insertion/extraction, which improves the cycling performance of the battery, (iii) the hollow conductive shell provides electrical conductivity and shows good elasticity to accommodate the effective strain of volume changes during Li^+ insertion/extraction, (iv) the large surface area and lower diffusion distance, and (v) the hollow shell protects the core from outside environmental changes and agglomeration. Thus, the YS provides a suitable and advanced electrode material for lithium battery applications, and exhibits improved cyclic capacity and energy density. As a result, many combinations have been tried as anode material for Li-ion batteries. For example, Si has been widely used as one of the promising candidates for Li-ion battery anode material because of its high theoretical specific capacity (4200 mA h g^{-1}). However, its low conductivity and fast decay in the cycling test due to large volume expansion (300%) during lithiation/delithiation limits its applicability in batteries. The unique Si/C YS structure with well-defined void space exhibits a high capacity (2800 mA h g^{-1} at C/10), a long cycle life (74% capacity retention after 1000 cycles), and a high columbic efficiency (99.84%) because of the presence of conductive carbon coating and freely expanding Si core.

Microwave Absorber

Microwave absorber materials have been used in military applications for several decades for EMI reduction, antenna pattern shaping, and radar cross reduction. More recently, with the demand for wireless electronics and the movement to higher frequencies, microwave absorbers are used to reduce electromagnetic interference (EMI) inside wireless electronic assemblies. Generally, microwave absorber materials either convert the EM energy into thermal energy or dissipate EM waves through interference. Mostly, magnetite (Fe_3O_4) NPs are used as microwave absorber materials because of their magnetic properties, low cost, and strong absorption characteristics. Recently, many efforts have been made to design YS NPs for microwave absorber application, which exhibit improved magnetic and dielectric loss properties. The YS NPs exhibit unique properties such as low density, high surface area, and synergistic effects of movable core and shell. Mostly, the magnetic core and dielectric shell YS NPs have been reported with excellent microwave absorption performance, which showed lower reflection loss and a wider absorption frequency range. For instance, Fe_3O_4/TiO_2 YS NPs have been reported as an attractive material for microwave absorption, where the maximum reflection loss value of YS can reach −37.6 db at 7 GHz with a thickness of 2 mm, compared to −10.5 db at 7 GHz with the same thickness of pure Fe_3O_4. The microwave absorption properties can also be tuned with different core sizes, interstitial void volumes, and shell thicknesses of YS NPs. Similarly, Fe_3O_4/SnO_2 YS NPs exhibited the maximum reflection loss value of −36.5 db at 7 GHz at a thickness of 2 mm.

Separation Processes

Magnetically driven YS NPs exhibit advantages in the field of adsorption and separation because of a high surface area and a unique structure. Generally, most adsorbent shells are reported with magnetic cores, where the high surface area shell exhibits very good adsorption behaviour and the magnetic core helps in the separation of YS NPs from the solution phase. In this case, YS with metal oxide and carbon shells are reported as good adsorbents. Compared to other shells, carbon exhibits much higher capability for adsorption application because of its better stability in acid or alkaline media. Additionally, the mobility of magnetic NPs enhances the recyclability efficiency of adsorbents. For example, a very high adsorption capacity of bilirubin (146.5 mg g^{-1}) was reported using iron oxide (Fe_3O_4/γ-Fe_2O_3)/C YS NPs. Similarly, Fe_3O_4/C YS NPs also exhibit excellent reusability for the adsorption of pyrene with high adsorption capacities (77.1 mg g^{-1}).

Gas Sensors

There has been increasing attention on gas sensors in recent years because of growing awareness on environmental pollution. It has been found that significant efforts are being paid to design YS NPs for gas sensor applications with ultra-high sensitivity, selectivity, response, stability, and reproducibility. The high surface area of NPs is advantageous for gas sensing and is widely used for industrial applications. Mostly, the metal oxide semiconductor NPs are used as gas sensors based on the working principle of the change in electrical conductivity after the adsorption or desorption of target gas molecules. The YS morphology provides more surface area for adsorption of target gases, which improve the sensitivity, selectivity, and response. The semiconductor metal oxide YS NP-based gas sensors can be categorized into two groups based on their sensing mechanism: (i) semiconductor YS gas sensors, where multishell layers of metal oxide provide a high surface area for the sensing of the target gas, (ii) catalytic driven oxide gas sensors, where metal NPs are encapsulated inside the oxide shell to provide additional sensitivity because of the catalytic effects of metal NPs.

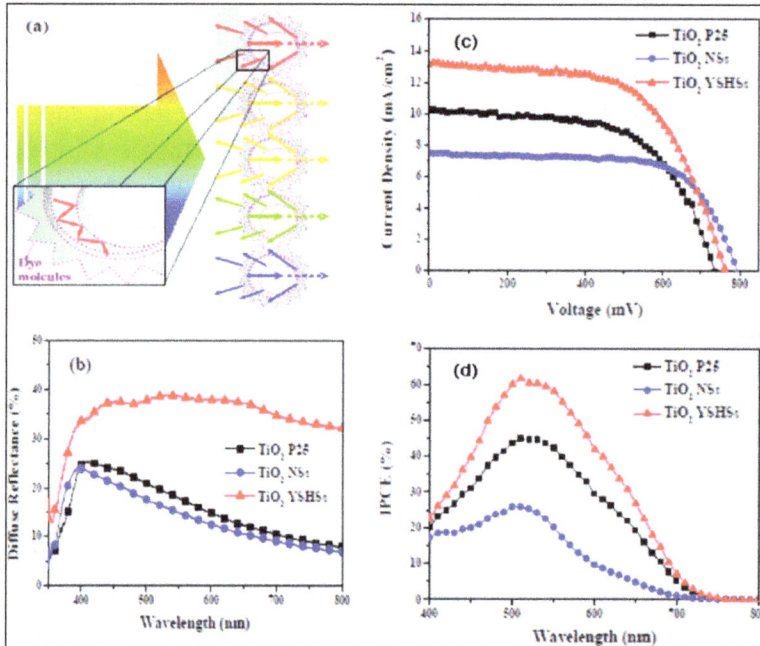

(a) Schematic illustration of multiple scattering and reflection of light inside void space, (b) diffuse reflectance spectra, (c) I–V characteistics, (d) IPCE spectra of DSSC photo-anode composed with YS, NPs and P25 TiO_2, respectively.

Among these sensors, only semiconductor oxide-based sensors are widely used for gas sensing applications, which are made from similar or different metal oxide materials. The multishell structure of YS provides more sites for target molecule adsorption, which enhances the sensitivity and response of the sensor.

The combination of metal with metal oxide gives new strategies for the design of advanced high-performance gas sensors with improved selectivity and sensitivity based on catalytic promotion. Particularly, the multishell layers of oxide prolong the retention time of the target gas molecule within the shell, which also helps in catalytic dissociation. The Pd/SnO_2 YS NPs are reported as useful methyl benzene sensors (xylene and toluene vapor), where Pd loading enhances the gas response either by chemical sensitization (catalytic promotion) or electronic sensitization (decreasing the electron concentration in the sensing material). Similarly, Au/SnO_2 YS NPs exhibit lower detection (3 ppm), faster response (0.3 s) and better selectivity towards CO sensing because of the catalytic effect of Au and enhanced electron depletion at the surface of the YS. Many other YS NP-based gas sensors are also reported for H_2S, CO, ethanol, H_2O_2, and acetone gas sensing.

Dye-sensitized Solar Cells (DSSC)

The DSSC has been an attractive field of research for a new generation of photovoltaic devices because of its low cost and easy fabrication since 1991, when it was first reported by Michael Gratzel. The DSSC is considered a potential alternative renewable energy source in place of expensive silicon-based solar cells. The DSSC is a third generation photovoltaic solar cell that converts visible light into electrical energy. In the DSSC, the dye molecules serve as a photo-sensitizer, in which a photoexcited electron is injected into a wide band-gap semiconductor and conducted away by the semiconductor to the electrolyte and finally the electron comes back to the dye. The movement of the electron in the circuit creates energy which can be harvested by the battery or the super-capacitor. A photoanode composed of semiconductor NPs is the heart of the DSSC, which requires a high surface area for adsorption of more dye molecules. Surface area, electron transport and light-harvesting properties are important parameters to achieve high efficiency of DSSC.

The unique YS structure helps to improve the efficiency of DSSC by enhancing both the surface area and light-harvesting properties. The multifunctional YS NP-based photo-anode offers two possible ways to enhance the efficiency of DSSC: (i) a high surface area for a higher adsorption capacity of dye molecules and (ii) enhanced scattering of incident light. For example, the improved light harvesting and scattering properties of anatase TiO_2 YS microspheres are attributed to the superior IPCE (60.4%) efficiency in visible light compared to simple TiO_2 NPs (24.4%) and P25 (44.6%) as shown in figure. Similarly, TiO_2 YS NPs also show higher overall photoconver-sion efficiency (6.01%) compared to those of TiO_2 NPs (4.01%) and P25 (4.46%).

Multifunctional Nanoparticles

Multifunctional or dynamic nanoplatforms (nanosomes) and tecto-dendrimers are comprised of interconnected nanomodules, each having been developed to fulfil a specific function. Some nanoparticles comprising a nanosome may carry drugs, others are molecules for targeted delivery, others may fulfil the role of biosensors (pH, redox potential, membrane potential, etc), and still others may carry nanoantennae made of gold nanocrystals that heat the nanosome when placed in an electromagnetic field with a certain frequency. Superparamagnetic nanoparticles attached to a nanosome enable nanosome visualisation using tomography. Fluorescence-based nanomodules that monitor the efficacy of nanomedical therapy, e.g. tumour cell death have been developed. Depending on a specific medical objective, nanosomes can be comprised of different functional modules and carry out several tasks, e.g. internal environment monitoring, visualisation of target cells, drug delivery and controlled drug release, treatment monitoring. Non-modular multifunctional nanoparticles include, for example, modified viral capsids. When these particles are assembling, both the content of the capsid (payload) and the composition of the capsid surface molecules responsible for targeted delivery and certain sensory functions may be changed. Nanosomes and other above-mentioned multifunctional nanodevices may be considered as prototypes of medical nanorobots.

A schematic polymer model of a multifunctional medical nanoparticle is shown on a figure. The solubilizing block (which can be represented by a polymer chain) allows the nanoparticle to operate in a biological medium (blood, lymph, etc.). The particle's hydrophilicity/hydrophobicity, electrostatic charge and its density effect the pharmacokinetic and pharmacodynamic properties of a drug. Polymer chains may differ in terms of stability, size, composition and the presence of specific domains (e.g., hydrophobic segments). Polymer molecular weight affects the membrane-penetrating ability of the drug (passing through the blood-brain barrier and endocytosis stimulation). An active pharmaceutical agent (pharmacon) may be bound to the polymer base (or enclosed in nanocontainer) via a biodegradable or stable bond, being an inactive drug precursor, or an active metabolite (active pharmaceutical ingredient). A "targeting device" acts as a vector (may be represented by antibodies, molecules appearing in the affected area, protein domains with specific sorption/binding properties, etc.) that guides a nanoparticle to a given segment of a tissue or target organ. In a biosystem a conjugate obtains the conformation that facilitates the assembly of a conjugate-based multifunctional therapeutic nanoparticle.

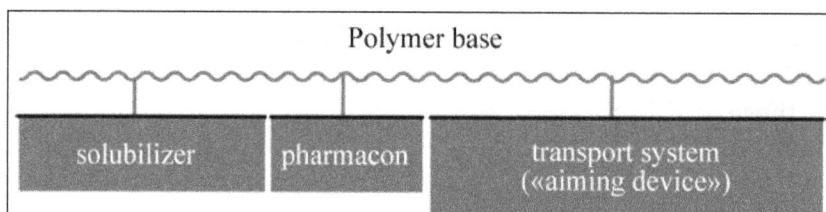

General schematic model of a medical polymer multifunctional nanoparticle.

Applications of Nanoparticles in Biology and Medicine

A list of some of the applications of nanomaterials to biology or medicine is given below:

- Fluorescent biological labels.

- Drug and gene delivery.

- Bio detection of pathogens.

- Detection of proteins.

- Probing of DNA structure.

- Tissue engineering.

- Tumour destruction via heating (hyperthermia).

- Separation and purification of biological molecules and cells.

- MRI contrast enhancement.

- Phagokinetic studies.

The fact that nanoparticles exist in the same size domain as proteins makes nanomaterials suitable for bio tagging or labelling. However, size is just one of many characteristics of nanoparticles that itself is rarely sufficient if one is to use nanoparticles as biological tags. In order to interact with biological target, a biological or molecular coating or layer acting as a bioinorganic interface should be attached to the nanoparticle. Examples of biological coatings may include antibodies, biopolymers like collagen, or monolayers of small molecules that make the nanoparticles biocompatible. In addition, as optical detection techniques are wide spread in biological research, nanoparticles should either fluoresce or change their optical properties.

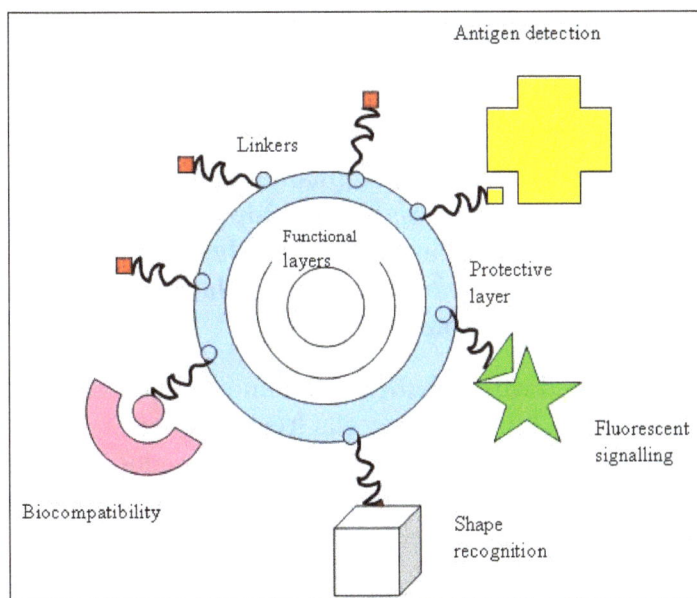

Typical configurations utilised in nano-bio materials
applied to medical or biological problems.

Nano-particle usually forms the core of nano-biomaterial. It can be used as a convenient surface for molecular assembly, and may be composed of inorganic or polymeric materials. It can also be in the form of nano-vesicle surrounded by a membrane or a layer. The shape is more often spherical but cylindrical, plate-like and other shapes are possible. The size and size distribution might be important in some cases, for example if penetration through a pore structure of a cellular membrane is required. The size and size distribution are becoming extremely critical when quantum-sized effects are used to control material properties. A tight control of the average particle size and a narrow distribution of sizes allow creating very efficient fluorescent probes that emit narrow light in a very wide range of wavelengths. This helps with creating biomarkers with many and well distinguished colours. The core itself might have several layers and be multifunctional. For example, combining magnetic and luminescent layers one can both detect and manipulate the particles.

The core particle is often protected by several monolayers of inert material, for example silica. Organic molecules that are adsorbed or chemisorbed on the surface of the particle are also used for this purpose. The same layer might act as a biocompatible material. However, more often an additional layer of linker molecules is required to proceed with further functionalisation. This linear linker molecule has reactive groups at both ends. One group is aimed at attaching the linker to the nanoparticle surface and the other is used to bind various moieties like biocompatibles (dextran), antibodies, fluorophores etc., depending on the function required by the application.

Tissue Engineering

Natural bone surface is quite often contains features that are about 100 nm across. If the surface of an artificial bone implant were left smooth, the body would try to reject it. Because of that smooth surface is likely to cause production of a fibrous tissue covering the surface of the implant. This layer reduces the bone-implant contact, which may result in loosening of the implant and further inflammation. It was demonstrated that by creating nano-sized features on the surface of the hip or knee prosthesis one could reduce the chances of rejection as well as to stimulate the production of osteoblasts. The osteoblasts are the cells responsible for the growth of the bone matrix and are found on the advancing surface of the developing bone.

The effect was demonstrated with polymeric, ceramic and, more recently, metal materials. More than 90% of the human bone cells from suspension adhered to the nano-structured metal surface, but only 50% in the control sample. In the end this findings would allow to design a more durable and longer lasting hip or knee replacements and to reduce the chances of the implant getting loose.

Titanium is a well-known bone repairing material widely used in orthopaedics and dentistry. It has a high fracture resistance, ductility and weight to strength ratio. Unfortunately, it suffers from the lack of bioactivity, as it does not support sell adhesion and growth well. Apatite coatings are known to be bioactive and to bond to the bone. Hence, several techniques were used in the past to produce an apatite coating on titanium. Those coatings suffer from thickness non-uniformity, poor adhesion and low mechanical strength. In addition, a stable porous structure is required to support the nutrients transport through the cell growth.

It was shown that using a biomimetic approach – a slow growth of nanostructured apatite film from the simulated body fluid – resulted in the formation of a strongly adherent, uniform nanoporous layer. The layer was found to be built of 60 nm crystallites, and possess a stable nanoporous structure and bioactivity.

A real bone is a nanocomposite material, composed of hydroxyapatite crystallites in the organic matrix, which is mainly composed of collagen. Thanks to that, the bone is mechanically tough and, at the same time, plastic, so it can recover from a mechanical damage. The actual nanoscale mechanism leading to this useful combination of properties is still debated.

An artificial hybrid material was prepared from 15–18 nm ceramic nanoparticles and poly (methyl methacrylate) copolymer. Using tribology approach, a viscoelastic behaviour (healing) of the human teeth was demonstrated. An investigated hybrid material, deposited as a coating on the tooth surface, improved scratch resistance as well as possessed a healing behaviour similar to that of the tooth.

Multicolour Optical Coding for Biological Assays

The ever increasing research in proteomics and genomic generates escalating number of sequence data and requires development of high throughput screening technologies. Realistically, various array technologies that are currently used in parallel analysis are likely to reach saturation when a number of array elements exceed several millions. A three-dimensional approach, based on optical "bar coding" of polymer particles in solution, is limited only by the number of unique tags one can reliably produce and detect.

Single quantum dots of compound semiconductors were successfully used as a replacement of organic dyes in various bio-tagging applications. This idea has been taken one step further by combining differently sized and hence having different fluorescent colours quantum dots, and combining them in polymeric microbeads. A precise control of quantum dot ratios has been achieved. The selection of nanoparticles used in those experiments had 6 different colours as well as 10 intensities. It is enough to encode over 1 million combinations. The uniformity and reproducibility of beads was high letting for the bead identification accuracies of 99.99%.

Manipulation of Cells and Biomolecules

Functionalised magnetic nanoparticles have found many applications including cell separation and probing. Most of the magnetic particles studied so far are spherical, which somewhat limits the possibilities to make these nanoparticles multifunctional. Alternative cylindrically shaped nanoparticles can be created by employing metal electrodeposition into nanoporous alumina template. Depending on the properties of the template, nanocylinder radius can be selected in the range of 5 to 500 nm while their length can be as big as 60 µm. By sequentially depositing various thicknesses of different metals, the structure and the magnetic properties of individual cylinders can be tuned widely.

As surface chemistry for functionalisation of metal surfaces is well developed, different ligands can be selectively attached to different segments. For example, porphyrins with thiol or carboxyl linkers were simultaneously attached to the gold or nickel segments respectively. Thus, it is possible to produce magnetic nanowires with spatially segregated fluorescent parts. In addition, because of the large aspect ratios, the residual magnetisation of these nanowires can be high. Hence, weaker magnetic field can be used to drive them. It has been shown that a self-assembly of magnetic nanowires in suspension can be controlled by weak external magnetic fields. This would potentially allow controlling cell assembly in different shapes and forms. Moreover, an external magnetic field can be combined with a lithographically defined magnetic pattern ("magnetic trapping").

Protein Detection

Proteins are the important part of the cell's language, machinery and structure, and understanding their functionalities is extremely important for further progress in human well being. Gold nanoparticles are widely used in immunohistochemistry to identify protein-protein interaction. However, the multiple simultaneous detection capabilities of this technique are fairly limited. Surface-enhanced Raman scattering spectroscopy is a well-established technique for detection and identification of single dye molecules. By combining both methods in a single nanoparticle probe one can drastically improve the multiplexing capabilities of protein probes. The group of Prof. Mirkin has designed a sophisticated multifunctional probe that is built around a 13 nm gold nanoparticle. The nanoparticles are coated with hydrophilic oligonucleotides containing a Raman dye at one end and terminally capped with a small molecule recognition element (e.g. biotin). Moreover, this molecule is catalytically active and will be coated with silver in the solution of Ag(I) and hydroquinone. After the probe is attached to a small molecule or an antigen it is designed to detect, the substrate is exposed to silver and hydroquinone solution. A silver-plating is happening close to the Raman dye, which allows for dye signature detection with a standard Raman microscope. Apart from being able to recognise small molecules this probe can be modified to contain antibodies on the surface to recognise proteins. When tested in the protein array format against both small molecules and proteins, the probe has shown no cross-reactivity.

Commercial Exploration

Some of the companies that are involved in the development and commercialisation of nanomaterials in biological and medical applications are listed. The majority of the companies are small recent spinouts of various research institutions. Although not exhausting, this is a representative selection reflecting current industrial trends. Most of the companies are developing pharmaceutical applications, mainly for drug delivery. Several companies exploit quantum size effects in semiconductor nanocrystals for tagging biomolecules, or use bio-conjugated gold nanoparticles for labelling various cellular parts. A number of companies are applying nano-ceramic materials to tissue engineering and orthopaedics.

Table: Examples of Companies commercialising nanomaterials for bio- and medical applications.

Company	Major area of activity	Technology
Advectus Life Sciences Inc.	Drug delivery	Polymeric nanoparticles engineered to carry anti-tumour drug across the blood-brain barrier
Alnis Biosciences, Inc.	Bio-pharmaceutical	Biodegradable polymeric nanoparticles for drug delivery
Argonide	Membrane filtration	Nanoporous ceramic materials for endotoxin filtration, orthopaedic and dental implants, DNA and protein separation
BASF	Toothpaste	Hydroxyapatite nanoparticles seems to improve dental surface
Biophan Technologies, Inc.	MRI shielding	Nanomagnetic/carbon composite materials to shield medical devices from RF fields
Capsulution Nano-Science AG	Pharmaceutical coatings to improve solubility of drugs	Layer-by-layer poly-electrolyte coatings, 8–50 nm
Dynal Biotech		Magnetic beads
Eiffel Technologies	Drug delivery	Reducing size of the drug particles to 50–100 nm
EnviroSystems, Inc.	Surface desinfectsant	Nanoemulsions
Evident Technologies	Luminescent biomarkers	Semiconductor quantum dots with amine or carboxyl groups on the surface, emission from 350 to 2500 nm
Immunicon	Tarcking and separation of different cell types	magnetic core surrounded by a polymeric layer coated with antibodies for capturing cells
KES Science and Technology, Inc.	AiroCide filters	Nano-TiO_2 to destroy airborne pathogens
NanoBio Cortporation	Pharmaceutical	Antimicrobal nano-emulsions
NanoCarrier Co., Ltd	Drug delivery	Micellar nanoparticles for encapsulation of drugs, proteins, DNA
NanoPharm AG	Drug delivery	Polybutilcyanoacrylate nanoparticles are coated with drugs and then with surfactant, can go across the blood-brain barrier
Nanoplex Technologies, Inc	Nanobarcodes for bioanalysis	
Nanoprobes, Inc.	Gold nanoparticles for biological markers	Gold nanoparticles bio-conjugates for TEM and fluorescent microscopy

Nanoshpere, Inc.	Gold biomarkers	DNA barcode attached to each nanoprobe for identification purposes, PCR is used to amplify the signal; also catalytic silver deposition to amplify the signal using surface plasmon resonance
NanoMed Pharmaceutical, Inc.	Drug delivery	Nanoparticles for drug delivery
Oxonica Ltd	Sunscreens	Doped transparent nanoparticles to effectively absorb harmful UV and convert it into heat
PSiVida Ltd	Tissue engineering, implants, drugs and gene delivery, bio-filtration	Exploiting material properties of nanostructured porous silicone
Smith & Nephew	Acticoat bandages	Nanocrystal silver is highly toxic to pathogenes
QuantumDot Corporation	Luminescent biomarkers	Bioconjugated semiconductor quantum dots

Most major and established pharmaceutical companies have internal research programs on drug delivery that are on formulations or dispersions containing components down to nano sizes. Colloidal silver is widely used in anti-microbial formulations and dressings. The high reactivity of titania nanoparticles, either on their own or then illuminated with UV light, is also used for bactericidal purposes in filters. Enhanced catalytic properties of surfaces of nano-ceramics or those of noble metals like platinum are used to destruct dangerous toxins and other hazardous organic materials.

As it stands now, the majority of commercial nanoparticle applications in medicine are geared towards drug delivery. In biosciences, nanoparticles are replacing organic dyes in the applications that require high photo-stability as well as high multiplexing capabilities. There are some developments in directing and remotely controlling the functions of nano-probes, for example driving magnetic nanoparticles to the tumour and then making them either to release the drug load or just heating them in order to destroy the surrounding tissue. The major trend in further development of nanomaterials is to make them multifunctional and controllable by external signals or by local environment thus essentially turning them into nano-devices.

Targeted Nanoparticles for Cancer Therapy

Using targeted nanoparticles to deliver chemotherapeutic agents in cancer therapy offers many advantages to improve drug/gene delivery and to overcome many problems associated with conventional chemotherapy. For example, nanoparticles via either passive targeting or active targeting have been shown to enhance the intracellular concentration of drugs/genes in cancer cells while avoiding toxicity in normal cells. In addition, the targeted nanoparticles can also be designed as either pH-sensitive or temperature-sensitive carriers. The pH-sensitive drug delivery system can deliver and release drugs within the more acidic microenvironment of the cancer cells and components within cancer cells. The temperature-sensitive system can carry and release drugs with changes in temperature locally in the tumor region provided by sources such as magnetic fields, ultrasound waves, and so on so that combined therapy such as chemotherapy and hyperthermia can be applied. The targeting of nanoparticles to tumors via cancer-specific features/moieties has also been shown to minimize the effects of composition, size, and molecular mass of nanoparticles on their efficacy. Targeted nanoparticles can be further modified or functionalized to reduce toxicity. For example, modifying nanoparticles surface chemistry could reduce their toxicity and immunotoxicity.

Challenges of Targeted NPs for Cancer Therapy

Although targeted nanoparticles have emerged as one strategy to overcome the lack of specificity of conventional chemotherapy, there are also potential risks and challenges associated with this novel strategy. For instance, some cancer cell types would develop drug resistance over the drug treatment course, thereby rendering drugs released from the targeted nanoparticles to be ineffective. Combined therapies, such as the use of targeted nanoparticles for delivering both chemotherapeutics and gene therapeutics, might be effectively delivered and specifically targeted to cancer cells and tissues to overcome this drug resistance and to stop the tumor growth. Another strategy to overcome this drug resistance is to develop multifunctional targeted nanoparticles.

Similar to other new technologies, targeted NPs for cancer therapy also face many challenges. One challenge of targeted NPs is that NPs might change the stability, solubility, and pharmacokinetic properties of the carried drugs. However, these aspects have not been extensively investigated. The shelf life, aggregation, leakage, and toxicity of materials used to make nanoparticles are other limitations for their use. Some materials used to make NPs such aspoly (lactic-co-glycolic acid) (PLGA) have low toxicity, but degrade quickly and do not circulate in tissues long enough for sustained drug/gene delivery. On the other hand, other materials such as carbon nanotubes and quantum dots are durable and can persist in the body for weeks, months, or even years, making them potentially toxic and limiting their use for repeated treatments. New materials to make targeted nanoparticles such as silicon/silica (solid, porous, and hollow silicon nanoparticles) have been developed; however, their use for drug delivery to cancer patients has taken off slowly due to the potential health risks associated with introducing new materials in the human body.

Besides developing new materials and selecting appropriate materials for each specific treatment, other factors need to be optimally selected in order to design better targeted nanoparticles. These factors include the particles size, shape, sedimentation, drug encapsulation efficacy, desired drug release profiles, distribution in the body, circulation, and cost. For instance, in the case of particle size, it has been well-known that the clearance rate of very small nanoparticles might be high, and most of these nanoparticles might end up in the liver and spleen, thus making the use of targeted nanoparticles impractical and ineffective. On the other hand, larger nanoparticles might be too big to go through small capillaries for drug delivery. Thus selecting the right materials and particle size is another important aspect in targeted NPs for cancer therapy.

Despite extensive research efforts to develop new targeted nanoparticles, only a few of them are in clinical use including Abraxane, Doxil, and Myocet that are approved by FDA. A major account forthe slow development of effective targeted nanoparticles has been due to the lack of knowledge about the distribution and location of targeted nanoparticles after either oral administration or injection. For example, most studies have not examined the targeting efficiency of nanoparticles real time in vivo, thus precise bio-distribution and subsequently therapeutic effects are not well-known. Therefore, detecting cancer (malignant) cells in the body and monitoring treatment effects on these cells in real time is another challenge needed to be overcome to develop efficient targeted nanoparticles.

References

- Nanoparticle, science-328326: britannica.com, Retrieved 17 August, 2019

- How-nanoparticles-are-made: nanowerk.com, Retrieved 12 March, 2019

- YolkShell-Nanoparticles-Classifications-Synthesis-Properties-and-Applications-283908469: researchgate.net, Retrieved 29 July, 2019

- The preparation of magnetic nanoparticles for applications in biomedicine. J Phys D: Appl Phys 2003, 36: R182-R197. 10.1088/0022-3727/36/13/202

- Targeted-nanoparticles-for-cancer-therapy-promises-and-challenges-2157-7439.1000103: longdom.org, Retrieved 09 June, 2019

- Kralj, Slavko; Makovec, Darko (27 October 2015). "Magnetic Assembly of Superparamagnetic Iron Oxide Nanoparticle Clusters into Nanochains and Nanobundles". ACS Nano. 9 (10): 9700–9707. doi:10.1021/acsnano.5b02328. PMID 26394039

- Cancer nanotechnology: opportunities and challenges. Nat Rev Cancer. 2005;5(3):161–171

- Superparamagnetic iron oxide nanoparticles (SPIONs): development, surface modification and applications in chemotherapy. Adv Drug Deliv Rev. 2011;63(1–2):24–46

Nanofibers: An Integrated Study

The fibers that are generated from different polymers and have radius in nanometer range are referred to as nanofibers. Electrospinning, self-assembly, template synthesis, and thermal-induced phase separation are some of the methods used to make nanofibers. This chapter closely examines the key concepts of nanofibers to provide an extensive understanding of the subject.

Nanofibers are an exciting new class of material used for several value added applications such as medical, filtration, barrier, wipes, personal care, composite, garments, insulation, and energy storage. Special properties of nanofibers make them suitable for a wide range of applications from medical to consumer products and industrial to high-tech applications for aerospace, capacitors, transistors, drug delivery systems, battery separators, energy storage, fuel cells, and information technology.

Generally, polymeric nanofibers are produced by an electrospinning process. Electrospinning is a process that spins fibers of diameters ranging from 10nm to several hundred nanometers. This method has been known since 1934 when the first patent on electrospinning was filed. Fiber properties depend on field uniformity, polymer viscosity, electric field strength and DCD (distance between nozzle and collector). Advancements in microscopy such as scanning electron microscopy has enabled us to better understand the structure and morphology of nanofibers. At present the production rate of this process is low and measured in grams per hour.

Another technique for producing nanofibers is spinning bi-component fibers such as Islands-In-The-Sea fibers in 1-3 denier filaments with from 240 to possibly as much as 1120 filaments surrounded by dissolvable polymer. Dissolving the polymer leaves the matrix of nanofibers, which can be further separated by stretching or mechanical agitation.

The most often used fibers in this technique are nylon, polystyrene, polyacrylonitrile, polycarbonate, PEO, PET and water-soluble polymers. The polymer ratio is generally 80% islands and 20% sea. The resulting nanofibers after dissolving the sea polymer component have a diameter of approximately 300 nm. Compared to electrospinning, nanofibers produced with this technique will have a very narrow diameter range but are coarser.

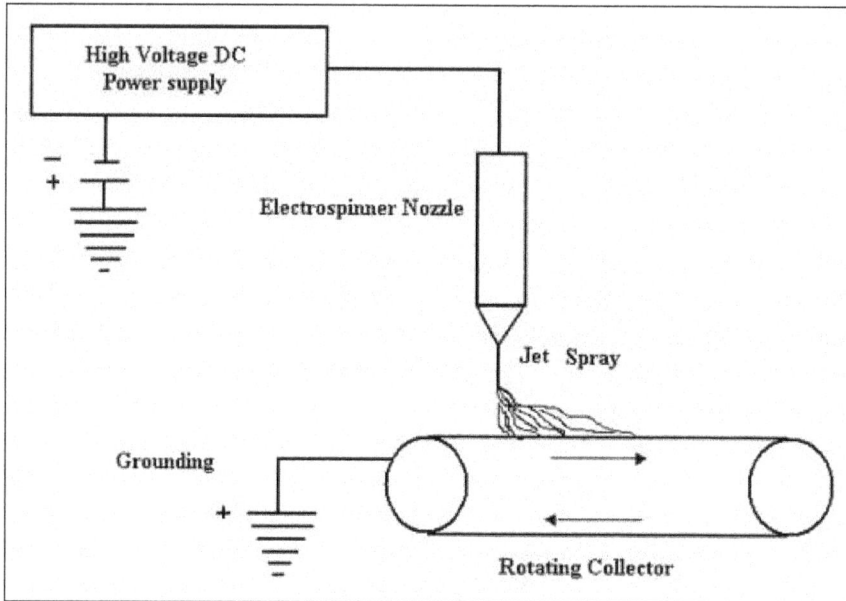

Schematic representation of electrospinning process.

Properties of Nanofibers

- Nanofibers exhibit special properties mainly due to extremely high surface to weight ratio compared to conventional nonwovens.

- Low density, large surface area to mass, high pore volume, and tight pore size make the nanofiber nonwoven appropriate for a wide range of filtration applications.

- Nanofibers are smaller compared to a human hair, which is 50-150 μm and the size of a pollen particle compared to nanofibers. The elastic modulus of polymeric nanofibers of less than 350 nm is found to be 1.0±0.2 Gpa.

Application of Nanofibers

Filtration

Nanofibers have significant applications in the area of filtration since their surface area is substantially greater and have smaller micropores than melt blown (MB) webs. High porous structure with high surface area makes them ideally suited for many filtration applications. Nanofibers are ideally suited for filtering submicron particles from air or water.

Medical Application

Electrospun fibers have diameters three or more times smaller than that of MB fibers. This leads to a corresponding increase in surface area and decrease in basis weight.

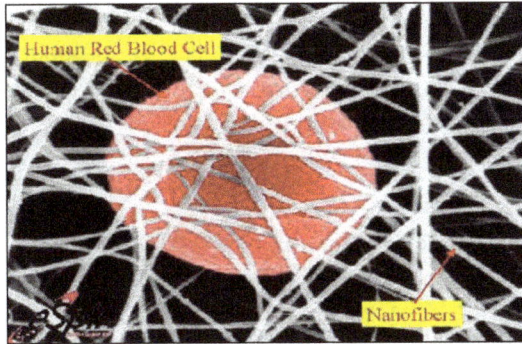

Comparison of red blood cell with nanofibers web.

Nanofibers and Electrospinning Method

Electrospinning is a simple and comprehensive process for generating an ultrafine fibre from varieties of materials which include polymer, composite and ceramic. The electrospinning setup consists of three major components namely, high voltage power supply, syringe with metal needle and a conductive collector. It is, in fact, very sophisticated, but a simple, processing mechanism of producing nanofiber. The electrospinning process can be classified into several techniques like vibration electrospinning, magneto-electrospinning, siro-electrospinning and bubble electrospinning. As the charge liquid jet moves from the syringe tip to the collector, the mode of current flow changes from ohmic to convective flow as the charge moves instead to the fibre surface.

A slow acceleration is a characteristic of the ohmic flow, since the geometry of the Taylor cone is controlled by the ratio of the surface tension to electrostatic repulsion. After successfully addressing the ohmic flow, the jet travels at a rapid acceleration, which includes the transition zone from liquid to dry solid. In the end, the jet penetrates the collector. The name 'Taylor Cone' simply represents the conical shape formed at the needle tip.

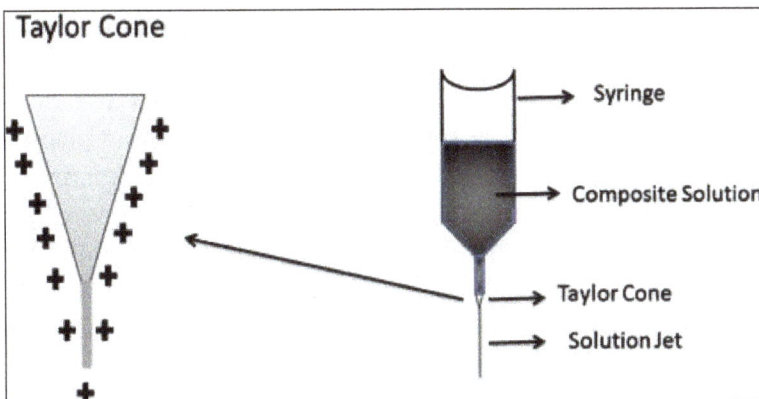

Illustration of Taylor Cone formations from the Syringe Needle Tip.

In 1964, Sir Geoffrey Ingram Taylor described this cone as a continuation of the study by Zeleny on the formation of a cone-jet of glycerine exposed to high electric fields. Several others including Wilson and Taylor, Nolan and Macky continued research in this area. However, it was Taylor who investigated further into the reactions between droplets and electric fields. Taylor's result is based on two assumptions: (1) that the surface of the cone is an equipotential surface and (2) that the cone exists in steady state equilibrium. Immediately after being discharged and the Taylor cone activated, the polymer jet goes through a whipping process in which the solvent evaporates, precipitating a charged polymer fibre, which lays itself at random on a grounded collecting metal screen. As far as the melt is concerned, the discharged jet solidifies when it travels in the air and is collected on the grounded metal screen. The terminal diameter of the 'whipping' jet (ht) is controlled by flow rate (Q), electric current (I) and fluid surface tension (γ) as given by the equation,

$$h_t = \left(\gamma \bar{\in} \frac{Q^2}{I^2} \frac{2}{\pi(2Inx-3)} \right)^{\frac{1}{3}}$$

where $\bar{\in}$ is the dielectric constant, (x) is the displacement, equation above offers a prediction that the terminal diameter of the whipping jet is controlled by the flow rate, electric current and the surface tension of the fluid, disregards the elastic effects and fluid evaporation, and also makes an assumption about the minimal jet thinning after the saturation of the whipping instability takes place.

Electrospinning Setups

Electrospinning Process

The electrospinning apparatus is really a simple idea, carrying only three main components: a high voltage power supply, a polymer solution reservoir (e.g., a syringe, with a small diameter needle) with or without a flow control pump, and a metal collecting screen. A high voltage power supply with adjustable control can well provide up to 50-kV DC output and, depending on the number of electrospinning jets, the multiple outputs that function independently, are necessitated. The polymeric solution is kept in a reservoir and connected to a power supply to establish a charged polymer jet. Charging the polymer solution could be done either with a syringe with a metal needle or a capillary with a metal tip in the polymer solution. If the syringe is not placed horizontally, polymer flow can be driven by gravity. However, to remove the experimental variables, a syringe pump is engaged to control the precise flow rate. The fibre collecting screen is expected to be conductive and it can either be a stationary plate or a rotating platform or substrate. The plate can produce non-woven fibres, whereas a rotating platform can produce both nonwoven and aligned fibres.

Presently, two standard electrospinning setups are available namely the vertical and horizontal, with three new electrospinning setups with different angles for the study of the effect of the gravity. As a result of the increasing interest in this technology, many research groups have developed sophisticated mechanisms by which more complex nanofibers structures, can be fabricated in a more controlled and efficient manner. For instance, motor-controlled multiple jets and fibre-collecting targets provide avenue for producing a single nanofibrous scaffold consisting of multiple layers, with each layer obtained from a different polymer type. Furthermore, this technology can be used to manufacture polymer composite scaffolds where the fibres of each layer represent a combination of various polymer types.

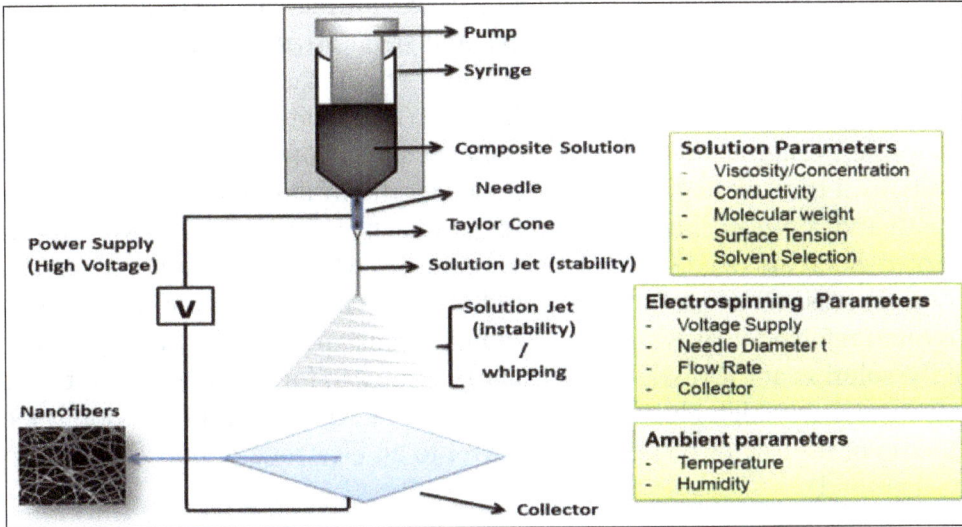

Electrospinning setup and controllable electrospinning process parameters.

Mechanism and Technique for Nanofiber Formation

Electrospinning process has been studied extensively. The mechanism takes place when the surface tension of the solution is overcome by an applied electric field, thereby ejecting tiny jets from the surface. Taylor identified a critical voltage at which this breakdown would occur:

$$V_c^2 = 4\frac{H^2}{L^2}\left(In\frac{2L}{R} - \frac{3}{2}\right)(0.117\pi\gamma R)$$

where Vc is the critical voltage, H is the separation between the capillary and the ground, L is the length of the capillary, R is the radius of the capillary and γ is the surface tension of the liquid. A similar relationship by Carson had progressed for the potential electrostatic spraying from a hemispherical drop pendant from a capillary tube:

$$V = 300\sqrt{20\gamma r}$$

where v is electric field, γ is the surface tension of the liquid and r is the radius of the pendant drop. As he investigated a small range of fluids, Taylor determined a 49.3°

equilibrium angle-balanced surface tension with electrostatic forces, and he had manipulated this value in his derivation. Taylor cones are required for electrospinning because they define the onset of subtle velocity gradients in the fibre forming process. When V > Vc, a thin jet of solution will explode from the cone surface and travel towards the nearest electrode of opposite polarity, or electrical ground. As a way to describe this, electrospinning jet is a string of charged elements connected by a viscoelastic medium, with one end attached to the point of origin and the other end is let free. When a polymer solution, held in a capillary by its surface tension, is subjected to an electrical field, a charge is induced on the liquid surface. Mutual charge repulsion occurs in a force opposing the surface tension forces and shear stresses are set up in the fluid. With increased intensity in the electrical field, ions in the like-polarity solution will be forced to aggregate at the surface of the drop. The length of the stable jet increases with increasing voltage. After the viscoelastic jet starts flowing away from the Taylor's cone, initially it traverses a linear path. The jet gradually starts to bend away from this linear path and complex shape changes may occur from repulsive forces generated in the charged elements inside the electrospinning jet. The jet may undergo substantial reductions in cross sectional area and spiralling loops may grow from it. This phenomenon is often referred to as whipping instability. Consequently, the hemispherical surface of the solution at the tip of the capillary elongates to form a cone, called the Taylor cone. At remarkably high electrical fields (V > Vc), a charged jet of solution is ejected from the tip of the Taylor cone and will travel to an electrode of opposite polarity (or electrical ground).

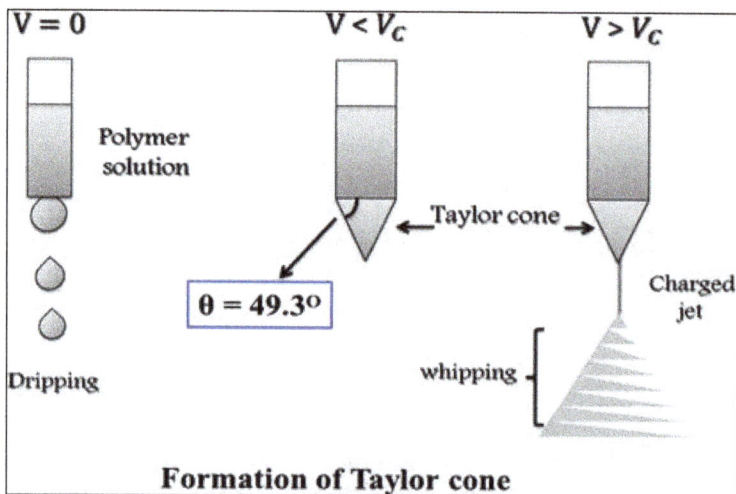

Formation of Taylor cone

Schematic illustration of the effects by electric field applied to a solution in a capillary.

A dimensionless parameter called the Berry number (Be) was used by various research groups as a processing index for controlling the diameter of fibres. Be number is defined as:

$$B_e = \eta C$$

where (η) is the intrinsic viscosity of the polymer which is the ratio of the specific viscosity to the concentration at an infinite dilution and C is the concentration of the polymer solution. Intrinsic viscosity is dependent on polymer molecular weight. B_e also describes the level of the entanglement of polymer chains in a solution. As far as highly diluted solutions are concerned, when B_e is less than unity, the polymer molecules are sparsely distributed in the solution. There is low probability for individual molecules to be entangled with each other.

When B_e is greater than one, the polymer concentration as well as the level of molecular entanglement are enhanced, leading to more favourable conditions for the fibre formation. The experiment implied that as the solution viscosity increased the fibre diameter also increased (approximately proportionally) to the jet length. Baumgarten detailed the relationship between fibre diameter and solution viscosity expressed by the following equation:

$$D = \sqrt[3]{\eta}$$

where D is the fibre diameter and η is the solution viscosity in poise. It is also proves that fibre diameter is highly dependent on the applied electric field. An increase in the applied voltage increases the electrostatic stress, which, in turn, produces smaller diameter fibres. According to Huang further increase in the electric field, is a critical value achieved when the electrostatic repulsive force deals with the surface tension and the charged jet of the fluid is ejected from the end of the Taylor cone. In the process, the solvent evaporates, contributing to the formation of charged polymer fibre. As for the melt, the discharged jet solidifies when it moves in the air. The shape of the base is dependent on the surface tension of the liquid and the force of the electric field; jets can be ejected from surfaces that are mostly flat if the electric field is high enough. The solvent in the polymer jet evaporates in the movement to the collecting screen, thereby increasing the surface charge on the jet. As it passes via the electric field, this increase in the surface charge induces instability in the polymer jet. The polymer jet divides geometrically, first into two jets, and then into many more as the process repeats itself in order to compensate for this instability. Nanofibers are formed from the spinning force action, given by the electrostatic force on the continuous splitting of the polymer droplets. They are deposited one layer after another on the metal target plate, forming a non-woven nanofibrous mat. The mechanisms, by which nanofibers are formed, much less controlled, have not been fully elucidated yet, although the electrospraying/ electrospinning technology has been used for such a long time. Not much of theoretical clarity has been achieved, although, several studies have been carried out to investigate the mechanism of fibre formation to reproducibly control scaffold design. A uniform fibrous structure is created only under optimised operating during the process electrospinning. The structural morphology of the nanofibers is affected by both extrinsic and intrinsic parameters. In order to produce inform nanofibers, external parameters, like environmental humidity and temperature, in addition to intrinsic parameters, including applied voltage, working distance and conductivity and viscosity of the polymer solution need to be optimised. The intrinsic parameters are more critical in determining the nanofiber structure in general sense.

Operating Parameters for Electrospinning

Three main parameters, which are solution parameters, process parameters and ambient parameters, tend to affect the electrospinning process. These operating parameters play a big role in determining the desired quality of the electrospun fibre produced. The most preferred in many applications is a fibre with diameter within 10–1000 nm in scale and a smooth surface morphology. The solution properties are difficult to alter, according to Lu and Ding, since the relationship between one parameter will drag the other parameters; and in addition, they are very difficult to isolate as one controllable parameter. Researchers discussed the effects of working parameters that govern the electrospinning process and the process, and discovered that these parameters could affect the fibre morphologies and diameters. In their study on the effects of the parameters on nanofiber diameter, Thompson found out that the jet radius could leave an impact to the production of the electrospun fibre. In their study, a number of parameters significantly affected the fibre formation compared to other parameters. For instance, the first electrospinning method by Formhals had had some technical disadvantages since it was not easy to dry the fibres entirely after electrospinning as the spinning and collection zones have a very short distance; this resulted in a less aggregated web structure. After half a decade, however Formhals, in his pioneering work, changed the distance between the nozzle and the collecting device in order to give more drying time for the electrospun fibres at a longer distance.

Polymer/Solution Parameters

Viscosity/Concentration

The most critical factor in controlling the structural morphology of the nanofibrous structure is polymer solution viscosity, a parameter that is directly proportional to the concentration of the polymer solution. For fibre formation, polymer viscosity should be in a particular range, depending on the type of polymer and solvent used. An electrospinning method by Zeng J was used to fabricate PLA nanofiber, with different concentrations or viscosities ranging from 1% to 5% by weight as shown in figure, and below 3% for beads or beaded fibres and 3% and above for continuous nanofibers. A bead that contains nanofibers structure was created below this range. Spherical beads had become longer and turned into spindle-shaped ones, with increased viscosity, and the number of beads in the structure was decreased. In the same manner, Liu also talked about that a different particular range of viscosity rendered appropriate for the formation of uniform nanofibers composed of cellulose. Recent studies done by Deitzel and Demir, in addition, have illustrated that a more viscous polymer solution can well create larger fibres. To sum up, these studies have been able to prove that there is a polymer-specific, optimal viscosity value for electrospinning.

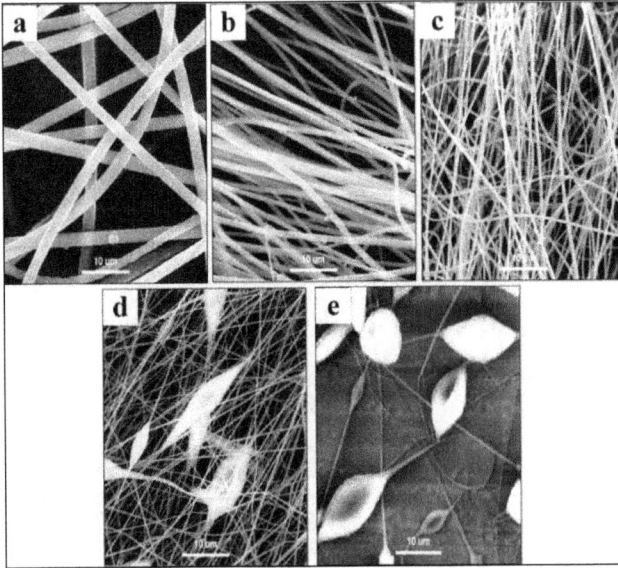

FESEM of electrospinning of nanofibers with different concentrations PLA.
(a) 5%, (b) 4%, (c) 3%, (d) 2%, and (e) 1%.

Polymer concentration determines the spinnability of a solution. For chain entanglements to occur, the solution must have high enough polymer concentration. However, the solution should not be either too diluted or too concentrated. Both, the viscosity and surface tension of the solution are affected by the polymer concentration.

Conductivity

The charge carrying capacity of polymer solutions with high conductivity is greater than solutions with low conductivity. Therefore, the fibre jet produced from a solution of high conductivity will tend to have higher tensile force when exposed to an applied voltage. Through observation, an increase in the solution conductivity brings about a substantial decrease in the nanofiber diameter; and also, evidently the radius of the nanofiber jet is inversely related to the cube root of the electrical conductivity of the solution.

The conductivity of a given cell has a connection with the molar conductivity following equation $\Lambda = \dfrac{k}{c}$ where k is the conductivity with units of mS/cm, c is the ion concentration with units of mol/L and therefore, the molar conductivity (Λ) has units of S cm2/mol.

$$\Lambda = \frac{k}{c}$$

Chitral and Shesha had published the results of a comprehensive investigation of the effects of change in the conductivity of polyethylene oxide (PEO)/water solution on the electrospinning process and fibre morphology. The effects of the conductivity of PEO solution on the jet current and jet path were elaborated further, with the addition of NaCl to the solution results in the formation of protrusion on the fibre surface, as shown in figure. The effects of the conductivity of polyethylene oxide (PEO) solution on the jet current and jet path were also considered.

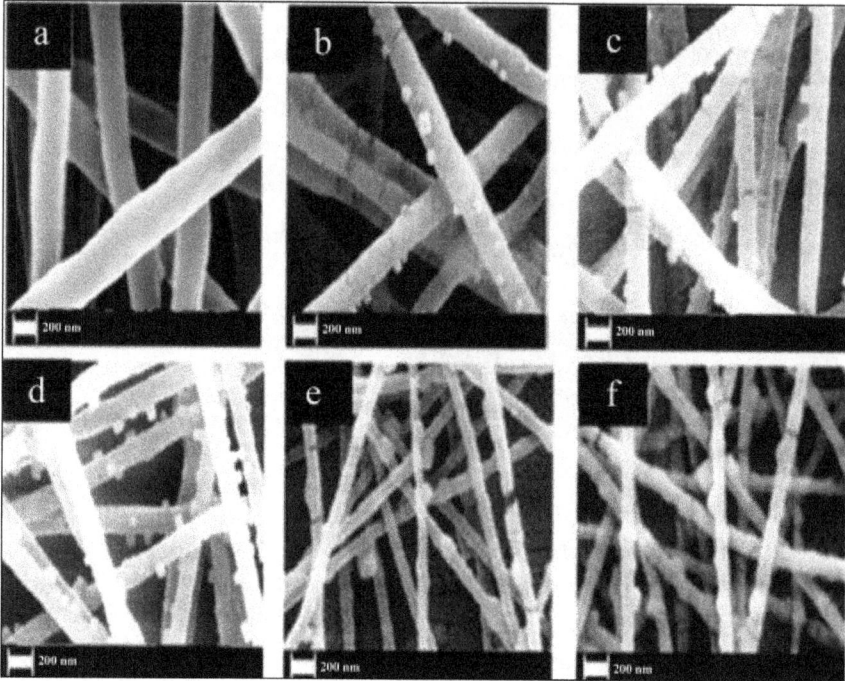

FESEM images of samples of electrospinning PEO/NaCl fibres for a range of conductivities. (a) 5 g/0 g, (b) 5 g/0.1 g, (c) 5 g/0.2 g, (d) 5 g/0.5 g, (e) 5 g/1.25 g, and (f) 5 g/2 g.

KH_2PO_4, NaH_2PO_4 and NaCl were studied, and each was added in separation at 1% W/V to PLLA solutions. The resulting electrospun nanofibers were smooth, bead-free and they also had smaller diameters than those of the nanofibers electrospun from solutions that did not have a salt. While KH_2PO_4 – which contains solutions produced nanofibers with the largest diameter, those containing NaCl produced nanofibers with the smallest diameter. The ion size was also found to ascertain the nanofiber diameter. Ions with smaller radii had higher charge density, and thus they gave greater forces of elongation on the electrospun nanofibers.

Molecular Weight

Molecular weight of the polymer also leaves a great effect on the morphologies of the electrospinning fibre. The entanglement of polymer chains in solutions, namely the solution viscosity, is principally a reflection of the molecular weight. Keeping the concentration fixed, and lowering the molecular weight of the polymer have the ability to form beads instead of the smooth fibre. Smooth fibre will be obtained by increasing the molecular weight. What is also worth noting is that too high molecular weight favours micro-ribbon formation even with low concentration. Çiğdem studied the impact of the molecular weight (MW) on the fibre structure of electrospun poly(vinylalcohol) (PVA) which has molecular weights that range from 89000 to 186,000 g/mol when dissolved in water.

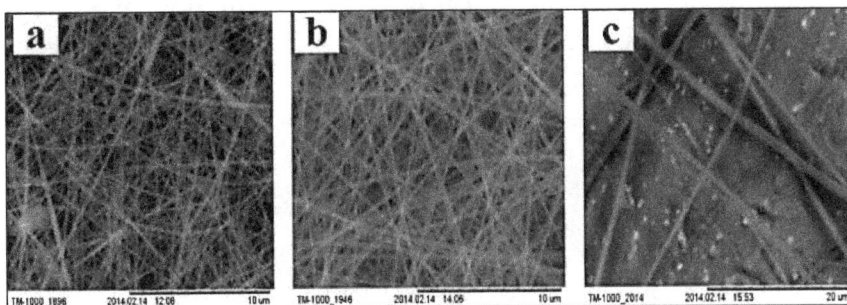

FESEM showing the typical structure in the electrospun PVA polymer for various molecular weights. (a) 89000–98,000 g/mol; (b) ~125,000 g/mol; and (c) 146,000–186,000 g/mol (solution concentration: 25 wt.%).

Surface Tension

As the function of solvent compositions of the solution, surface tension is an important factor in electrospinning. Yang conducted a study on the influence of surface tensions on the morphologies of electrospun products with PVP as model with ethanol, N,N-dimethylformamide (DMF) and dichloromethane (MC) as solvents. In the process, it was discovered that different solutions may contribute different surface tensions. With the concentration fixed, and the surface tension of the solution, reduced beaded fibres can be converted into smooth fibres.

Solvent Selection

For a particular polymer to solubilise and be transformed into nanofibers via the process of electrospinning, the choice of solvent is very important. The solubility of the polymer in the solvent and the boiling point of the solvent, which altogether indicate its volatility, are two major aspects worth considering when it comes to choosing a solvent. Volatile solvents are the more favourable choice as they assist the dehydration of the nanofibers during trajectory from the capillary tip to the collector surface, because of their lower boiling point, and thus causing a rapid evaporation rate. Nonetheless, highly volatile solvents that have very low boiling points should be prevented as they may evaporate at the capillary tip and further leading to the clogging and the obstruction of the flow-rate of the polymer solution. Solvents that have high boiling points may not dehydrate entirely before reaching the collector, thus resulting in ribbon-like, flat, nanofiber morphologies or conglutination of nanofibers at the boundaries. The ability of the electrospinning polyvinylpyrrolidone (PVP) by Yang and Coworkers, was investigated with different solvents. The solvents examined were MC, ethanol and DMF, while the beaded nanofibers were formed from DCM and DMF solutions of PVP, the use of ethanol produced PVP nanofibers. Nanofibers electrospun from an integration of the ethanol and DMF had small diameters of 20 nm, while a combination of ethanol and DCM resulted in the formation of nanofibers with diameters as large as 300 nm. It is therefore conclusive that nanofiber morphology and porosity may be regulated by the defensible use of solvents or a combination of solvents.

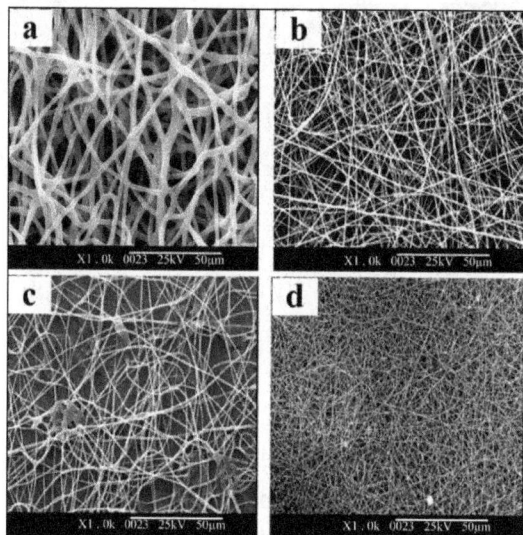

TEM images of PEO nanofibers electrospun different solvent. (a) chloroform (3%), (b) Ethanol (4%), (c) DMC (5%), (d) Water (7%).

Electrospinning Parameters

Voltage Supply

The amount of charge per unit surface area of the polymer droplet which constitutes charge density is determined by the applied voltage, working distance and the conductivity of the polymer solution. Applied voltage is used to provide the driving force to spin fibres by imparting charge to the polymer droplet. The working distance which is the distance between the tip of syringe and the collecting plate, in addition to the applied voltage, can influence the structural morphology of nanofibers. Demir suggested that when higher voltages are applied, more polymer is ejected to form a larger diameter fibre. Similarly, high voltage conditions also created a rougher fibre structure. To reduce bead formation, Zong proposed an approach to increase charge density on the surface of the droplet by adding salt particles. However, they concluded that high-charge density produced thinner fibres, a finding not corroborated by Demir. In the state of low voltages or field strengths, typically, a drop is suspended at the needle tip, and a jet will originate from the Taylor cone producing bead-free spinning (under the assumption that the force of the electric field is sufficient to address the surface tension). Hao Shao studied the effect of high voltage, and observed change in the morphology as a result of change in high voltage as shown in figure.

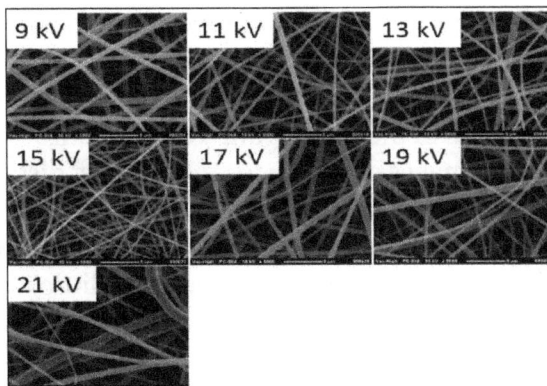

SEM images of the PVDF nanofibers prepared at different applied voltages.

Needle Diameter (Nozzle)

The size of needle has a certain effect on the nanofibers diameters. It was discovered that a reduction in the diameters of the electrospun nanofibers was caused by a decrease in the internal diameter of the needle. The nozzle (usually the syringe needle set up) determines the amount of polymer melt that comes out, which, in turn, affects the size of the drop being formed and also the pressure or the amount of force required by the pump to push the melt out. If the polymer melt is less viscous, it can easily flow out of the nozzle. The polymer melt is usually a thick highly viscous fluid. So, if the nozzle is too small, and the melt is too viscous, the melt cannot be forced out. Therefore, an appropriate nozzle should be used. Different types of nozzles or spinnerets have been used over the years. The effect of needle diameters on the resulting electrospun poly(methyl methacrylate) (PMMA) average nanofiber diameters was evaluated for three different needle gauges by Javier Macossay. These fibres presented regular surface morphologies, with a few nanofiber bundles being evident in figure.

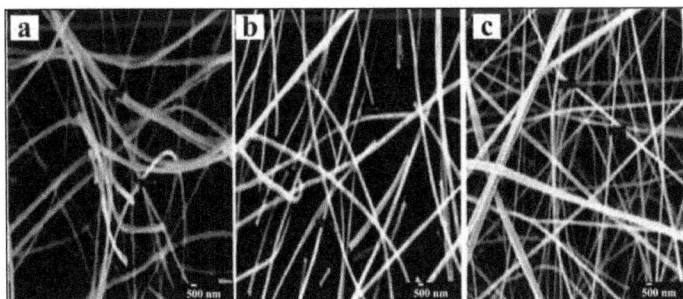

SEM of PMMA nanofibers, utilizing internal diameter needle.
(a) 0.83 mm, (b) a 0.4 mm, (c) a 0.1 mm.

Distance between Tip and Collector

The distance from the needle to the collector is very important, because by decreasing it, the electrical field increases instead; and also the stretching force does and the time at which the fibre undergoes the field is lower, causing sufficient evaporation of the solvent of the fibres. The result is that when there is reduction in the distance between the needle and the collector the fibres grow and may be subject to structural deformities like blobs. The high voltage was determined at 15 kV and the distance from the tip of the needle to the collector is in the range from 9 cm to 21 cm. The fibres' morphology was assessed from the SEM images of figure; for the shortest distance (9 cm), the fibres came together at their intersections following the incomplete evaporation of the solvent before the jet arrived at the collector. For the other four distances used, the fibres appeared similar and the mean fibre diameter increased a little with the distance to the collector. The distance established between the tip and the collector exerted a direct influence in flight time and electric field strength. For the fibres to form, the electrospinning jet must be given ample time for most of the solvents to be evaporated. The electric field strength will increase at the same time and this will increase the acceleration of the jet to the collector. As a result, there may not be enough time for solvents to evaporate when they reach the collector.

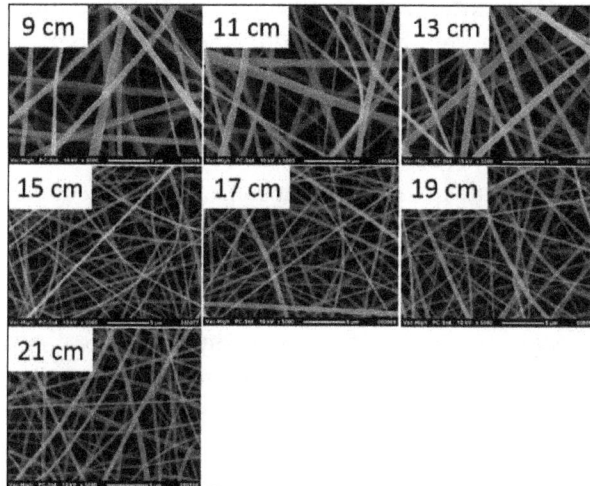

SEM images of the PVDF nanofibers
prepared at different spinning distances.

Flow Rate

Another important parameter process is the flow rate of the polymer solution within the syringe. For the polymer solution to have enough time for polarization, lower flow rate is more preferred. If the flow rate is very high, bead fibres with thick diameter will form instead of smooth fibres with thin diameters owing to the short drying time before reaching the collector, and also due to low stretching forces. There is a corresponding rise in the fiber diameter or blobs size, as a result of greater volume of solution ejected from the needle tip, when there is increase in the feed rate. Shamim Z indicated that when the flow rate is decreased with other parameters kept constant; there is a decrease in the blobs size and an increase in nanofiber diameter. The inference here is that, with the decrease in flow rate, blobs size could get smaller until the non-beaded structure is obtained.

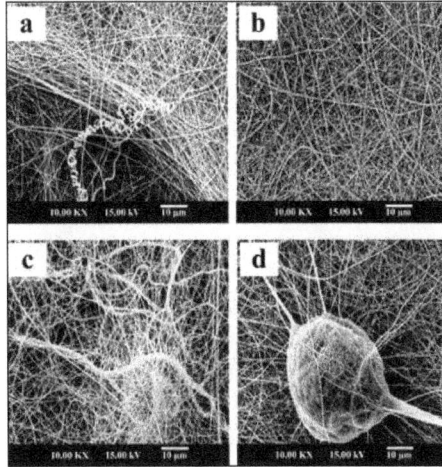

SEM images of PVA different flow rate. (a) 0.1 ml/h,
(b) 0.5 ml/h, (c) 1 ml/h, (d) 1.5 ml/h.

Collector

The formation of nanofibers can be classified into woven and non-woven nanofibers. The type of collector used plays a big role in differentiating the type or nanofiber alignments. The use of oriented collector, as well as static double grounded collector, constitutes the methods adopted in developing aligned woven nanofibers. The rotating drum collector is used for collecting the aligned arrays nanofibers, while the rotating disk is used for collecting uniaxially aligned nanofibers. The alignment fibres obtained from the rotating drum correspond to the rotational speed applied on the drum. This type of electrospinning method is more complex because the speed of the rotation needs to be very properly controlled to produce nanofibers with such a good alignment. The rotating disk collector can also serve to collect continuous nanofibers, since they can very much attract the large electrical field applied at the edge of the disk.

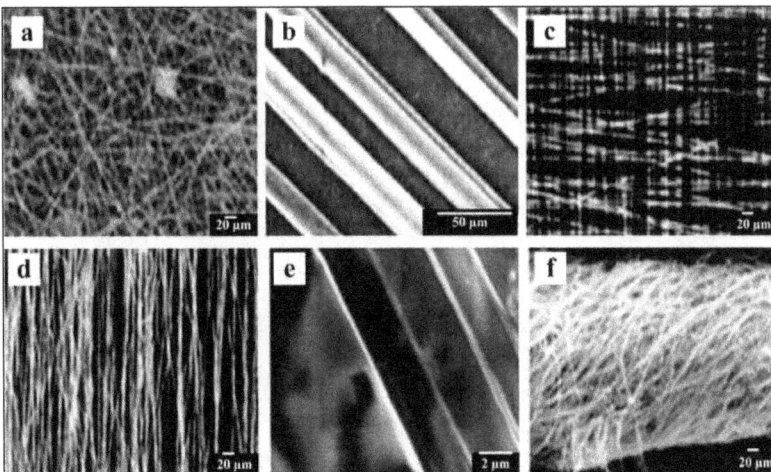

SEM images of the different electrospun products with various types of collectors. (a) Wire Screen, (b) Pin, (c) Gridded Bar, (d) Parallel Bar, (e) Rotating Wheel, and (f) Liquid Bath.

SEM images of diverse collectors for many reports, have been developed including the wire mesh studied by Wang X, pin studied by Sundaray B, grids studied by Li D parallel or gridded bar and rotating rods or wheel studied by Xu CY, and liquid bath studied by Ki CS. Kim proved that the different types of composition used in the collector affected the structure of the poly (L-lactide) (PLLA) and poly (lactide-co-glycolide) (PLA$_{50}$GA$_{50}$) fibres.

Ambient Parameters

Fibre diameters and morphologies such as humidity and temperature could also be affected by ambient parameters. Increasing temperature, as noted by Mituppatham for instance, favours the thinner fibre diameter with polyamide-6 fibres for the inverse relationship between the solution viscosity and the temperature. With regards to humidity, low humidity could dry the solvent totally and increase the velocity of the solvent evaporation. On the contrary, high humidity will lead to thick fibre diameter because the charges on the jet can be neutralised and the stretching forces become small.

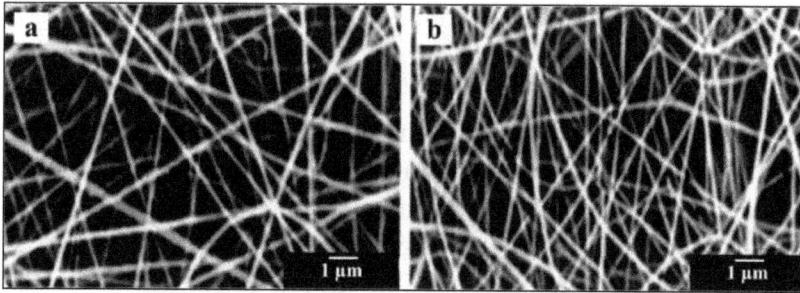

SEM images of the electrospun PA-6-32 fibers under
different temperatures. (a) 30° and (b) 60°.

The variety of humidity can also affect the surface morphologies of electrospun PS fibres, as recently show by Casper. Nezarati observed that low humidity (5% RH) resulted in beads connected by thin fibres, but increasing the RH (20–75% RH) resulted in smooth, uniform fibres for poly(ethylene glycol) (PEG). In addition as relative humidity was increased from 50–75%, fibre density decreased.

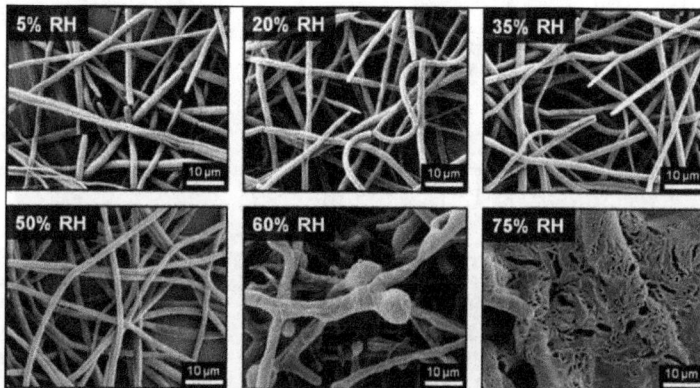

SEM images of poly(ethylene glycol) (PEG) electrospun
at relative humidity (RH) ranging from 5 to 75%.

Carbon Nanofibers

Carbon nanofibers have received growing interests due to their unique chemical and physical properties, depending upon their size, surface area, and shape. Indeed the attractive structural, electrical, and mechanical properties of carbon nanotubes (CNTs) make it an ideal supporting material for various applications. Particularly, the CNTs can be used as an efficient support for the decoration of catalytic active materials. Carbon nanofibers (CNFs) also have the similar physicochemical properties to CNTs and the diameter varying from some tens of nanometers to 500 nm and are also suitable to be used as catalyst support. Electrospinning is a very simple but powerful method for the fabrication of high-quality carbon nanofibers. In general, the CNFs with sub-micrometer diameters as well as some tens of nanometers to 500 nm are prepared by carbonization of electrospun polymer nanofibers under inert atmosphere at high temperature. Undoubtedly, polyacrylonitrile (PAN) is a well-known and efficient precursor for the fabrication of carbon fibers; therefore, several attempts were made to prepare the electrospun-derived carbon nanofibers from PAN. Several approaches, including wet-chemical synthesis, electrodeposition, and dry synthesis, are developed to obtain various multifunctional active materials loaded with carbon nanocomposites. By using these techniques, various types of metal or metal oxide nanoparticles (NPs), such as Au, Co, Ru, Pt, Pd, Ag, Co, Rh, Ti and Cu, have been decorated or immobilized on/into the carbon nanofibers. These metal NP-supported CNF nanocomposites have shown great promises in catalysis, fuel cells, and highly sensitive chemical/biological sensing applications. In particularly, the CNF composites showed excellent results in various catalytic systems such as in photocatalytic activity, water gas shift reactions (WGS), enzyme immobilization or biocatalysts and direct oxidation of alcohols. So far, TiO_2-deposited CNFs have gained much attention in the photocatalytic reactions. Alike, Pd NP-supported CNFs are often preferred for the catalytic organic conversions such as hydrogenation reaction and Heck coupling reaction. It is proven that the CNFs are one of the highly suitable supports for the decoration of Pd NPs and the resultant Pd/CNF composite often demonstrated an enhanced catalytic activity. In fact, the unique structure, high conductivity, huge surface area, and chemical inertness of CNFs often help to obtain high dispersion of metal nanoparticles on CNFs. Most of the Pd NP-supported CNFs showed better activity than the conventional Pd/C catalysts.

Preparation and Characterization of Carbon Nanofiber Composites

Electrospun Carbon Nanofibers

Electrospinning is a straightforward method to obtain the nanofibers. Figure shows the fundamental electrospinning setup and the list of important parameters to be controlled. The nanofibers can be produced by applying high voltage to a polymer solution which could create electrostatically repulsive force and an electric field between two electrodes, so that the nanofibers can be formed. Obviously, the formation of nanofibers is highly dependent on the viscosity and electric conductivity of the polymeric fluids, humidity, and applied voltage. To date, over 100 kinds of polymers have been employed to produce their nanofibers via electrospinning. However, a very limited number of polymers such as polyacrylonitrile (PAN), polyimide (PI), poly(vinyl alcohol) (PVA), poly(vinylidene fluoride) (PVDF), cellulose acetate, and pitch have been successfully used to obtain carbon nanofibers. The carbon nanofibers are often characterized by various techniques. Scanning electron microscope (SEM) is one of the very common methods to characterize the CNFs. The surface morphology, particularly the fiber diameter, uniformity, and surface smoothness are often studied by SEM analysis. Alike, transmission electron microscopy (TEM) and atomic force microscopy (AFM) are also employed for the detail surface analysis. The crystalline and amorphous nature of the CNFs is often investigated by means of X-ray diffraction (XRD) analysis. Raman spectroscope is a very useful technique for the analysis of G-band and D-band of CNFs. X-ray photoemission spectroscopy (XPS) was also effectively used to analyze the CNFs. The specific surface area and textural properties such as pore volume and pore size of CNFs are evaluated by using the Brunauer-Emmett-Teller (BET) method.

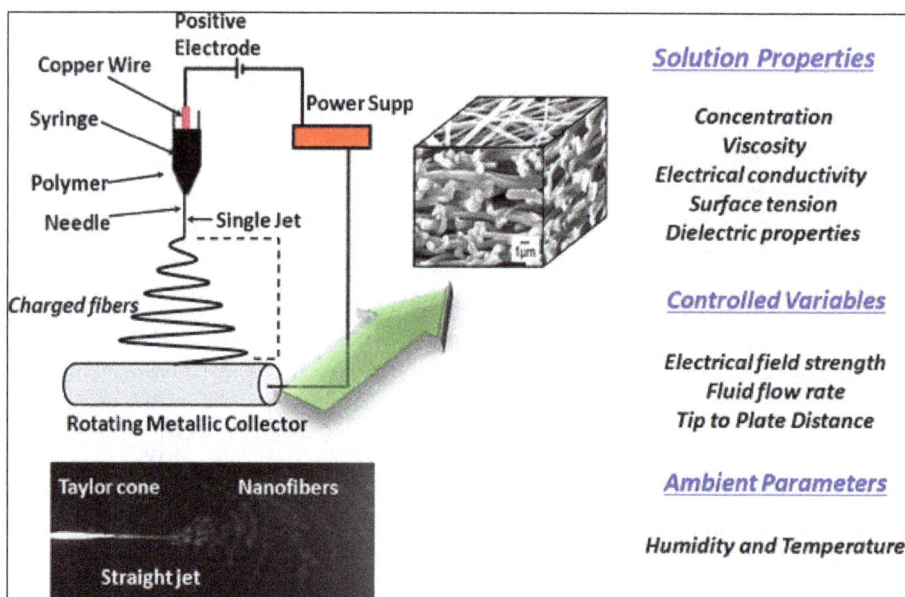

Scheme of fundamental setup for electrospinning and electrospinning parameters.

Kim prepared CNFs via electrospinning by using PVA and DMF as precursor and solvent, respectively. In a typical preparation method, PVA was dissolved in DMF and the polymer mixture was electrospun. In the first step, the resultant nanofiber webs were oxidatively stabilized at 280 °C under air flow (heating at 1°C/min). Then the stabilized nanofiber web was activated by steam resulting in activated carbon nanofibers. The stabilized nanofiber webs were heated at a rate of 5 °C/min up to 700, 750, and 800 °C and activated for 30 min by supplying 30 vol.% of steam in a carrier gas of N_2. They confirmed that the resultant CNFs have well-developed mesopores and the CNFs demonstrated excellent specific capacitance (173 F/g at 10 mA/g).

Kuzmenko and co-workers prepared nitrogen-doped carbon nanofibrous mats from regenerated cellulose impregnated with ammonium chloride. The ammonium chloride provided the thermal stabilization of incompletely regenerated cellulose fibers. In a typical preparation, cellulose acetate solution was prepared in acetone/DMAc mixture which was subsequently electrospun. The voltage was 25 kV, distance between needle and collector was 25 cm, and the process was performed at temperature around 20 ± 2 °C and relative humidity 45–60%. Aluminum foil was used for collecting the nanofibers. The prepared cellulose acetate nanofibers were deacetylated by using dilute NaOH solution. Then the regenerated cellulose webs were impregnated with NH_4Cl by immersion in 0.3 M aqueous solution of NH_4Cl for 24 h at 20 ± 2°C. The NH_4Cl-treated regenerated cellulose samples were carbonized in a quartz tube furnace for general annealing in N_2 flow (1 L/min) by heating up to 800 °C with the heating rate of 5 °C/min. Figure shows the SEM images of the CNFs synthesized from the regenerated cellulose.

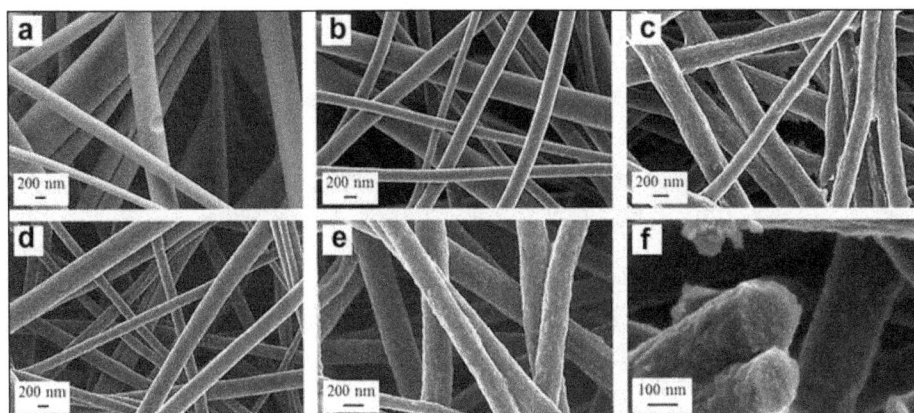

(a-f) SEM images of the CNFs synthesized from the differently regenerated cellulose with additional NH_4Cl impregnation.

Porous Carbon Nanofibers with Hollow Cores

Kim successfully prepared porous CNFs with hollow cores through the thermal treatment of electrospun copolymeric nanofiber webs. Figure shows the schematic diagram for producing porous CNFs with hollow cores. For the preparation of pores CNFs with hollow cores, PAN and poly(methyl methacrylate) (PMMA) polymers were chosen. The PAN is a widely used precursor for the preparation of CNFs, and the PMMA can be thermally decomposed at elevated temperatures. Dissolving these two polymers (PAN and PMMA) in a solvent would create phase separation (continuous phase (sea) changes into pore walls (or skeletons of nanofibers) and the discontinuous phase (islands) changes into many hollow pores), which results in the sea-islands feature. It is well known that the low-surface-tension polymer (PAN) would occupy the continuous phase of the solution (sea), while the high-surface-tension polymer (PMMA) forms the discontinuous phase (islands). In fact the two separate phases are due to the intrinsic properties (e.g., interfacial tension, viscosity, elasticity) of the polymers. The electrospinning technique was used to obtain the PAN/PMMA nanofibers containing two separate phases. The thermal treatment of PAN/PAMM nanofibers at over 1000 °C would eventually form the porous carbon nanofibers with hollow cores. The complete removal of PMMA phase and the transformation continuous PAN phase would result in the formation of porous CNFs with hollow cores.

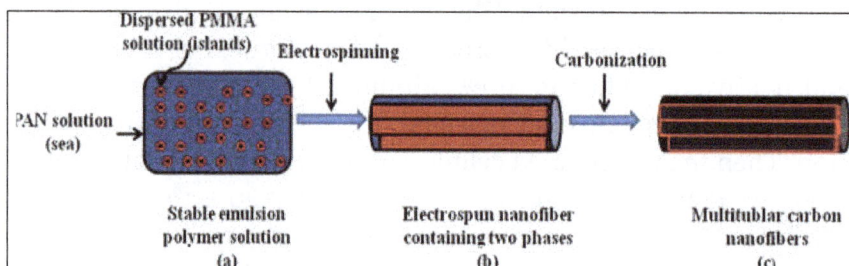

Schematic diagram for the preparation of porous CNFs with hollow cores: (a) Preparation of stable polymer solutions from two separate phases; nanoscale phase separation occurs due to their different molecular weights; PMMA forms the discontinuous phase and PAN forms the continuous phase; (b) nanofiber formation (with two phases) by electrospinning; (c) removal of the PMMA phase at elevated temperatures.

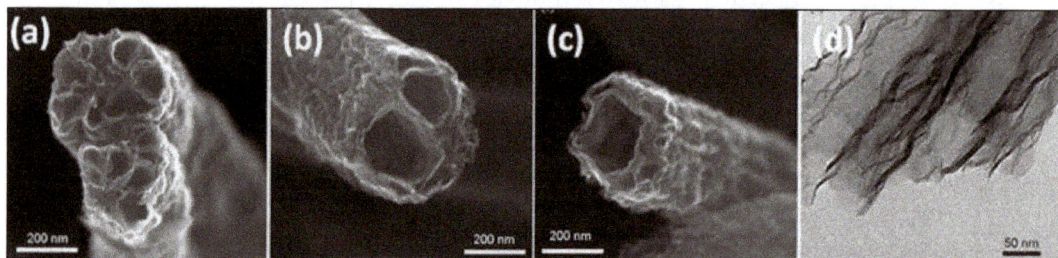

(a–c) Cross-sectional TEM images of CNFs thermally treated at 2800 °C [PAN:PMMA (a) 5:5, (b) 7:3, and (c) 9:1] and (d) TEM image of CNFs showing structurally developed core walls after thermal treatment.

Highly flexible N- and O-containing porous ultrafine CNFs were prepared by Wei and co-workers. The ultrafine porous CNFs were obtained by simply varying the PAN/PMMA ratios (10/0, 7:3, 5:5, and 3:7). Briefly, PAN/PMMA solutions with different ratios (10:0, 7:3, 5:5, and 3:7) are prepared in DMF. For better dispersion, the PAN/PMMA solution was sonicated followed by stirring at 60 °C for 2h. The polymer blend was electrospun under an electric field of 9 kV at a tip-to-collector distance of 15 cm. The resultant electrospun PAN/PMMA nanofibers were stabilized under air flow at 300 °C with the heating rate of 1 °C/min for 1h. Subsequently, the stabilized nanofibers were carbonized under N_2 atmosphere at 900 °C with heating rate of 5 °C/min for 1h. It was proven that the increasing the ratio of PMMA would result in the formation of ultrafine CNFs. Figure shows the FE-SEM images of CNFs; CNFs, 7:3; CNFs, 5:5; and u-CNFs, 7:3. The FE-SEM images show that the morphology of CNFs is homogeneous, continuous, and a typical cylindrical shape. The FE-SEM images of CNFs, 7:3, and CNFs, 5:5, showed that the CNFs have several hollow cores along the fiber axis. Notably, the morphology of the u-CNFs (3:7) was completely changed. The morphology of CNFs (3:7) fibers was homogenous, continuous, and a cylindrical shape with an average diameter of ~50 nm. In fact, the complete decomposition of PMMA during the thermal treatment is the main reason. The BET specific surface area of the u-CNFs-3,7 was determined to be 467.57 m²/g and pore volume of 1.15 cm³ g⁻¹ and an average pore size of 9.48 nm.

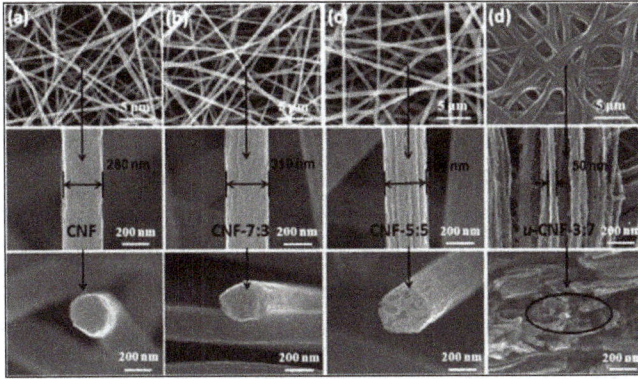

FE-SEM images of electrospun PAN/PMMA nanofibers with different mixing ratios. (a) CNFs; (b) PAN/PMMA, 7:3; (c) PAN/PMMA, 5:5; and (d) PAN/PMMA, 3:7.

Chang introduced a novel technique of centrifuged-electrospinning for the preparation of ultrathin carbon fibers. Figure shows the preparation diagram of the ultrathin porous CNFs by centrifuged-electrospinning. In a typical procedure, PAN/PMMA polymer blend was prepared in DMF at different weight ratios of 80/20 (PAN80/PMMA20) and 10/90 (PAN10/PMMA90). The polymer blends were used for the preparation of PAN/PMMA nanofibers by centrifuged electrospinning. The centrifuged-electrospinning conditions were as follows: an applied positive voltage of 45 kV, a three-phase induction motor spinning at 4000 rpm, a syringe feed rate of 1.5 mL/min, and a stainless steel ring with a diameter of 50 cm as the collector. Finally, the resultant PAN/PMMA nanofibers were stabilized at 280 °C for 2 h at a heating rate 0.5 °C/min in air atmosphere and then carbonized at 800°C for 4 h under argon atmosphere at a heating rate of 5 °C/min. Figure (b, c, and d) shows the TEM images of ultrathin micro-/mesoporous CNFs.

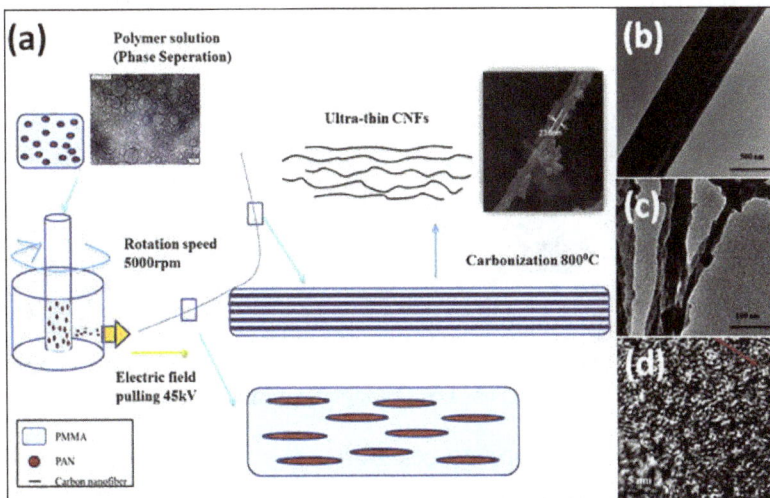

(a) Schematic diagram for the preparation of ultrathin micro-/mesoporous CNFs by centrifuged-electrospinning followed by carbonization and (b, c, and d) TEM images of ultrathin micro-/mesoporous CNFs.

Carbon Nanofiber Composites

Recently, preparation of metal oxide-supported carbon nanofiber composites via electrospinning has been extensively studied. The carbon nanocomposites are used in various applications such as energy conversion and storage, capacitive deionization, catalysis, adsorption/separation, and in the field of biomedicine. In order to achieve higher activity, various synthetic routes were developed to achieve porous carbon nanofibers composites with high surface area and tunable pore size distribution. Most of the preparation methods involve carbonization process at elevated temperatures of typically above 1200 °C. So far, various metal or metal oxide nanoparticle (Pd, Pt, Ti, Ag, Au, Cu, Ni, Zn, and Ru)-supported CNF nanocomposites were reported.

Atchison prepared metal carbide-supported carbon nanocomposites through carbothermal reduction process. Zirconium carbide/carbon nanocomposite (ZrC/C), titanium carbide/carbon nanocomposite (TiC/C), and niobium carbide/carbon nanocomposite (NbC/C) were prepared by electrospinning followed by carbothermal reduction at elevated temperatures. Cellulose acetate and PVP were used as precursor.

Chen prepared Pd nanoparticle-supported carbon nanofibers (Pd-NP/CEPFs: Pd-NP/CEPFs) through the electrospinning process. Shortly, electrospinning solution was prepared by using 10 wt% PAN and 3.3 wt% $Pd(OAc)_2$ in DMF. The electrospinning process was performed in an electric field of 30 kV and the tip-to-collector distance of 30 cm. Then the electrospun PAN/$Pd(OAc)_2$ nanofiber involved three steps as follows: (1) 210 °C annealing for 1 h under air flow for the oxidation of PAN, (2) heating up to 400 °C at a rate of 5 °C/min and annealing for 2 h in H_2 and Ar mixture (H_2/Ar = 1/3) atmosphere for the reduction of Pd^{2+}, and (3) heating up to 550 °C at a rate of 5 °C/min and annealing for 1 h in Ar for the formation of metal nanoparticles on/in the carbonized nanofibers.

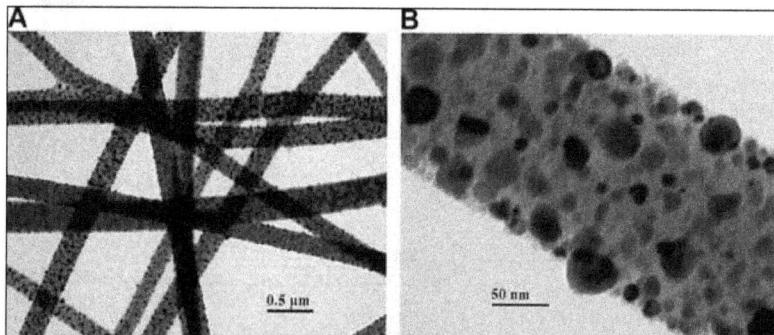

TEM images of Pd-NP/CENFs. (A) Lower magnification; (B) higher magnification.

Zhang et al. obtained AgNP-immobilized carbon nanocomposite by a two-step preparation: electrospinning followed by the hydrothermal growth of the AgNPs on the CNFs.

In a typical procedure, the electrospinning solution of PAN was prepared in DMF, and it was electrospun at an applied electric voltage of 10 kV. The PAN nanofibers were then stabilized under air flow at 270 °C for 1 h and subsequently carbonized under N_2 atmosphere at 1000 °C for 1 h at the rate of 5 °C/min. After the preparation of CNFs from PAN nanofibers, the CNFs were treated with HNO3, centrifuged, and washed with water for several times. Finally an aqueous mixture of glucose, $Ag(NH_3)_2OH$, and CNFs was stirred for 5 min. After being stirred for more than 5 min, the mixture was transferred into a Teflon-lined autoclave, and it was sealed in a stainless steel tank and heated at 180 °C for 3 h. Finally the AgNP-immobilized CNFs (Ag/CNFs) were obtained.

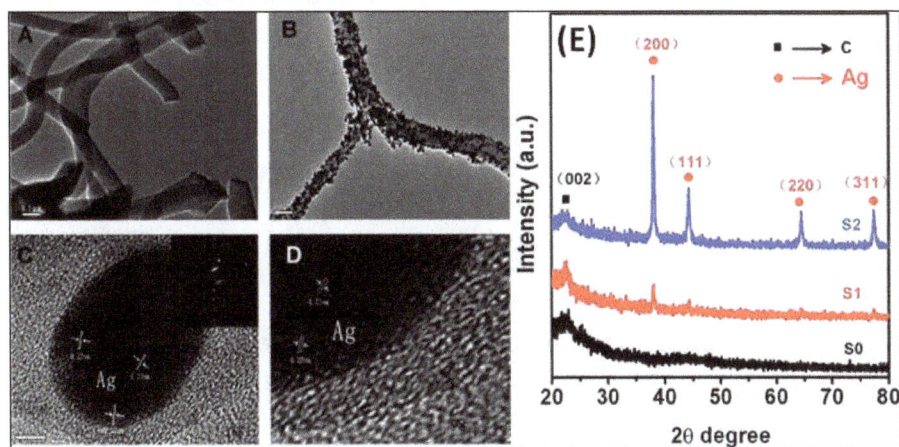

TEM images of sample (A) CNFs and (B) CNFs/AgNPs, (C) HRTEM images of CNFs/AgNPs, and (E) XRD patterns of CNFs/AgNPs and CNFs.

Yu and co-workers prepared electrospun $Ag/g-C_3N_4$-loaded composite carbon nanofibers ($Ag/g-C_3N_4/CNFs$) through combing the electrospinning technology and carbonization treatment. The microstructure of $Ag/g-C_3N_4/CNFs$ was characterized by XRD, FE-SEM, EDS, TEM, and XPS.

Ghouri et al. achieved Co/CeO_2-decorated carbon nanofibers ($Co/CeO_2/CNFs$) by calcination of electrospun nanofibers composed of cerium (III) acetate hydrate, cobalt (II) acetate tetrahydrate, and poly(vinyl alcohol) in nitrogen environment at 700 °C. PVA was used as carbon source due to its high carbon content. In a typical preparation, CoAc and CeAc aqueous solutions were prepared in distilled water. The resultant aqueous solutions were mixed with PVA aqueous solution. After stirring for 6 h, the mixture was electrospun at high voltage of 22 kV using DC power supply at room temperature with 65% relative humidity. The tip-to-collector distance of 22 cm was fixed. Finally, the dried nanofiber mats were calcined at 700 °C for 6 h in N_2 flow with a heating rate of 2.0 °C/min. The physicochemical properties of the $Co/CeO_2/CNFs$ were characterized by XRD, FE-SEM, EDS, TEM, XPS, and Raman.

The utilization of noble metals in green technologies has garnered an increasing level of research interest. Particularly, the Pt-based nanocomposites are often preferred as the anode because of their excellent performance in catalyzing the dehydrogenation of methanol. For example, Formo achieved Pt nanostructure-supported CNF nanofibers through electrospinning followed by calcination in air at 510 °C for 6 h.

Applications of Carbon Nanofiber Composites

Electrospun carbon nanofibers have proven to be efficient catalytic supports owing to the high porosity and large surface areas. The high porosity in a nonwoven mat of nanofibers enables direct growth of catalytic nanostructures. Till date, there are number of applications found for the electrospun carbon nanofibers and its composites.

Carbon Nanocomposites in Organic Transformations

Owing to high surface area, porosity, stability, metal-support interaction, smaller particle size, and high dispersion in reaction medium, the metal nanoparticle-supported carbon nanocomposites demonstrated excellent activity in organic reactions. They can be highly reusable due its stability which is one of the hallmarks of the carbon nanocomposites.

Palladium-catalyzed Sonogashira coupling reaction is the most straightforward and powerful method used for the construction of $C(sp_2)$–$C(sp)$ bond, drugs, and polymeric materials. The conventional protocols of the Sonogashira reactions are carried out in the homogeneous phase, using soluble palladium (Pd) composites such as $Pd(PPh_3)_4$, $Pd(PPh_3)_2Cl_2$, and $Pd(OAc)_2$ as catalysts in the presence of CuI as co-catalyst. Even with the high reaction rate and high turnover numbers, homogeneous catalysis has a number of disadvantages, in particular the lack of reuse of the catalyst. Chen developed Pd-supported CNF catalytic system for the Sonogashira reaction. Figure shows Pd-NP/CENF catalyzed Sonogashira reaction of iodobenzene and phenylacetylene in liquid phase. The catalyst showed superior catalytic activity toward the Sonogashira reaction. In addition, the catalyst was found to be highly reusable, at least for 10 runs without any significant loss in its catalytic activity.

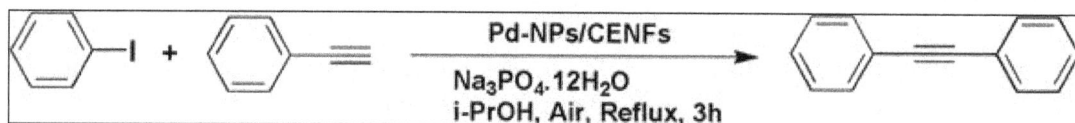

The Sonogashira reaction equation of iodobenzene and phenylacetylene in liquid phase catalyzed by Pd-NP-supported CNFs.

Alike, electrospun $Ag/g\text{-}C_3N_4$-loaded composite carbon nanofibers ($Ag/g\text{-}C_3N_4/CNFs$) were used for the conversion of 4-nitrophenol to 4-aminophenol and benzylamine to N-benzylbenzaldimine. The $Ag/g\text{-}C_3N_4/CNFs$ offered the significant advantages, such as low catalyst use, high activity, easy recycling, and excellent stability. In fact, the synergistic effect between catalytic activity of Ag nanoparticles (NPs) and $g\text{-}C_3N_4$ and excellent adsorption capacity of carbon nanofibers are the main reason for the excellent catalytic activity.

Carbon Nanocomposites in Catalytic Reduction of 4-Nitrophenol

Catalytic transformation of 4-nitrophenol to 4-aminophenol is one of the very significant reactions in green chemistry. It is well known that the nitrophenols are harmful to the environment, and therefore the US Environmental Protection Agency has listed it as 114th organic pollutant. In recent days, catalytic conversion of nitrophenols to aminophenols is widely studied. In fact, the catalytic product, aminophenols, can be used as an excellent intermediate in synthesizing various drugs and reducing agent. The 4-nitrophenols are often employed as a photographic developer, corrosion inhibitor, anticorrosion lubricant, and hair-dyeing agent. Although the reaction is very simple, greener, and most efficient, the reaction without metal catalysts is not achievable. To perform this reaction, various metal catalysts (based on graphene oxide, silica, alumina, activated carbon, CNTs, fullerenes, and so on) are developed and proven to be an excellent candidate for the reduction of 4-nitrophenols to 4-aminophenol. For example, RGO-ZnWO$_4$-Fe$_3$O$_4$, AgNPs-rGO, PdNiP/RGO, and NiNPs/silica are reported for the reduction of nitrophenol.

Carbon nanofibers/silver nanoparticle (CNFs/AgNPs) composite nanofibers were used for the reduction of 4-nitrophenol (4-NP) with NaBH4. The reaction was tracked by time-dependent UV-visible spectroscopy. It was found that the CNF/AgNP composite demonstrated an excellent catalytic activity in the reduction of 4-nitrophenol. The catalytic efficiency was found to be enhanced with the increasing of the content of silver on the CNF/AgNP catalyst. Reaction kinetic was studied for the CNF/AgNP reduction of 4-nitrophenol. It was reported that the rate constant of 6.2×10^{-3} s^{-1} was determined for the reduction of 4-nitrophenol over CNF/AgNP catalyst. The excellent active might be attributed to the high surface areas of Ag NPs and synergistic effect on delivery of electrons between CNFs and AgNPs. The CNF/AgNP composite nanofibers can be easily recycled and reused without any significant loss in its catalytic activity.

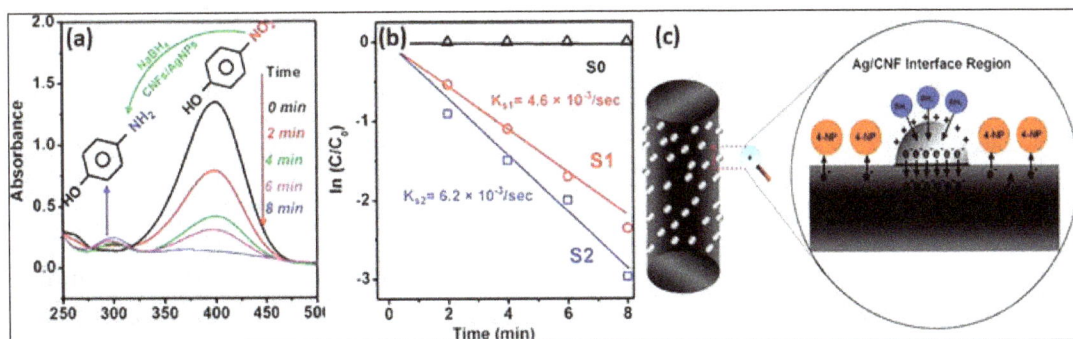

Catalytic evolution of CNFs/AgNPs: (a) UV-vis absorption spectra during the reduction of 4-nitrophenol over CNFs/AgNPs; (b) ln(C/Co) and C/Co vs. reaction time for the reduction of 4-nitrophenol, So = fresh CNFs, S1 = CNFs/AgNPs, and S2 = CNFs/AgNPs; (c) proposed mechanism of the catalytic reduction of 4-nitrophenol with the CNF/AgNP composite nanofibers.

Carbon Nanofiber Composites in Energy Applications

Due to the excellent properties such as high surface area, conductivity, and porosity, the CNF-based nanocomposites are widely used for the energy applications. Without a doubt, the development of electrochemical energy storage systems (EES) is largely focused due to its vital demand for clean and sustainable energy. Mainly three types of devices are very important and most commercialized energy storage systems such as batteries, electrochemical capacitors (ECs), and fuel cells.

The ultrafine CNFs prepared via electrospinning of PAN/PMMA blend followed by thermal treatment in inert atmosphere were used as a flexible electrode material for the supercapacitor applications. Figure shows the cyclic voltammetry results for the ultrafine CNFs at different scan rates in 1.0 mol/L H_2SO_4 and specific capacitance for the ultrafine CNFs at different scan rates in 1.0 mol/L H_2SO_4. The ultrafine CNFs demonstrated an enhanced specific capacitance of 86 F g^{-1} in 1 mol/L H_2SO_4. Being a flexible electrode material, this is the highest specific capacitance for the CNFs reported so far. The excellent specific capacitance of ultrafine CNFs is due to its unique properties. The results proved that the fiber diameter of ultrafine CNFs was about 50 nm. The XPS and Raman studies confirmed the presence of N and O in the form of various functional groups such as pyridinic, benzenoid amine, graphitic N, and N-oxides. High specific surface area of 467.57 m²/g with an excellent pore volume (1.15 cm³ g^{-1}) and pore size (9.48 nm) was determined for the ultrafine CNFs. The BET results confirmed the interconnected micro-/meso-/macropores on the surface of the ultrafine CNFs.

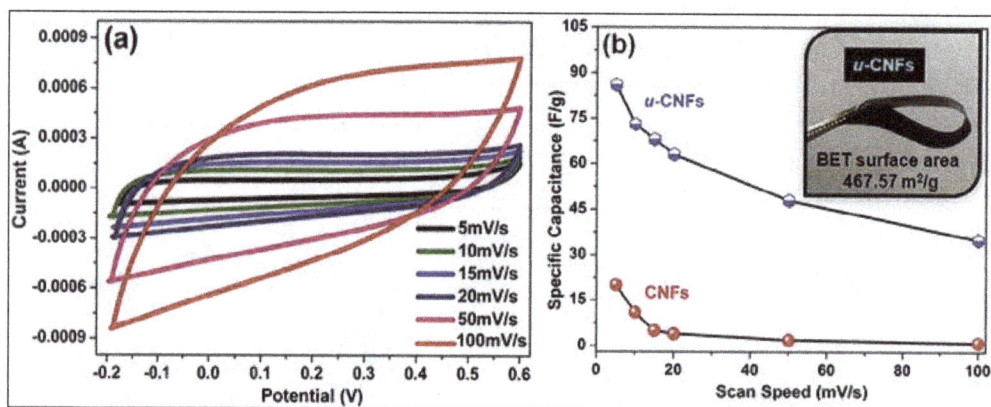

(a) Cyclic voltammetry results for the ultrafine CNFs at different scan rates in 1.0 mol/L H_2SO_4 and; (b) specific capacitance for the u-CNFs (3:7) and CNFs at different scan rates in 1.0 mol/L H_2SO_4.

The Pt nanostructure-supported CNF nanocomposite was employed for the direct oxidation of methanol. It was found that the Pt nanostructure-supported CNF nanocomposite showed better activity than the commercial Pt/C which may be due to the synergistic effect of the underlying anatase surface and the Pt nanostructures with well-defined facets. Alike, Co/CeO$_2$-decorated carbon nanofibers were developed for the methanol oxidation. The results showed that the electrocatalytic activity of the Co/CeO_2-decorated carbon nanofibers toward methanol oxidation was excellent. Interestingly, the introduced catalyst revealed negative onset potential (50 mV vs. Ag/AgCl) which is a superior value among the reported non-precious electrocatalyst.

Applications of Nanofibers

Healthcare

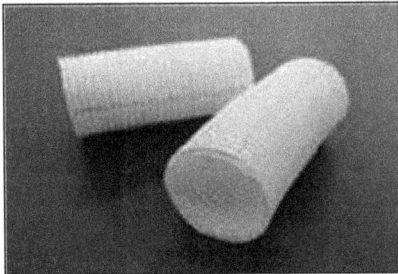

Artificial blood vessel made of cloth.

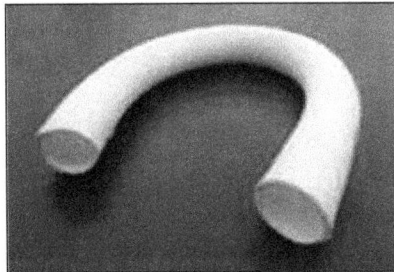

Artificial blood vessel made from PTFE.

Nanofibers are used in healthcare sector for regenerative medicine, wound treatment, drug delivery system. Nanofibers is thought to be advantageous in manufacturing 3D structures and scaffold materials of cells.

Other than the field of regenerative medicine applications are being developed in wound treatment, drug delivery system and dosage form.

Environmental Engineering

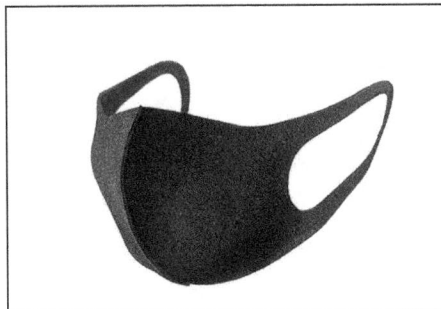

A face mask.

In the field of environmental engineering nanofibers are well known with products such as filters or face masks.

Targeted molecules can be eliminated at high efficiency with water processing filters made of nanofibers making use of their high performance of filtering.

As one of the applications filters to eliminate cesium are under development. Extremely small particles of cesium, which were not collected only with meltblown nonwovens, can be collected by integrating nanofiber nonwoven layers to meltblown nonwovens.

Functional Goods

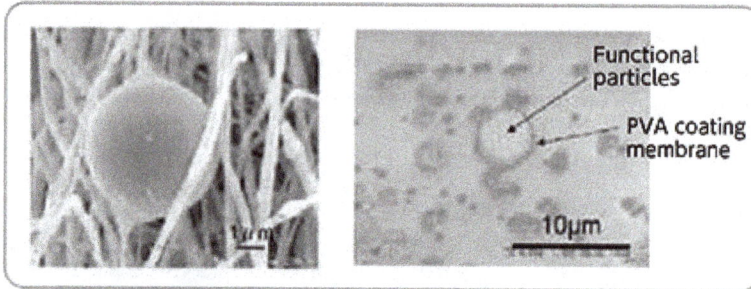

Nanofiber nonwovens containing functional particles.

In recent years some research results report that a variety of textiles can be spun using not only a single and simple nanofiber but also composite materials of textiles, composite materials with particles or coating materials of functional objects.

That is, nanofibers spun with the electrospinning method are expected to be used for food delivery system to protect nutritions between processing and storage or during transfer to another location inside a body.

Electronic Materials

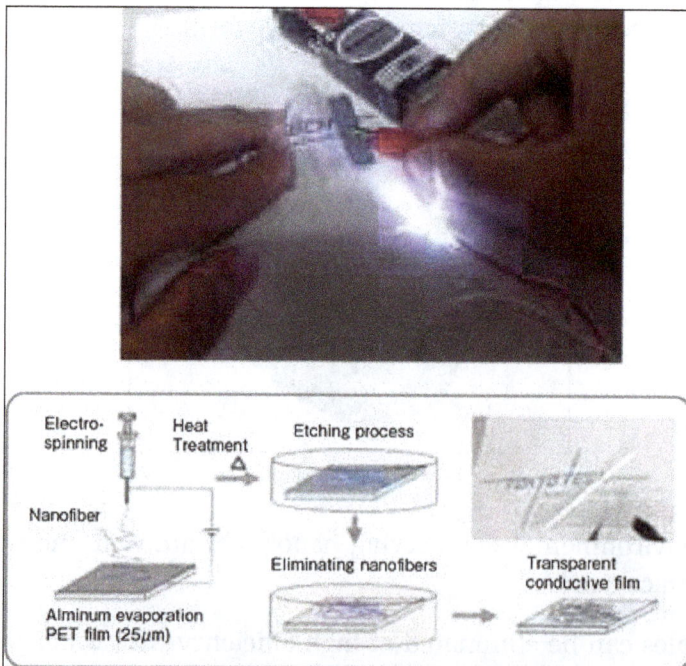

Transparent conductive film.

In the field of electronic materials nanofibers are expected for applications such as:

- Electrodes or separators for high efficiency solar batteries, fuel cells, and secondary batteries.

- Transparent conductive filters (electrodes) for displays, touch panels and functional glass.

Transparent conductive films feature that they have as high visible light transmission rate as that of ITO at 80% and high conductivity at $45\Omega/sq$ of surface resistance, and they are very thin, light, flexible and unbreakable.

Understanding Nanocomposites

Nanocomposites are multiphase soild materials that have nano-scale repeat distances between their different phases. They improve the properties of mechanical strength, toughness and electrical, and thermal conductivity. Ceramic Matrix nanocomposites, metal matrix nanocomposites and polymer matrix nanocomposites are a few of its types. The topics elaborated in this chapter will help in gaining a better perspective about the subject of nanocomposites.

Nanocomposites are materials that incorporate nanosized particles into a matrix of standard material. The result of the addition of nanoparticles is a drastic improvement in properties that can include mechanical strength, toughness and electrical or thermal conductivity. The effectiveness of the nanoparticles is such that the amount of material added is normally only between 0.5 and 5% by weight.

Improved Properties

Nanocomposites can dramatically improve properties like:

- Mechanical properties including strength, modulus and dimensional stability.
- Electrical conductivity.
- Decreased gas, water and hydrocarbon permeability.
- Flame retardancy.
- Thermal stability.
- Chemical resistance.
- Surface appearance.
- Optical clarity.

How Nanocomposites Work

Nanoparticles have an extremely high surface to volume ratio which dramatically changes their properties when compared with their bulk sized equivalents. It also changes the way in which the nanoparticles bond with the bulk material. The result is that the composite can be many times improved with respect to the component parts. Some nanocomposite materials have been shown to be 1000 times tougher than the bulk component materials.

Applications

Nanocomposites are currently being used in a number of fields and new applications are being continuously developed. Applications for nanocomposites include:

- Thin-film capacitors for computer chips.

- Solid polymer electrolyes for batteries.

- Automotive engine parts and fuel tanks.

- Impellers and blades.

- Oxygen and gas barriers.

- Food packaging.

Preparation of Nanocomposites

It explained that the preparation of nanocomposites can be done by three routes, which are "solution blending, the molten state, and in situ polymerization." They pointed out that the latter consists in placing the monomer and the catalyst between the clay layers, and polymerization takes place in the gap, so as polymerization progresses the spacing between the clay's layers increases gradually and the dispersion state of the clays changes from intercalated (the ordered of layered silicate gallery is retained) to exfoliated (delamination with destruction of the clay sheet order). The advantages of this method are 1) the one step synthesis of the metallocene polymer nanocomposites; 2) improved compatibility of the clay and the polymer matrix; and 3) enhanced clay dispersity.

Nanocomposites can also be prepared by dispersing a Nanomer nanoclay into a host polymer, generally at less than 5wt% levels. This process is also termed exfoliation. When a nanoclay is substantially dispersed it is said to be exfoliated. Exfoliation is facilitated by surface compatibilization chemistry, which expands the nanoclay platelets to the point where individual platelets can be separated from another by mechanical shear or heat of polymerization. Nanocomposites can be created using both thermoplastic and thermoset polymers, and the specific compatibilization chemistries designed and employed are necessarily a function of the host polymer's unique chemical and physical characteristics. In some cases, the final nanocomposite will be prepared in the reactor during the polymerization stage. For other polymer systems, processes have been developed to incorporate Nanomer nanoclays into a hot-melt compounding operation. Figure 1 shows a Nanomer particles protruding from a plasma-etched polymer matrix.

Nanomer particles protruding from a plasma-etched polymer matrix.

The first step in the preparation of the nanoclay involves purifying approximately 99 percent of the montmorillonite. The second step involves surface treatment of the clay. montmorillonite is hydrophilic and relatively incompatible with most hydrophobic polymers, so it must be chemically modified to make its surface more receptive to dispersion. After the clays are chemically treated, they are dispersed in the polymer. the clays are incorporated into the polymer matrix by one of two approaches: during polymerization or by melt compounding (This is the difficult part of the technology and may limit the use of nanocomposites. The dispersion process requires a custom solution for each polymer used, so developing polymer nanocomposites becomes a capital intensive research and development project. very few compounding firms have this kind of capability, and this leaves resin producers or well-funded startup companies to develop these materials.

In general, nanocomposites exhibit gains in barrier, flame resistance, structural, and thermal properties yet without significant loss in impact or clarity. Because of the nanometer-sized dimensions of the individual platelets in one direction, exfoliated Nanomer nanoclays are transparent in most polymer systems. However, with surface dimensions extending to one micron, the tightly bound structure in a polymer matrix is impermeable to gases and liquids, and offers superior barrier properties over the neat polymer. Nanocomposites also demonstrate enhanced fire resistant properties and are finding increasing use in engineering plastics.

Nanomer nanoclays provide plastics product development teams with exciting new polymer enhancement and modification options. With the proper choice of compatibilizing chemistries, the nanometer-sized clay platelets interact with polymers in unique ways. Application possibilities for packaging include food and non-food films and rigid containers. In the engineering plastics arena, a host of automotive and industrial components can be considered, making use of lightweight, impact, scratch-resistant and higher heat distortion performance characteristics.

Classification of Nanocomposites

The general class of nanocomposite organic/inorganic materials is a fast growing area of research. Significant effort is focused on the ability to obtain control of the nanoscale structures via innovative synthetic approaches. The properties of nanocomposite materials depend not only on the properties of their individual parents, but also on their morphology and interfacial characteristics.

It pointed out that "most nanocomposites that have been developed and that have demonstrated technological importance have been composed of two phases, and can be microstructurally classified in three principal types: (a) Nanolayered composites composed of alternating layers of nanoscale dimension; (b) nanofilamentory composites composed of a matrix with embedded (and generally aligned) nanoscale diameter filaments; (c) nanoparticulate composites composed of a matrix with embedded nanoscale particles.

The general class of nanocomposite organic/inorganic materials is a fast growing area of research. Significant effort is focused on the ability to obtain control of the nanoscale structures via innovative synthetic approaches. The properties of nanocomposite materials depend not only on the properties of their individual parents, but also on their morphology and interfacial characteristics.

There are basically two modes of classification for nanocomposites. They are the organic and inorganic nanocomposites. So many efforts are taken by the researchers to take control over nanostructures by synthetic approaches. The properties of the nanocomposites not only depend upon the individual parent compositions but also on their morphology and interfacial characteristic.

In the classification of the nanocomposites the inorganic components can be 3 dimensional framework systems such as zeolites ; two dimensional layered materials such as clays, metal oxides, metal phosphates, chalcogenides and even one-dimensional and zero-dimensional materials, such as (Mo3Se3-)n, chains and clusters. Experimental work has generally shown that virtually all types and classes of nanocomposite materials lead to new and improved properties, when compared to their macrocomposite counterparts. Therefore, nanocomposites promise new applications in many fields such as mechanically-reinforced lightweight components, non-linear optics, battery cathodes and ionics, nanowires, sensors and other systems.

The general class of organic/inorganic nanocomposites may also be of relevance to issues of bioceramics and biomineralization, in which in-situ growth and polymerization of biopolymer and inorganic matrix is occurring. Finally, lamellar nanocomposites represent an extreme case of a composite in which interface interactions between the two phases are maximized.

Nanocomposites can generally be divided into two types: multilayer structures and inorganic/organic composites. Multilayer structures are typically formed by gas phase deposition or from the self-assembly of monolayers. Inorganic/organic composites can be formed by sol-gel techniques, bridging between clusters, or by coating nanoparticles, in polymer layers, for example.

Properties of Nanocomposites

Nano-composites have gained much interest recently. Significant efforts are underway to control the nano-structures via innovative synthetic approaches. The properties of nano-composite materials depend not only on the properties of their individual parents but also on their morphology and interfacial characteristics.

The physical, chemical and biological properties of nano materials differ from the properties of individual atoms and molecules or bulk matter. By creating nano particles, it is possible to control the fundamental properties of materials, such as their melting temperature, magnetic properties, charge capacity and even their colour without changing the materials' chemical compositions.

Nano-particles and nano-layers have very high surface-to-volume and aspect ratios and this makes them ideal for use in polymeric materials. Such structures combine the best properties of each component to possess enhanced mechanical and superconducting properties for advanced applications.

The properties of nano-composite materials depend not only on the properties of their individual parents but also on their morphology and interfacial characteristics. Some nanocomposite materials could be 1000 times tougher than the bulk component. The general class of nanocomposite organic/inorganic materials is a fast growing area of research.

The inorganic components can be three-dimensional framework systems such as zeolites, twodimensional layered materials such as clays, metal oxides, metal phosphates, chalcogenides, and even onedimensional and zero-dimensional materials such as $(Mo_3Se_3-)n$ chains and clusters. Thus, nanocomposites promise new applications in many fields such as mechanically reinforced lightweight components, non-linear optics, battery cathodes, nano-wires, sensors and other systems.

Inorganic layered materials exist in many varieties. They possess well defined, ordered intralamellar space potentially accessible by foreign species. This ability enables them to act as matrices for polymers yielding hybrid nano-composites. Lamellar nanocomposites represent an extreme case of a composite in which interface interactions between the two phases are maximized. By engineering the polymer-host interactions, nanocomposites could be produced with the broad range of properties. Lamellar nano-composites can be divided into two distinct classes viz. intercalated and exfoliated. In the former, the polymer chains are alternately present with the inorganic layers in a fixed compositional ratio and have a well-defined number of polymer layers in the intralamellar space.

In exfoliated nano-composites, the number of polymer chains between the layers is almost continuously variable and the layers stand >100 Å apart. The intercalated nano-composites are useful for electronic and charge transport properties. On the other hand, exfoliated nano-composites possess superior mechanical properties. For example, the electronics industry utilizes materials that have high dielectric constants and that are also flexible, easy to process, and strong. Finding single component materials possessing all these properties is difficult.

The most commonly used ceramic materials with high dielectric constant are found to be brittle and are processed at high temperatures, while polymer materials, which are very easy to process have low dielectric constants. Composite materials having micron-scale ferroelectric ceramic particles as the filler in liquid crystal polymer, fluoropolymer, or thermoplastic polymer matrices do not possess ideal processing characteristics and are difficult to form into the thin uniform films used for many microelectronics applications. Here comes the necessity of utilizing nanocomposite materials having a wide range of materials mixed at the nanometer scale.

By optimized fabrication process and controlled nano-sized second phase dispersion, thermal stability and mechanical properties such as adhesion resistance, flexural strength, toughness and hardness can be enhanced which can result into improved nano-dispersion.

The possibilities of producing materials with tailored physical and electronic properties at low cost could result in interesting applications ranging from drug delivery to corrosion prevention to electronic/automotive parts to industrial equipment and several others.

Nanocomposites are materials that incorporate nanosized particles into a matrix of standard material. The result of the addition of nanoparticles is a drastic improvement in properties that can include mechanical strength, toughness and electrical or thermal conductivity. The effectiveness of the nanoparticles is such that the amount of material added is normally only between 0.5 and 5% by weight.

Typically, nanocomposites are clay, polymer or carbon, or a combination of these materials with nanoparticle building blocks. They have an extremely high surface to volume ratio which dramatically changes their properties when compared with their bulk sized equivalents. It also changes the way in which the nanoparticles bond with the bulk material. The result is that the composite can be many times improved with respect to the component parts. Some nanocomposite materials have been shown to be 1000 times tougher than the bulk component materials.

Nanocomposites can dramatically improve properties like:

- Mechanical properties including strength, modulus and dimensional stability.
- Electrical conductivity.
- Decreased gas, water and hydrocarbon permeability.
- Flame retardancy.
- Thermal stability.
- Chemical resistance.

Benefits of Nanocomposites

In general, nanocomposites exhibit gains in barrier, flame resistance, structural, and thermal properties yet without significant loss in impact or clarity. Because of the nanometer-sized dimensions of the individual platelets in one direction, exfoliated Nanomer nanoclays are transparent in most polymer systems. However, with surface dimensions extending to 1 micron, the tightly bound structure in a polymer matrix is impermeable to gases and liquids, and offers superior barrier properties over the neat polymer. Nanocomposites also demonstrate enhanced fire resistant properties and are finding increasing use in engineering plastics.

Recent efforts have focused upon polymer-layered silica nanocomposites and other polymer-clay composites. These materials have improved mechanical properties without the large loading required by traditional particulate fillers. Increased mechanical stability in polymer-clay nanocomposites also contributes to an increased heat deflection temperature. These composites have a large reduction gas and liquid permeability and solvent uptake. Traditional polymer composites often have a marked reduction in optical clarity; however, nanoparticles cause little scattering in the optical spectrum and very little UV scattering.

Although flame retardant additives to polymers typically reduce their mechanical properties, polymerclay nanocomposites have enhanced barrier and mechanical properties and are less flammable. Compression injection molding, melt-intercalation, and co-extrusion of the polymer with ceramic nanopowders can form nanocomposites. Often no solvent or mechanical shear is needed to promote intercalation.

The Nanocomposites 2000 conference has revealed clearly the property advantages that nanomaterial additives can provide in comparison to both their conventional filler counterparts and base polymer. Properties which have been shown to undergo substantial improvements include:

- Mechanical properties e.g. strength, modulus and dimensional stability.

- Decreased permeability to gases, water and hydrocarbons.

- Thermal stability and heat distortion temperature.

- Flame retardancy and reduced smoke emissions.

- Chemical resistance.

- Surface appearance.

- Electrical conductivity.

- Optical clarity in comparison to conventionally filled polymers.

Other benefits from nanocomposites include improvement in modulus, flexural strength, heat distortion temperature, barrier properties, and other benefits and, unlike typical mineral reinforced systems, they are without the conventional trade-offs in impact and clarity.

In plastics the advantages of nanocomposites over conventional ones don't stop at strength. The high heat resistance and low flammability of some nanocomposites also make them good choices to use as insulators and wire coverings. Another important property of nanocomposites is that they are less porous than regular plastics, making them ideal to use in the packaging of foods and drinks, vacuum packs, and to protect medical instruments, film, and other products from outside contamination.

Particle Loadings

In addition it is important to recognize that nanoparticulate/fibrous loading confers significant property improvements with very low loading levels, traditional microparticle additives requiring much higher loading levels to achieve similar performance. This in turn can result in significant weight reductions (of obvious importance for various military and aerospace applications) for similar performance, greater strength for similar structural dimensions and, for barrier applications, increased barrier performance for similar material thickness.

Types of Nanocomposites

According to the matrix materials, nanocomposites are classified as:

- Ceramic Matrix Nanocomposites(CMNC).

- Metal Matrix Nanocomposites(MMNC).

- Polymer Matrix Nanocomposites(PMNC).

Ceramic matrix nanocomposites mainly has Al_2O_3 or SiC system. Most studies reported so far have confirmed the noticeable strengthening of the Al_2O_3 matrix after addithithe on of a low (i.e appro: 10%)volume fraction of SiC particles of suitable size and hot pressing of the resulting mixture.

Metal matrix nanocomposites (MMNC) refer to materials consisting of a ductile metal or alloy matrix in which some nanosized reinforcement material is implanted. These materials combine metal and ceramic features.

Polymer matrix nanocomposites are widely used in industry to their ease of production, lightweight and ductile nature. They have some disadvantages such as low modulus and strength compared to metals and ceramics.

Among the structural applications, nanocomposites have been the most studied and used materials because of their application in all sectors from domestic to aerospace.

Here applications like fuel cell applications, biomedical applications, and electronic applications and in aerospace are reviewed.

Table: Different types of nanocomposites.

Class	Examples
Metal	$Fe\text{-}Cr/Al_2O_3, Ni/Al_2O_3, Co/Cr, Fe/MgO$ $AI/CNT, Mg/CNT$
Ceramics	$Al_2O_3/SiO_2, \ SiO_2/Ni, Al_2O_3/TiO_2$ $Al_2O_3/SiC, Al_2O_3/CNT$
Polymer	Thermoplastic/thermoset polymer/layered silicates, polyester/TiO_2, polymer/CNT, polymer/layered double hydroxides

Polymer Nanocomposites

Polymer nanocomposites are materials in which nanoscopic inorganic particles, typically 10-100 A in at least one dimension, are dispersed in an organic polymer matrix in order to dramatically improve the performance properties of the polymer. Systems in which the inorganic particles are the individual layers of a lamellar compound - most typically a smectite clay or nanocomposites of a polymer (such as nylon) embedded among layers of silicates - exhibit dramatically altered physical properties relative to the pristine polymer. For instance, the layer orientation, polymer-silicate nanocomposites exhibit stiffness, strength and dimensional stability in two dimensions (rather than one). Due to nanometer length scale which minimizes scattering oflight, nanocomposites are usually transparent.

Polymer nanocomposites represent a new alternative to conventionally filled polymers. Because of their nanometer sizes, filler dispersion nanocomposites exhibit markedly improved properties when compared to the pure polymers or their traditional composites. These include increased modulus and strength, outstanding barrier properties, improved solvent and heat resistance and decreased flammability.

Polymers that contain transition metal complexes either attached to or directly in a π-conjugated backbone are an exciting and a promising class of modern materials. These macromolecules are hybrid of π-conjugated organic and transition metalcontaining polymers. π-conjugated organic polymers, such as polyacetylene, poly thiophene, and polypyrrole, as well as oligomers and derivatives of these materials have been extensively explored. These materials are endowed with many important properties such as nonlinear optical properties, electronic conductivity and luminescence, and have been proposed for their use in various applications including chemical sensors, electroluminescent devices, electro catalysis, batteries, smart windows and memory devices.

Layered silicate/polymer nanocomposites exhibit superior mechanical characteristics (e.g. 40% increase of room temperature tensile strength), heat resistance (e.g. 100% increase in the heat distortion temperature) and chemical resistance (e.g.~ 10 fold decrease in O_2 and H_2O permeability) compared to the neat or traditionally filled resins. These property improvements result from only a 0.1-10 vol. % addition of the dispersed nanophase. Polyimide-clay hybrids represent another example of polymer nanocomposites. These nanocomposites have been prepared by intercalation of the organoclay with a polyamic acid. The claypolyimide hybrid composite films exhibit greatly improved CO2 barrier properties at low clay content; less than 8.0 vol. % clay results in almost a ten-fold decrease in permeability. Adding nanoscale ceramic powders to commercial products can produce another class of polymer nanocomposites. The addition of reinforcing agents is widely used in the production of commodities (packaging films and tyres). It is expected that the reduction of the added particle size down to nanometric scale could enhance the performance of these materials, even though not to the extent as layer addition. These new materials are aimed at being a substitute for more expensive technical parts (gear systems in wood drilling machines, wear resistance materials) and in the production of barrier plastic film for food industry. Besides structural applications, polymer nanoparticle compounds have very interesting functional applications. For instance, $\gamma-Fe_2O_3$ polymer nanocomposites are used as advanced toner materials for high quality colour copiers and printers and as contrast agents in NMR analysis, memory devices. The key to forming such novel materials is understanding and manipulating the guest-host chemistry occurring between the polymer and the layered compounds or the nanoparticles, in order to obtain a homogenous dispersion and a good contact between polymer and added particle surfaces. There have been major advances in solid state and materials chemistry in the last two decades and the subject is growing rapidly.

The coatings of magnetic particles are of special interest because of their important applications viz. technological energy transformation, magnetic recording, magnetic fluids and magnetic refrigeration system. Polymer materials have been filled with several inorganic compounds in order to increase properties like heat resistance, mechanical strength and impact resistance and to decrease other properties like electrical conductivity, dielectric constant thereby increasing the permeability for gases like oxygen and water vapor.

In recent years considerable efforts have been devoted to the development of methods for the preparation of composite particles consisting of polymer cores covered with shells of different chemical composition. In several of these powders, particles covered with magnetic materials have been used as beads for gas separation, or as pigments, catalysts, coatings, flocculents, toners, raw materials recovery, drug delivery and anticorrosion protection.

Polymer composites containing ferrites are increasingly replacing conventional ceramic magnetic materials because of their mouldability and reduction in cost. They are also potential materials for microwave absorbers, sensors and other aerospace applications. These flexible magnets or rubber ferrite composites are possible by the incorporation of magnetic powders in various elastomer matrices. This modifies the physical properties of the polymer matrix considerably.

Solvent casting method is one of the easiest methods for the preparation of polymer nanocomposites. It needs simple equipment and is less time consuming. Figures (a-c) show the optical micrograph images of $\gamma-Fe_2O_3$ dispersed natural rubber $(\gamma-Fe_2O_3-NR), \gamma-Fe_2O_3$ dispersed polyethylene glycol $(\gamma-Fe_2O_3-PEG)$ and $\gamma-Fe_2O_3$ dispersed polycarbonate $(\gamma-Fe_2O_3-PC)$, nanocomposite films. The fine dispersion of the magnetite inside the polymer matrix makes it a magnetic polymer. Figure led) shows the structure of the polymer polyisoprene (natural rubber).

Optical micrograph images of $\gamma-Fe_2O_3$ dispersed (a) natural rubber.
(b) polyethylene glycol (c) polycarbonate composite films.

Characterization

The vigorous development of polymeric science and extensive utilization of polymeric materials in technology has led in recent years to the increased interest in the preparation and characterization of polymer and its composite films. Characterization is an essential part of all investigations dealing with materials. The important aspects of characterization are chemical composition and compositional homogeneity (chemical homogeneity), structure (including crystal system where possible atomic coordinates, bonding and ultra structure) and identification and analysis of defects and impurities influencing the properties of the materials. Characterization, therefore, describes all those features of composition and structure of a material that would suffice for reproducing the material. The advances made in the last few years in characterization techniques, especially in the structure elucidation, have been stupendous and have opened new vistas in solid state materials. Among the several characterization techniques, X-ray diffraction (XRD), scanning electron micrography (SEM) and infrared (IR) spectroscopy are the three important techniques.

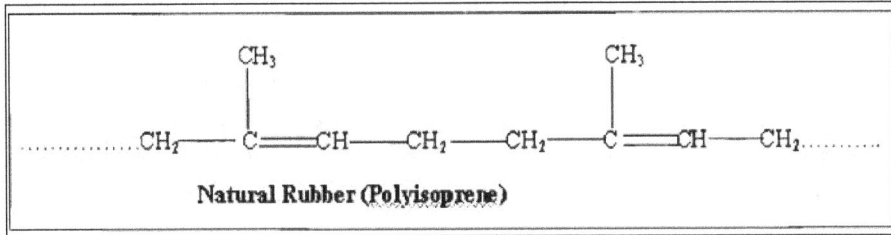

Structure of natural rubber. Polyisoprene.

X-ray Diffraction

X-ray diffraction 1 has played a central role in identifying and characterizing solids since the early part of this century. The June 2005. nature of bonding and the working criteria for distinguishing between short-range and long-range order of crystalline arrangements from the amorphous substances are largely derived from X-ray diffraction and thus it remains as a useful tool to obtain structural information.

X-ray diffraction pattern of amorphous polymer will not show any sharp and highly intensed peaks whereas the nanocomposites of amorphous polymer show sharp and highly intensed peaks. This is due to the development of crystallinity in the amorphous polymer. The XRD pattern of pure natural rubber and $\gamma - Fe_2O_3$ dispersed natural rubber composite. Highly intensed peaks occur in the pattern due to the presence of inserted gamma iron oxide materials in the rubber matrix. X-ray diffraction has been most commonly used for routine characterization as well as for detailed structural elucidation. In order to obtain detailed structural information, knowledge of X-ray diffraction intensities is also essential, the intensities being related to the structure factor.

Scanning Electron Micrograph

Structural phenomena play an important role in determining the properties of a polymer. Mechanical properties are determined not only by the changes in shape confirmation and by motion of individual molecules of the polymers, but by the behavior of larger and more complex structural formations as well. The interface boundaries of these formations, known as super molecular structures, are the sites where chemical reactions in the polymer are most likely to begin and centers of crack formation and incipient destruction are likely to arise. It has been found that extensive occurrences of ordered structures are typical not only of crystalline, but also of amorphous polymers. Despite the complex morphology of structural formation in polymers it should not be forgotten that all these structures are built up of separate polymeric molecules. At a glance it seems self-evident that direct relations must exist between the properties of macromolecules and their ability to form super molecular structures. The shapes of most polymer molecules may vary within wide limits when studying the simplest phenomenon of structure formation. Quite a long time ago it was found that there are two ways by which structures can form. Sufficiently flexible molecules roll up into spherical coil globules, which form in very much the same way as the drops of a liquid under the action of surface tension, But if the macromolecules are sufficiently rigid, the simple.stlinear structures result. No separate linear polymer molecules have been observed so far. Evidently in majority of cases, they aggregate into chain bunches usually containing several dozen molecules. The phenomenon of structural transformations occurring during deformation is very typical of polymers. A classic example of structural transformation is the formation of a 'neck' on deformation, described for the case of crystalline polymers some time ago by Kakina. It is also observed in the case of amorphous polymers with developed structures and it is firmly described as a phase transformation. The nature of this phenomenon remained obscure for a long time, but electron microscopy revealed that 'neck' formation is actually a jump wise transition from one super molecular structure to another with a sharp interfacial boundary which is also observed on a microscopic scale.

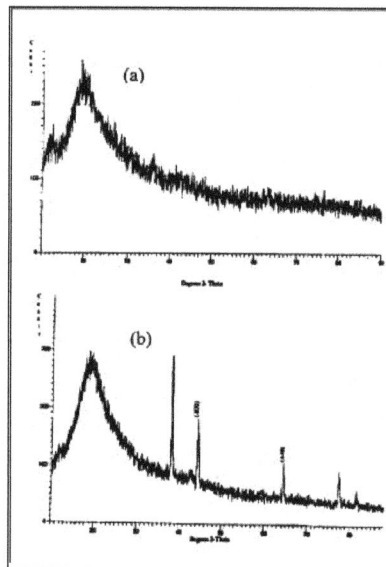

XRD pattern of (a) pure natural rubber (b) $\gamma-Fe_2O_3$ dispersed natural rubber.

A well-known example is that of poly(methyl methacrylate) where a sharp boundary can be seen between the isotropic and the oriented parts of the specimen. The formation of a 'neck' on deformation of a large spherulite of isotactic poly (methyl methacrylate) takes place and a sharp boundary can be discerned between the unchanged and the oriented portions of a spherulite. In addition, secondary formations can be seen which have resulted from recrystallisation of the oriented parts, and these are also separated by sharp boundary lines.

The SEM images of $\gamma-Fe_2O_3$ dispersed natural rubber composite. From the figure one can observe the fine dispersion of iron oxide particles in the rubber matrix. The dispersed particles have irregular shape and show agglomeration.

Infrared Spectroscopy

Infrared spectroscopy is one of the most powerful analytical techniques, which offers the possibility of chemical identification. This technique when coupled with intensity measurements may be used for quantitative analysis. One of the important advantages of infrared spectroscopy over' the other usual methods of structural analysis (X-ray diffraction, electron spin resonance, etc.) is that it provides information about the structure of a molecule quickly, without tiresome evaluation methods. This method can solve many problems in organic chemistry (polymeric materials) and coordination chemistry, and also advantageously complements the results obtained by other methods. This technique is based upon the simple fact that a chemical substance shows marked selective absorption in the infrared region giving rise to dose-packed absorption bands called an IR absorption spectrum, over a wide wavelength range. Various bands will be present in the IR spectrum, which will correspond to the characteristic functional groups and bonds present in a chemical substance. Thus an IR spectrum of a chemical substance is a fingerprint for its identification. IR spectrum of polymer nanocomposite shows the presence of both nanomaterials and polymers (depending upon the polymer chain) at various frequencies.

SEM image of $\gamma - Fe_2O_3$ dispersed natural rubber.

Thermal Analysis

Thermal analysis may be defined as the measurement of physical and chemical properties of materials as a function of temperature. The two main thermal analysis techniques are thermogravimetric analysis (TGA) which automatically records the change in weight of a sample as a function of either temperature or time, and the differential thermal analysis (DT A), which measures the difference in temperature, .1T, between a sample and an inert reference material as a function of temperature; DT A therefore detects change in heat content. A technique closely related but modified to DTA is differential scanning calorimetry (DSC). In DSC, the equipment is designed to allow a quantitative measure of the enthalpy changes, (.1H), that occurs in a sample as a function of either temperature or time.

DSC is an analytical tool which helps to understand the thermal behavior of polymer nanocomposites. It helps in finding glass transition temperature (T_g) of polymer and its polymer composites. The increase in T_g values shows the presence of inorganic materials in the polymer matrix.

Exciting developments can be expected in the area of polymer nanocomposites and structures in the near future. Investigations on polymer nanocomposite can thus pay rich dividends.

Non-polymer based Nanocomposites

Metal/Metal Nanocomposite

Bimetallic nanoparticles either in the form of alloy or core-shell structures or being investigated in some depth because of their improved catalytic properties and changes in the electronic/optical properties related to individual, separate metals. It is postulated their interesting Physico-chemical properties, result from the combination of two kinds of metals and their fine structures.

Metal/Ceramic Nanocomposites

In these types of composites, the electric, magnetic, chemical, optical and mechanical properties of both phases are combined. Size reduction of the components to the nanoscale causes improvement of the above mentioned properties and leads to new application. The polymer precursors techniques offers an attractive rough to such composites proving a chemically inert and hard ceramic matrix.

Ceramic/Ceramic Nanocomposites

Ceramic Nano composites could solve the problem of fracture failures in artificial joint implants; these would extend patient's mobility and eliminate the high cost of surgery. The use of Zirconia-toughened alumina nanocomposite to form Ceramic/ceramic implants with potential life spends of more than 30 years.

Applications of Nanocomposites

Strengths and Limitations of Nanocomposites

Nanoparticles have an extremely high surface typical not only of crystalline, but also of amorphous polymers to volume ratio which dramatically changes their properties when compared to their bulk sized equivalents. It also changes the way in which the nanoparticles bond with the bulk material. The result is that the composite can be many times improved with respect to the component parts. Some nanocomposite materials have been shown to be 1000 times tougher than the bulk component materials. These explain its adoption and why it is playing a prominent role in manufacturing. There are a few disadvantages associated with using nanoparticle viz. toughness and impact performance. Some researches have shown that nanoclay modification of polymers such as polyamides could even reduce impact performance.

There is a need for better understanding of formulation/structure/property relationships to platelet exfoliation and dispersion etc. The improved properties vis-à-vis the disadvantages of the nanoparticles and resultant composites are shown in table:

Improved properties	Disadvantages
Mechanical properties (tensile strength, stiffness, toughness).	Viscosity increase (limits process ability).
Gas barrier.	Dispersion difficulties.
Synergistic flame retardant additive.	Optical issues.
Dimensional stability.	Sedimentation.
Thermal expansion.	Black color when different carbon containing nanoparticles are used.
Thermal conductivity.	
Ablation resistance.	
Chemical resistance.	
Reinforcement.	

Real world applications for nanocomposites are coming slowly. Nanocomposites prove there appears to be a reluctance to embrace this new technology due to cost and variability in the quality of some of the products. They pointed out that John Jones, a market development specialist at Honeywell, believes that manufacturers of these materials have to prove to the market that these new materials can meet their performance expectations.

Another challenge nanocomposite producers face is the production of the nanocomposite itself. Both methods, pre-polymerization and post polymerization, for preparing nanocomposites have drawbacks. The Nanoparticle News explained that "Pre-polymerization production can disrupt the polymerization process, which is often critical and requires much developmental time and expense to achieve good yields and controllability, and post polymerization often requires a lot of time to achieve a good dispersion of the nanoparticles in the composite". This then becomes an expensive and low cost-competitive initiative.

Commenting on the limitations, Demetrakakes, stated that "another concern deals with equipment conversion that accepts new material through recalibration; this is a big investment for converters to make." He explained that it is a complicated process to go from plastic pellets to a blown bottle, as it requires heating and blowing that form to the shape of the bottle. This is expensive equipment, very high-speed equipment, designed for the material that you're going to run. You can't just take another material with different flow characteristics, crystallization rate, and those kinds of things, throw that in there and make it run.

To date one of the few disadvantages associated with nanoparticle incorporation has concerned toughness and impact performance. Some of the data presented has suggested that nanoclay modification of polymers such as polyamides could reduce impact performance. Clearly this is an issue which would require consideration for applications where impact loading events are likely. In addition, further research will be necessary to, for example, develop a better understanding of formulation/structure/property relationships, better routes to platelet exfoliation and dispersion etc.

Applications of Nanocomposites

Experimental work has generally shown that virtually all types and classes of nanocomposite materials lead to new and improved properties, when compared to their macrocomposite counterparts. Therefore, nanocomposites promise new applications in many fields such as mechanically-reinforced lightweight components, non-linear optics, battery cathodes and ionics, nanowires, sensors and other systems.

Such mechanical property improvements have resulted in major interest in nanocomposite materials in numerous automotive and general/industrial applications. These include potential for utilisation as mirror housings on various vehicle types, door handles, engine covers and intake manifolds and timing belt covers. More general applications currently being considered include usage as impellers and blades for vacuum cleaners, power tool housings, mower hoods and covers for portable electronic equipment such as mobile phones, pagers etc.

Nanomer nanoclays provide plastics product development teams with exciting new polymer enhancement and modification options. With the proper choice of compatibilizing chemistries, the nanometer-sized clay platelets interact with polymers in unique ways. Application possibilities for packaging include food and non-food films and rigid containers. In the engineering plastics arena, a host of automotive and industrial components can be considered, making use of lightweight, impact, scratch-resistant and higher heat distortion performance characteristics.

Oil and Gas Pipelines

Corrosion has a costly and deleterious effect on aging infrastructure throughout the world. As such, considerable attention has been focused on innovative techniques to arrest corrosion in the carbon steel found in pipelines, bridges, and water systems. In the United States, the annual cost associated with corrosion damage of structural components is greater than the combined annual cost of natural disasters, including hurricanes, storms, floods, fires and earthquakes. Similar findings have also been made by studies conducted in the United Kingdom, Germany, and Japan.

According to the U.S. Department of Transportation Office of Pipeline Safety, "between 1989 and 2008 pipeline corrosion incidents resulted in over $582M in property damages, 28 fatalities and 94 injuries." The need to manage and mitigate corrosion damage has therefore rapidly increased, as materials are placed in more extreme environments and pushed beyond their original design life.

Corrosion damage and failure are not always considered in the design and construction of pipeline systems. Even if corrosion is considered, unanticipated changes in the environment in which the structure operates can result in unexpected corrosion damage. Moreover, combined effects of corrosion and mechanical damage, and environmentally assisted material damage can result in unexpected failures due to the reduced load carrying capacity of pipelines.

The application of nanocomposites is becoming increasingly important in the manufacture and structural repair of damaged pipelines. In their work, Kessler and Goertzen pointed out that nanocomposites offer more advantages in pipelines manufacture, while its overwraps are used to repair corroded steel pipelines, as the repair can be completed in a relatively short period and the fluid transmission in the piping system can remain undisrupted while the repair is being made.

Fiber-reinforced nanocomposite pipelines are emerging as a feasible alternative to steel pipelines with regard to performance and cost. The pipeline is typically constructed including an inner non- permeable barrier tube that transports the fluid (pressurized gas or liquid), a protective layer over the barrier tube, an interface layer over the protective layer, multiple glass or carbon fiber composite layers, an outer pressure barrier layer, and an outer protective layer. It is a nanocomposite structure in the purest engineering sense of the term, as each of the several components provides a distinct function and the interaction between the components produces a structure with exceptional performance characteristics.

The pipeline has improved burst and collapse pressure ratings, increased tensile strength, compression strength, and load carrying capacity, compared to non-reinforced, non-metallic pipelines. The ability of re-inforced nanocomposite piping to withstand large strains allows the piping to be coiled such that long lengths can be spooled onto a reel in an open bore configuration.

In addition, the pipe can be manufactured with fiber optics, copper signal wires, power cables or capillary tubes installed directly into the structural wall of the piping. This offers the option of manufacturing the pipe with embedded sensors and operating it as a so- called smart structure. Sensors embedded in the pipe can be powered via copper wire from remote locations and real-time data from the sensors can be returned through fiber optics. This provides the unique advantage of lifetime performance monitoring of the pipe.

The application of re- inforced nanocomposites in oil and gas pipelines has the following advantages:

- Anisotropic characteristics of nanocomposite piping provide extraordinary burst and collapse pressure ratings, increased tensile and compressive strengths, and increased load carrying capacities.

- No welding and minimal joining - many miles of continuous pipeline can be emplaced as a seamless monolith.

- Emplacement requirements should be dramatically less than those for metal pipe, enabling the pipe to be installed in areas where right-of-way restrictions are severe.

- Structurally integrated sensors provide real-time structural health monitoring and could reduce need for pigging.

- Corrosion resistant and damage tolerant.

- Meets or exceeds published and consensus standards for pipeline in oil and gas applications.

The basic idea in implementing the nanocomposites in mechanical stream is the resistant to fracture and the often occurrence of wear and tear of the machine parts. Nanocomposites used as a blend against plastics can be used for strengthening the portions of the automobiles where higher efficiency is required. As the world is affected by the pollution the automobile manufacturers are working towards developing a technology which controls the same cost effectively, this led to the acceptance of polymeric nanocomposites. Owing to their polymeric nature, polymer nanocomposites fit this description. Because of their nanometer size features, polymeric nanocomposites possess unique properties, such as enhanced mechanical, impact, barrier and heat resistant properties, compared to other composites.

Combining the unique properties of nanocomposite and recyclable polymers to produce light-weight recyclable and biodegradable polymer/nanocomposite is a great challenge. These compositions were widely used for making the body parts of the automobiles. The industry was mainly concerned over the following aspects:

- Weight reduction.

- Improved performance.

- Aesthetics.

- And recyclability.

Nanotechnology is already driving changes throughout this industry at nearly every level involving material, components, and systems. This is because most cars produced in the United States contain some nanocomposite material, most typically carbon nano-tube in nylon blend for the use of the fuel system to protect against static electricity.

A plastic nanocomposite is being used for "step assists" in the General Motors Safari and Astro vans as it is scratch- resistant, light-weight, and rust-proof, and generates improvements in strength and reductions in weight, which lead to fuel savings and increased longevity. And in 2001, Toyota started using nanocomposites in a bumper that makes it 60% lighter and twice as resistant to denting and scratching.

Nanocrystals of various metals have been shown to be 100 percent, 200 percent and even as much as 300 percent harder than the same materials in bulk form. Because wear resistance often is dictated by the hardness of a metal, parts made from nanocrystals will significantly last longer than conventional parts.

Aircrafts

Researchers have made relatively awesome discoveries on nanocomposites over the last decade, ever since the pioneering work on nanocaly by the company Toyota. The dispersion of the silicate nanolayer with its high aspect ratio, large surface area, and high stiffness within a polymer matrix results in significant improvement of the properties of polymeric materials, including mechanical properties, barrier properties, resistance to solvent swelling, ablation performance, thermal stability, fire retardancy, controlled release of drugs, anisotropic electrical conductivity, and photo activity.

Layered-silicate nanocomposites have great applications, ranging from automotive and aerospace to food packaging and tissue engineering. Epoxy materials are widely used in adhesives, coatings, composites and electronics. These are also used in designing of aircraft parts too.

This epoxy system has a high glass transition temperature (Tg), good mechanical and physical performance characteristics, and low viscosity. In addition, epoxy nanocomposites as primer layer for aircraft coatings for improved anticorrosion properties are used.

High performance nanocomposites are also used in fuselage sinks in aircrafts.

Electronics

Conductive nanocomposites are capable of conducting electric current well owing to the electric charges in their structure. Polycarbonates which is an insulator can be made conductive Polycarbonates, the inexpensive plastics known for their excellent optical and mechanical properties, could in future, find applications into newer and more important horizons. Polycarbonates are tagged as poor electrical conductors, but a research team has altered this very property by adding carbon nanotubes to them thereby resulting in highly conductive nanocomposites. The team has come up with a strategy to achieve higher conductivities using carbon nanotubes in plastic hosts than what has been currently achieved. By combining nanotubes with polycarbonates, the team was able to reach a milestone of creating nanocomposites with ultra-high conductive properties. Shay Curran, associate professor of physics at UH demonstrated ultra-high electrical conductive properties in these plastics by mixing them with just the right amount and type of carbon nanotubes. As a result, the inexpensive plastic used to make optical discs will feature in high-end military aircrafts to shield them against build up of electrical charges and pulses which can lead to significant failures.

Additionally, by modifying the amount of carbon nanotubes added to the polycarbonate-nanotube mix, the electrical conductivity of the nanocomposite could be changed from that of silicon to a few orders below what is achieved by metals.

Films

The presence of filler incorporation at nano-levels has also been shown to have significant effects on the transparency and haze characteristics of films. In comparison to conventionally filled polymers, nanoclay incorporation has been shown to significantly enhance transparency and reduce haze. With polyamide based composites, this effect has been shown to be due to modifications in the crystallisation behaviour brought about by the nanoclay particles; sperilitic domain dimensions being considerably smaller. Similarly, nano-modified polymers have been shown, when employed to coat polymeric transparency materials, to enhance both toughness and hardness of these materials without interfering with light transmission characteristics.

Environmental Protection

Water laden atmospheres have long been regarded as one of the most damaging environments, which polymeric materials can encounter. Thus an ability to minimize the extent to which water is absorbed can be a major advantage.

Available data indicate that significant reduction of water absorption in a polymer could be achieved by nanoclay incorporation. Similar effects could also be achieved with polyamide-based nanocomposites.

Specifically, increasing aspect ratio diminishes substantially the amount of water absorbed, thus indicating the beneficial effects likely from nanoparticle incorporation compared to microparticle loading.

Hydrophobicity enhancement would clearly promote both improved nanocomposite properties and diminish the extent to which water would be transmitted through to an underlying substrate. Thus applications in which contact with water or moist environments is likely could clearly benefit from materials incorporating nanoclay particles.

Food Packaging

The gaseous barrier property improvement that can result from incorporation of relatively small quantities of nanoclay materials has been shown to be substantial. Data provided from various sources indicate oxygen transmission rates for polyamide- organoclay composites, which are usually less than half of the unmodified polymer.

Further data reveals the extent to which both the amount of clay incorporated in the polymer, and the aspect ratio of the filler contributes to overall barrier performance. In particular, aspect ratio has been shown to have a major effect, with high ratios (and hence tendencies towards filler incorporation at the nano-level) quite dramatically enhancing gaseous barrier properties.

Development of a combined active/passive oxygen barrier system for polyamide-6 materials is underway at various laboratories across the world. Passive barrier characteristics are provided by nanoclay particles incorporated via melt processing techniques whilst the active contribution comes from an oxygen- scavenging ingredient.

Oxygen transmission results reveal substantial benefits provided by nanoclay incorporation in comparison to the base polymer (rates approximately 15-20% of the bulk polymer value, with further benefits provided by the combined active/passive system).

Increased tortuosity provided by the nanoclay particles essentially slows transmission of oxygen through the composite and drives molecules to the active scavenging species resulting in near zero oxygen transmission for a considerable period of time.

Such excellent barrier characteristics have resulted in considerable interest in nanoclay composites in food packaging applications, both flexible and rigid.

Specific examples include packaging for processed meats, cheese, confectionery, cereals and boil-in-the- bag foods, also extrusion-coating applications in association with paperboard for fruit juice and dairy products, together with co-extrusion processes for the manufacture of beer and carbonated drinks bottles. The use of nanocomposite packaging would be expected to enhance considerably the shelf life of many types of food.

Fuel Tanks

The ability of nanoclay incorporation to reduce solvent transmission through polymers such as polyamides has been demonstrated. Available data reveals significant reductions in fuel transmission through polyamide–6/66 polymers by incorporation of nanoclay filler.

As a result, considerable interest is now being seen in these materials as both fuel tank and fuel line components for cars. Of further interest for this type of application, the reduced fuel transmission characteristics are accompanied by significant material cost reductions.

Nanocomposites are currently being used in a number of fields and new applications are being continuously developed. Other applications for nanocomposites include:

- Thin-film capacitors for computer chips.

- Solid polymer electrolyes for batteries.

- Automotive engine parts and fuel tanks.

- Impellers and blades.

- Oxygen and gas barriers.

References

- Nanocomposites_and_its_Applications-Review-3038972: academia.edu, Retrieved 28 January, 2019

- Polymeric Nanocomposites: A review Journal of Polymer-plastics technology and engineering vol. 43, no 2, pp. 427-443

- Nanocomposites, Synthesis, Structure, properties and new application opportunities, Pedro Henrique Cury Camargo, Kesturn Gundappa Satyanarayana, Fernanado Wypych, Curitiba –PR, Brazil

- Polymer-clay nanocomposites, John Wiley and Sons, 2001

Permissions

We would like to thank the editorial team for lending their expertise to make the book truly unique. They have played a crucial role in the development of this book. Without their invaluable contributions this book wouldn't have been possible. They have made vital efforts to compile up to date information on the varied aspects of this subject to make this book a valuable addition to the collection of many professionals and students.

This book was conceptualized with the vision of imparting up-to-date and integrated information in this field. To ensure the same, a matchless editorial board was set up. Every individual on the board went through rigorous rounds of assessment to prove their worth. After which they invested a large part of their time researching and compiling the most relevant data for our readers.

The editorial board has been involved in producing this book since its inception. They have spent rigorous hours researching and exploring the diverse topics which have resulted in the successful publishing of this book. They have passed on their knowledge of decades through this book. To expedite this challenging task, the publisher supported the team at every step. A small team of assistant editors was also appointed to further simplify the editing procedure and attain best results for the readers.

Apart from the editorial board, the designing team has also invested a significant amount of their time in understanding the subject and creating the most relevant covers. They scrutinized every image to scout for the most suitable representation of the subject and create an appropriate cover for the book.

The publishing team has been an ardent support to the editorial, designing and production team. Their endless efforts to recruit the best for this project, has resulted in the accomplishment of this book. They are a veteran in the field of academics and their pool of knowledge is as vast as their experience in printing. Their expertise and guidance has proved useful at every step. Their uncompromising quality standards have made this book an exceptional effort. Their encouragement from time to time has been an inspiration for everyone.

The publisher and the editorial board hope that this book will prove to be a valuable piece of knowledge for students, practitioners and scholars across the globe.

Index

www.ingramcontent.com/pod-product-compliance
Lightning Source LLC
Chambersburg PA
CBHW061933190326
41458CB00009B/2732